WILLIAM F. MAAG LIBRARY
YOUNGSTOWN STATE UNIVERSITY

A Specialist Periodical Report

Electrochemistry
Volume 6

A Review of the Literature Published up to the End of 1976

Senior Reporter
H. R. Thirsk, *Department of Physical Chemistry University of Newcastle upon Tyne*

Reporters
R. D. Armstrong, *University of Newcastle upon Tyne*
M. F. Bell, *University of Newcastle upon Tyne*
O. R. Brown, *University of Newcastle upon Tyne*
A. A. Metcalfe, *University of Newcastle upon Tyne*
I. Morcos, *Institut de recherche de l'Hydro-Québec, Canada*
P. P. Schmidt, *Oakland University, Rochester, Michigan, U.S.A.*

The Chemical Society
Burlington House, London, W1V 0BN

British Library Cataloguing in Publication Data

Electrochemistry. – (Chemical Society. Specialist Periodical Reports).
  Vol. 6
  1. Electrochemistry – Collected works
  I. Thirsk, Harold Reginald II. Series
  541'.37   QD552   (72-23822)

  ISBN 0-85186-057-5
  ISSN 0305-9979

Copyright © 1978
The Chemical Society

All Rights Reserved
No part of this book may be reproduced or transmitted
in any form or by any means – graphic, electronic,
including photocopying, recording, taping or
information storage and retrieval systems – without
written permission from The Chemical Society

Printed in Great Britain
at the Alden Press, Oxford

Preface

The responsibility for compiling these Specialist Periodical Reports leaves one very much indebted to the considerable efforts of the authors, who invariably are active scientists with a heavy commitment. At times pressure of work is so great that a deadline cannot be met and there is a resultant imbalance in the final volume.

The Senior Reporter apologizes for the fact that this has happened in the present volume, which is more heavily weighted towards theory and technique than was the original intention. Hopefully the balance may be redressed in subsequent volumes.

The interest in organic electrochemistry is still well sustained, and there would seem to be a continuing need for an annual review (Chapter 1). Professor Schmidt contributes the second part of the review on electron-transfer reactions, the first part having appeared in Volume 5 of this series (Chapter 4).

The interfacial tension of solid electrodes is a matter of considerable interest to electrochemists and a review of progress in this field seemed both useful and timely but of necessity covering a fairly long time-scale (Chapter 2). It is also hoped that the review of a.c. impedance methods (Chapter 3) will prove helpful to those electrochemists who are interested in the application of the technique to complex reactions and reactions involving intermediates at the interface, being supplementary to existing extensive reviews of studies more related to reactions in the solution.

JUNE 1977   H.R.T.

Contents

Chapter 1 Organic Electrochemistry–Synthetic Aspects   1
By O. R. Brown

1 Reductions   2
    Hydrocarbons   2
    Activated Olefins   4
    $\alpha\beta$-Unsaturated Carbonyl Compounds   6
    Other Carbonyl Compounds   8
    Quaternary Salts   11
    Nitro-compounds   13
    Other Nitrogen-containing Compounds   15
    Halides   18
    Carboxylic Acids and Derivatives   25
    Sulphur Compounds   26
    Oxygen Compounds   29
    Organometallic Compounds   31
    Simple Inorganic Materials   33

2 Oxidations   34
    Hydrocarbons and Unsaturated Ethers   34
    Alkyl Iodides   41
    Phenols   43
    Esters   45
    Carboxylic Acids   45
    Amines   48
    Other Nitrogen-containing Compounds   52
    Sulphur Compounds   58
    Halogenation   59
    Oxygen Compounds   62
    Miscellaneous   64

Chapter 2 The Interfacial Tension of Solid Electrodes   65
By I. Morcos

1 Introduction   65

2 Principles and Experimental Results   70
    General Criteria and Classification of Methods   70
    Indirect Measurement of Surface Stress   73

Direct Measurement of Surface Stress by a Piezoelectric Element	81
Direct Measurement of Contact Angles	84
Indirect Measurement of Contact Angles by Meniscus-rise Techniques	85

Chapter 3 The A.C. Impedance of Complex Electrochemical Reactions 98
By R. D. Armstrong, M. F. Bell, and A. A. Metcalfe

1 Introduction	98
2 Historical Summary	98
3 Generalized One-step Reactions	104
4 Adsorption	105
Adsorption of Neutral Molecules	105
Adsorption Coupled with a Homogeneous Reaction in Solution	108
5 Two-step Reactions	109
Reactions in which there is an Adsorbed Intermediate	109
Reactions in which there is a Solution-soluble Intermediate	113
6 Passive Behaviour of Metals	115
The Influence of a Catalyst/Inhibitor on an Electrode Reaction	115
The Behaviour of Metals in the Passive and Transpassive Regions	118
7 Special Cases Due to Electrocrystallization Effects	118
Adatom Model	119
Two-dimensional Nucleation and Growth: Potentiostatic Case	121

Chapter 4 Electron-transfer Reactions: Part II 128
By P. P. Schmidt

1 Introduction	128
2 The General Treatment of Electron-transfer Reactions	129
Introduction	129
Summary of the Outer-sphere Theory	130
The Linear-response Theory of Chemical Reactivity	134
Activated-complex Theory	139
The Linear-response Analysis of the Levich–Dogonadze Theory	149

3 Generalizations of the Polar Continuum Theory of Outer-sphere Reactions: the Consideration of the First Solvation or Co-ordination Shell — 154
Refinements in the Model Representation and Associated Hamiltonian Operator — 154
The Polar-solvent Model and Models of the Source Charge Densities — 159
Molecular Vibrations of the Inner Solvation or Co-ordination shell — 174

4 The Collective-continuum Treatment of the Inner-sphere Complex — 185
The Collective Model of Highly Excited Molecular Vibrational States — 186

5 The Evaluation of the Expression for the Transition Probability — 208
The Dogonadze–Kuznetsov–Vorotyntsev Analysis — 209
Generating-function Techniques from the Theory of Molecular Radiationless Transitions — 217

6 Inner-sphere Reactions and Bridge-assisted Outer-sphere Transfers — 225

7 Conclusion — 237

Author Index — 243

1
Organic Electrochemistry – Synthetic Aspects

BY O. R. BROWN

During 1974 the number of publications in this area of research was similar to that of the previous year. The incidence of completely new reactions being reported is quite small, and the rapid progress of the past decade may be giving way to a period of consolidation, emphasis being placed upon increased understanding of detailed reaction mechanisms. One major area for future work will clearly be concerned with transition-metal organic compounds; great interest and activity already exist in the application of electrochemical methods to various porphyrin compounds.

The material included here and the organization of the coverage will be the same as in Volume 5. New textbooks on organic electrochemistry continue to appear.[1—3] A particularly important collection of papers, too numerous to review here individually, has been published in the Soviet Union.[4] Several useful review articles should be mentioned. Morris has given a timely survey of organometallic electrochemistry,[5] and the physical parameters involved in the control of organic electrode processes have been discussed.[6] Intramolecular cyclizations, an important class of electrode reactions, have been reviewed,[7] as has the electrochemistry of carbohydrates and their derivatives.[8] An omission from previous Reports is Lund's survey of the electrochemistry of the hydroxy-group.[9] Other aspects of organic electrochemistry to have been reviewed include the use of redox systems in reductions,[10] amino-acid syntheses,[11] electrochemistry of dithiolylium salts,[12] and the polarography of heterocyclic compounds.[13]

---

[1] F. Beck, 'Principles and Applications of Electro-organic Chemistry', Verlag Chemie, Weinheim, 1974.
[2] 'Techniques of Electro-organic Synthesis', ed. N. L. Weinberg, Wiley, New York, 1973.
[3] M. Rifi and F. H. Covitz, 'Introduction to Organic Electrochemistry', Dekker, New York, 1974.
[4] Novosti Elektrokhim. Org. Soedin. Tezisy. Dokl. Vses. Soveshch, Elektrokhim. Org. Soedin. 8th, ed. L. G. Feoktistov, 1973, 'Zinatne', Riga, 1973.
[5] M. D. Morris, *Electroanalyt. Chem.*, 1974, **7**, 79.
[6] M. Fleischmann and D. Pletcher, *Adv. Phys. Org. Chem.*, 1973, **10**, 155.
[7] M. Lacan and I. Tabakovic, *Kem. Ind.*, 1974, **23**, 225.
[8] M. Fedoronko, *Adv. Carbohyd. Chem. Biochem.*, 1974, **29**, 107.
[9] H. Lund, in 'Chemistry of the Hydroxyl Group', ed. S. Patai, Wiley, London, 1971, Part 1, p. 253.
[10] P. N. Anantharaman and H. V. K. Udupa, *Trans. Soc. Adv. Electrochem. Sci. Technol.*, 1974, **9**, 108.
[11] I. A. Avrutskaya and M. Ya. Fioshin, *Itogi Nauki Tekh. Electrokhim.*, 1974, **9**, 228.
[12] C. T. Pederson, *Angew. Chem. Internat. Edn*, 1974, **13**, 349.
[13] J. Stradins, V. Kadis, and S. Hiller, *Khim. geterotsikl. Soedinenii*, 1974, 147.

An educational experiment combining organic synthesis and electrochemistry has been devised.[14]

Abbreviations used throughout this chapter are as follows: DMF, NN-dimethylformamide; DMSO, dimethyl sulphoxide; THF, tetrahydrofuran; HMPT, hexamethylphosphoramide; TFA, trifluoroacetic acid; MeCN, acetonitrile; HOAc, acetic acid; TEAB, tetraethylammonium bromide; TBAP, tetra-n-butylammonium perchlorate; TMAF, tetramethylammonium fluoroborate; TPAT, tetrapropylammonium tosylate.

1 Reductions

Hydrocarbons.—A further spate of patents has been taken out by Japanese workers for the reduction of benzenoids to the 1,4-dihydro-compounds. Various solvent systems have been suggested for use with quaternary ammonium salt electrolytes and either mercury or amalgam cathodes.[15—20]

It has been shown that rigorous removal of electrophiles from various aprotic solvents (DMF, THF, HMPT, pyridine and, to a lesser extent, MeCN) enables the dianions derived from anthracene, 9,10-diphenylanthracene, benzophenone, or nitrobenzene to remain stable on the time-scale of cyclic voltammetry.[21] These conditions were achieved by the addition of neutral alumina to the solution. Quaternary ammonium ions Et_4N^+ and Bu_4N^+ are not rapid proton donors for the anthracene dianion at -30 °C.

Electrochemical methods have been used to determine the rates of protonation of polynuclear aromatic radical anions generated in DMSO.[22] Substituted phenols and 9-phenylfluorene were used as donors. Protonation rates correlated with optimum electron densities in the anions.

The reduction of 9,10-diethylideneacenaphthene, in DMF containing phenol, consumes more than 4 F mol^{-1}, indicating that there is reduction of the olefinic bonds and of the naphthalene system.[23] The thermodyamics and kinetics of disproportionation and protonation of radical anions generated electrochemically from tetraphenylethylene, triphenylethylene, 1,1-diphenylethylene, α-methylstilbene, and *trans*-stilbene have been studied in DMF and in HMPT.[24]

Reductions of the cycloheptatrienyl and 1,2,3-triphenylcyclopropenyl cations in MeCN each give two amperometric peaks, corresponding to the formation of the radicals and anions respectively.[25] The products resulting from controlled-potential

[14] P. E. Iverson, *J. Chem. Educ.*, 1974, **51**, 489.
[15] A. Misono and T. Osa, Ger. P. 1 668 471 (*Chem. Abs.*, 1974, **80**, 140 596).
[16] T. Hatayama and T. Yamamoto, Japan. P. 74 47 740 (*Chem. Abs.*, 1975, **82**, 155 576).
[17] T. Hatayama, Y. Hamano, K. Udo, and T. Yamamoto, Japan. P. 74 41 192 (*Chem. Abs.*, 1975, **82**, 147 142).
[18] T. Hatayama, K. Udo, Y. Hamano, and G. Inoue, Japan. P. 74 33 955 (*Chem. Abs.*, 1975, **82**, 117 890).
[19] M. Fujii, M. Moritake, and T. Okazaki, Japan. Kokai 74 56 952 (*Chem. Abs.*, 1974, **81**, 135 682).
[20] M. Fujii, M. Moritake, and T. Okazaki, Japan. Kokai 74 100 050 (*Chem. Abs.*, 1975, **82**, 86 181).
[21] B. S. Jensen and V. D. Parker, *J.C.S. Chem. Comm.*, 1974, 367.
[22] A. J. Fry and A. Schuttenberg, *J. Org. Chem.*, 1974, **39**, 2452.
[23] S. Valcher and A. M. Ghe, *J. Electroanalyt. Chem.*, 1974, **55**, 407.
[24] T. Troll and M. M. Baizer, *Electrochim. Acta*, 1974, **19**, 951.
[25] R. Breslow and R. F. Drury, *J. Amer. Chem. Soc.*, 1974, **96**, 4702.

electrolyses at each wave are the dimers. Even in the presence of hydroxylic solvents, *e.g.* HOAc, the anions react to form the dimer. The anion can in part be protonated, however, by the use of guanidinium perchlorate as supporting electrolyte (Scheme 1). The extraordinary appearance of the second cyclic voltammetric peak will clearly attract further investigation of this system.

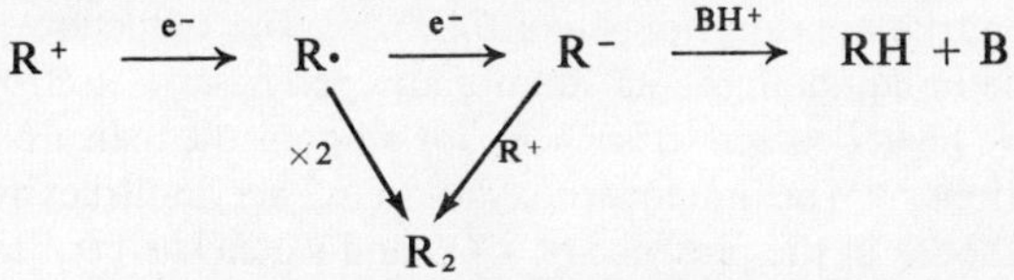

$$R^+ \xrightarrow{e^-} R\cdot \xrightarrow{e^-} R^- \xrightarrow{BH^+} RH + B$$

$$\times 2 \searrow \swarrow R^+$$

$$R_2$$

Scheme 1

Alicyclic polyene compounds can be reduced in HMPT–alcohol solvents containing inorganic electrolytes[26] (*e.g.* see Scheme 2). Cyclododeca-1,5,9-triene yields 2% diene, 94% cyclododecene, and 4% cyclododecane.

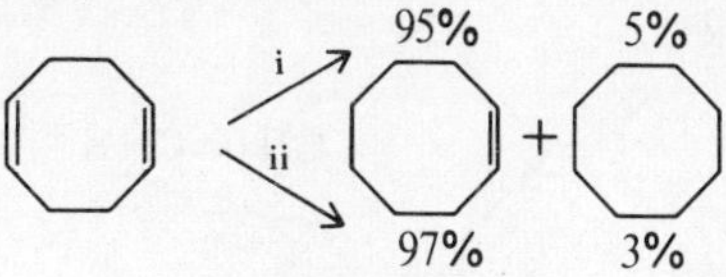

Reagents: i, EtOH–HMPT–LiCl; ii, PrOH–HMPT–LiCl

Scheme 2

Cathodic hydrogenation of olefinic bonds of several steroids has been carried out at palladized cathodes in acidic ethanol.[27] Even isolated olefinic bonds were reduced, but carbonyl groups were unaffected. Some of the reductions were stereoselective;

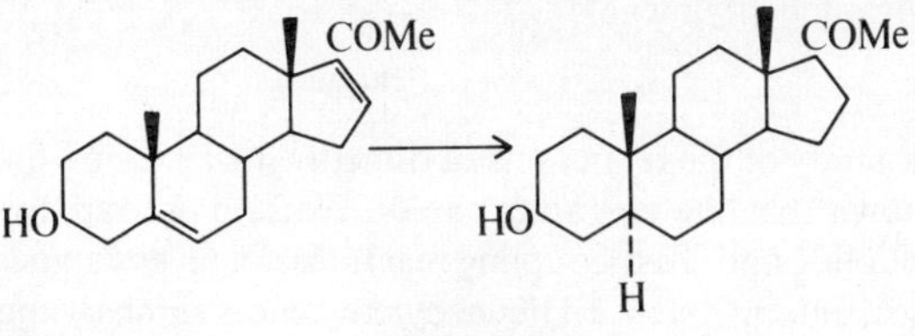

Scheme 3

for example, 3β-hydroxypregna-5,16-dien-20-one gave 3β-hydroxy-5α-pregnan-20-one in 90% yield (33% when purified) (Scheme 3). The olefinic bond in 1-chloroacenaphthylene-2-carboxylic acid can be reduced polarographically in aqueous alcoholic solution; functional groups are left intact.[28]

Rates of reaction of electrogenerated polycyclic hydrocarbon radical anions with organic halides, azides, tosylates, and sulphonamides have been studied by the polarographic method of catalytic currents.[29]

[26] S. Senoo and K. Saotome, Japan. P. 74 16 413 (*Chem. Abs.*, 1974, **81**, 169 220).
[27] K. Junghans, *Chem. Ber.*, 1974, **107**, 3191.
[28] A. M. Ghe and S. Valcher, *J. Electroanalyt. Chem.*, 1974, **55**, 417.
[29] H. Lund, M.-A. Michel, and J. Simonet, *Acta Chem. Scand.* (*B*), 1974, **28**, 900.

Activated Olefins.—The reduction of fumaric and maleic acids to succinic acid with 100% current yield at a rotating Pb cylinder in 5% H_2SO_4 has been patented.[30]

In 0.1N-H_2SO_4, acrylonitrile is converted into amines at a platinum electrode held at the reversible hydrogen potential.[31] At more positive potentials, propionaldehyde is produced in 56—7% yield.

The usual crop of patents concerned with electrohydrodimerization of acrylonitrile to adiponitrile appeared during 1974.[32—35] The reduction performed at a mercury cathode in liquid ammonia–ammonium perchorate electrolyte at -78 °C is claimed to give higher current efficiencies for adiponitrile than are obtained under aqueous conditions.[36] The monocarboxylation of acrylonitrile by its reduction at cadmium cathodes in the presence of CO_2 and water has been carried out with 48% current efficiency[37] (see below).

A crossed hydrocoupling of 1,1-dicyano-2-methylprop-1-ene and its analogues with acrylonitrile or methyl acrylate in a divided cell has been reported[38] (Scheme 4). Increased length and branching of the alkylidene chain caused current yields to fall.

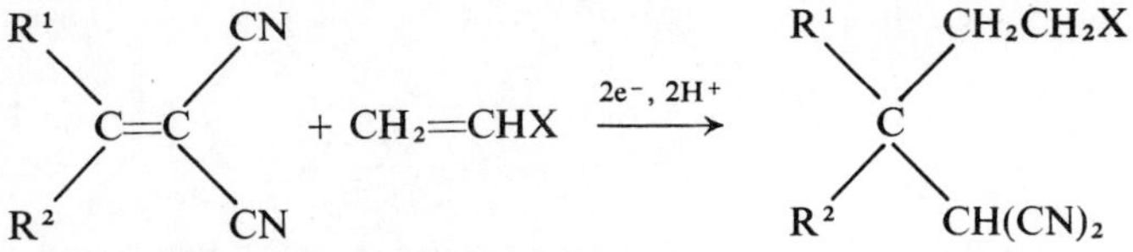

$$\begin{array}{c}R^1\\ \\R^2\end{array}\!\!\!C\!=\!C\!\!\!\begin{array}{c}CN\\ \\CN\end{array} + CH_2\!=\!CHX \xrightarrow{2e^-,\,2H^+} \begin{array}{c}R^1\\ \\R^2\end{array}\!\!\!C\!\!\!\begin{array}{c}CH_2CH_2X\\ \\CH(CN)_2\end{array}$$

(X = CN or CO_2Et)

(R^1 = Me, R^2 = Me, Et, Me_2CH, or Me_3C;
$R^1 = R^2$ = Et; or
$R^1\overset{\|}{C}R^2$ = cyclopentylidene)

Scheme 4

A mechanistic study of the reductions of dimethyl maleate and dimethyl fumarate in DMF has shown that the *cis* radical anion reacts more rapidly than the *trans*-isomer in self-coupling and cross-coupling reactions, and it spontaneously isomerizes to the *trans* radical anion.[39] *cis*-1,2-Dibenzoylethylene is similarly more reactive than the *trans*-isomer. The effect of ion-pairing upon the dimerization rate of diethyl

---

[30] H. V. Udupa, M. S. Venkatachalapathy, and R. I. Kanakam, Indian P. 102 485 (*Chem. Abs.*, 1974, **81**, 130 199).
[31] N. N. Gudeleva, N. A. Zakarina, and G. D. Zakumbaeva, *Izvest. Akad. Nauk Kazakh. S.S.R., Ser. khim.*, 1974, **24**, 75.
[32] C. R. Campbell, D. E. Danly, and W. H. Müller, Ger. Offen. 2 343 138 (*Chem. Abs.*, 1974, **81**, 49 287).
[33] D. L. Sadler and W. A. Heckel, Ger. Offen. 2 343 137 (*Chem. Abs.*, 1974, **81**, 49 288).
[34] F. N. Ruehlen, Ger. Offen. 2 338 341 (*Chem. Abs.*, 1974, **81**, 4347).
[35] J. H. Lester and J. S. Stewart, Ger. Offen. 2 344 294 (*Chem. Abs.*, 1974, **80**, 140 595).
[36] T. Chiba, Y. Takata, and A. Suzuki, *Chem. Letters*, 1974, 1241.
[37] D. A. Tyssee, Ger. Offen. 2 356 657 (*Chem. Abs.*, 1974, **81**, 44 737).
[38] Yu. D. Smirnov, S. K. Smirnov, and A. P. Tomilov, *Zhur. org. Khim.*, 1974, **10**, 1597.
[39] A. J. Bard, V. J. Puglisi, J. V. Kenkel, and A. Lomax, *Faraday Discuss. Chem. Soc.*, 1973, **56**, 353.

fumarate radical anions has been examined.[40] In MeCN the anion radicals are predominantly paired with sodium ions from the supporting electrolyte, but in DMSO a substantial number of the dimerizing anion radicals are unpaired. Radical anions derived from dimethyl fumarate, *trans*-stilbene, ethyl cinnamate, and anthracene all react with butyl bromide, in DMSO–TBAP.[41] In dry solvents, dialkylation occurs, but with added water monoalkylation takes place.

Stereoselective cathodic reductions of some compounds containing exocyclic double bonds have been investigated.[42] The less stable axial isomers are favoured,

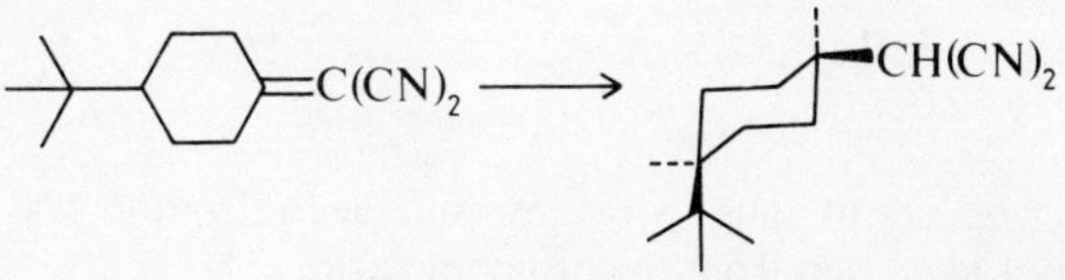

[Scheme 5 shows: a 4-t-butyl cyclohexylidene compound with =C(CN)$_2$ reducing to give axial CH(CN)$_2$ product: 67% axial, 33% equatorial (relative yields). A 4-t-butyl cyclohexanone reducing to give axial OH product: >98% axial (relative)]

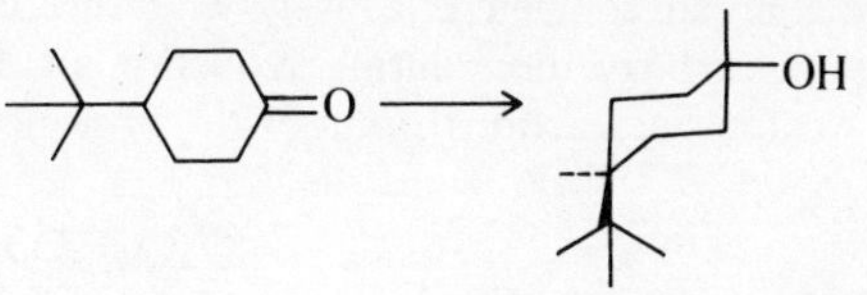

Scheme 5

but to different extents, according to the electrolytic medium (Scheme 5). Mercury and vitreous carbon cathodes gave similar results. Stereoselectivity was enhanced by the use of LiCl instead of TBAI and by the use of HOAc–DMF in place of ethanol.

The reduction of activated olefins at a mercury cathode in the presence of excess carbon dioxide has been examined in detail.[43] In MeCN–TEAT, disubstituted succinic acids were obtained, provided that the potential was held at a value corresponding to a two-electron wave (Scheme 6). A similar result is obtained even if

$$CH_2=CHX \xrightarrow{e^-} (CH_2=CHX)^{\cdot -} \xrightarrow{CO_2} {}^-O_2CCH_2\dot{C}HX$$
$$[X = CO_2Me, CN, \text{ or } C(O)Me] \qquad \downarrow e^-$$
$$^-O_2CCH_2CHXCO_2{}^- \xleftarrow{CO_2} {}^-O_2CCH_2\bar{C}HX$$

Scheme 6

carbon dioxide is reduced in preference to the olefin (Scheme 7). Monocarboxylation of the olefins to give β-substituted propionic acids proved to be possible in solutions containing water (~ 2.8 mol l^{-1}):

$$CH_2=CHCN + CO_2 + H_2O \xrightarrow{2e^-} OH^- + NCCH_2CH_2CO_2{}^-$$

[40] M. D. Ryan and D. H. Evans, *J. Electrochem. Soc.*, 1974, **121**, 881.
[41] S. Margel and M. Levy, *J. Electroanalyt. Chem.*, 1974, **56**, 259.
[42] R. J. Holman and J. H. P. Utley, *Tetrahedron Letters*, 1974, 1553.
[43] D. A. Tyssee and M. M. Baizer, *J. Org. Chem.*, 1974, **39**, 2819.

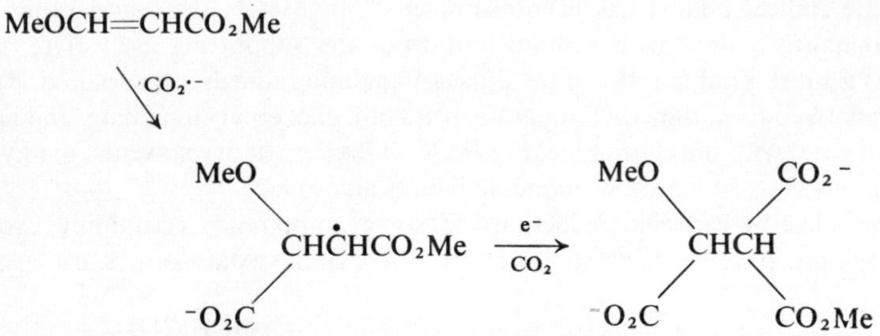

MeOCH=CHCO₂Me
  ↓ CO₂•⁻

$$\underset{^-O_2C}{\overset{MeO}{\diagdown}}CH\dot{C}HCO_2Me \xrightarrow[CO_2]{e^-} \underset{^-O_2C\quad\quad CO_2Me}{\overset{MeO\quad\quad CO_2^-}{\diagdown\quad\quad\diagup}}CHCH$$

Scheme 7

A flow system with continuous extraction and neutralization was developed to avoid evolution of hydrogen from the propionic acid.

The competitive reactions of radical anions formed by the reductions of activated olefins at potentials on their first (1e⁻) reduction wave have been studied.[44] Reaction with a second activated olefin can compete effectively with carboxylation. Thus dimerization can accompany carboxylation during reduction in MeCN–0.25M-TEAT (*e.g.* Scheme 8). Analogous reactions occur with those activated olefins

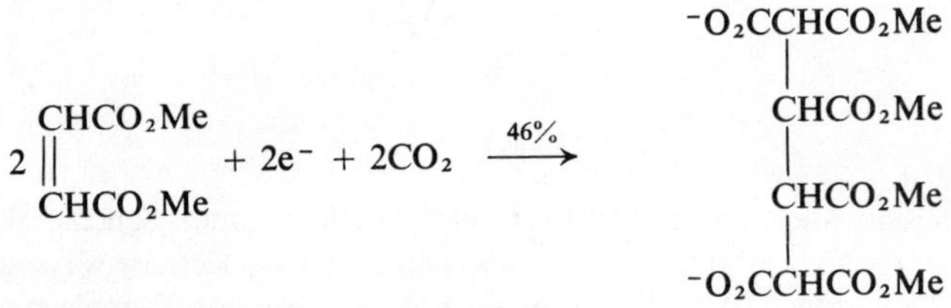

$$2 \underset{CHCO_2Me}{\overset{CHCO_2Me}{\|}} + 2e^- + 2CO_2 \xrightarrow{46\%} \begin{array}{c} ^-O_2CCHCO_2Me \\ | \\ CHCO_2Me \\ | \\ CHCO_2Me \\ | \\ ^-O_2CCHCO_2Me \end{array}$$

Scheme 8

which show only one two-electron reduction wave, provided that the olefin concentration is high. Bis-activated olefins, under similar conditions, can cyclize as they are carboxylated. Products obtained after treatment with methyl iodide indicated that reactions such as those shown in Scheme 9 are occurring.

αβ-Unsaturated Carbonyl Compounds.—Cathodic hydrogenations carried out at platinized platinum have included conversion of benzalacetone into benzylacetone and of chalcone into dihydrochalcone.[45, 46] On a mercury-poisoned electrode in H_2SO_4–aq. dioxan, the hydro-dimer was obtained from chalcone.[47] Chalcone in media of low acidity yields radical anions which dimerize.[48] Cinnamaldehyde, cinnamate esters, isophorone, *etc.* behave similarly.[49] The stability of radical anions

---

[44] D. A. Tyssee and M. M. Baizer, *J. Org. Chem.*, 1974, **39**, 2823.
[45] L. A. Taran, S. I. Berezina, L. G. Smolentseva, and V. A. Likhachev, *Soviet Electrochem.*, 1974, **10**, 749.
[46] L. A. Taran, S. I. Berezina, L. G. Smolentseva, and V. A. Likhachev, *Soviet Electrochem.*, 1973, **9**, 756.
[47] L. A. Taran, S. I. Berezina, L. G. Smolentseva, and Yu. P. Kitaev, *Izvest. Akad. Nauk. S.S.S.R., Ser. khim.*, 1974, 2611.
[48] V. N. Nikulin, N. M. Kargina, and Yu. M. Kargin, *Russ. J. Gen. Chem.*, 1974, **44**, 2479.
[49] E. Lamy, L. Nadjo, and J. M. Saveant, *J. Electroanalyt. Chem.*, 1974, **50**, 141.

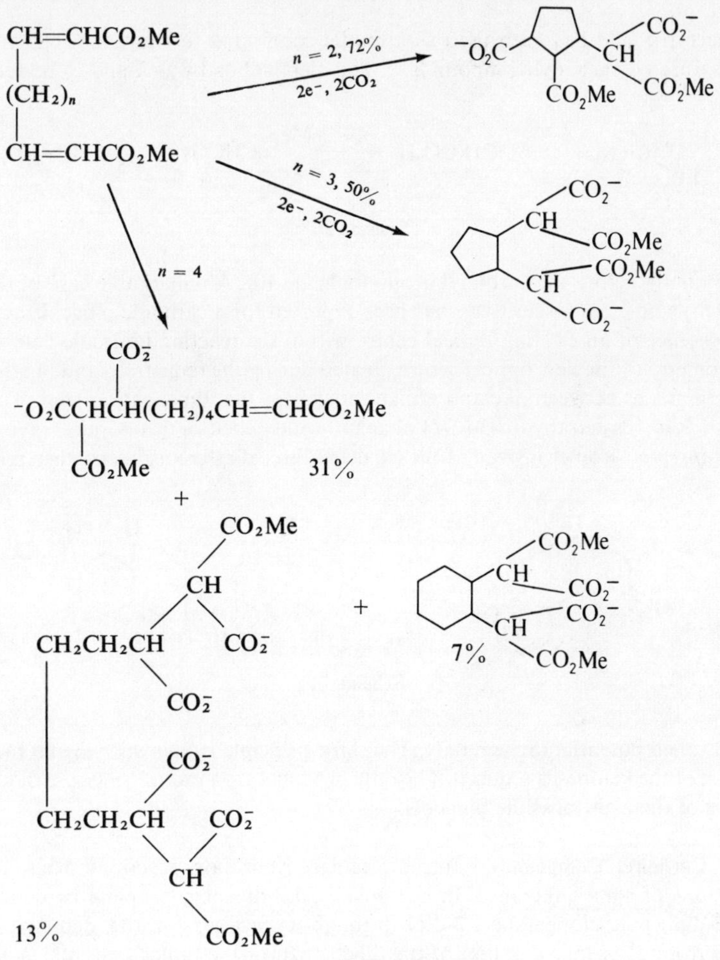

Scheme 9

derived from furan-substituted chalcones depends upon the cation of the supporting electrolyte.[50]

4,5-Di(2-furyl)octane-2,7-dione has been prepared by hydrodimerization of 4-(2-furyl)but-3-en-2-one in aqueous organic solvents containing H_2SO_4 electrolyte.[51] In DMSO–TBAP incorporating 0.05M-H_2O, trans-4,4-dimethyl-1-phenylpent-1-en-3-one radical anions formed at a mercury cathode dimerize at the 1- and 1'-positions.[52]

---

[50] A. Rusina, J. Volke, J. Cernak, J. Kovac, and V. Kollar, *J. Electroanalyt. Chem.*, 1974, **50**, 351.
[51] H. Satonaka, Z. Saito, and T. Shimura, *Kanagawa-Ken Kogyo Shikensho Kenkyu Hokoku*, 1974, 29.
[52] S. C. Rifkin and D. H. Evans, *J. Electrochem. Soc.*, 1974, **121**, 769.

When the water content is increased, competition from a first-order reaction becomes significant.

Ethacrynic acid has been used as a model compound to study the reduction of α,β-unsaturated carbonyl compounds.[53] The olefinic bond was found to be reduced

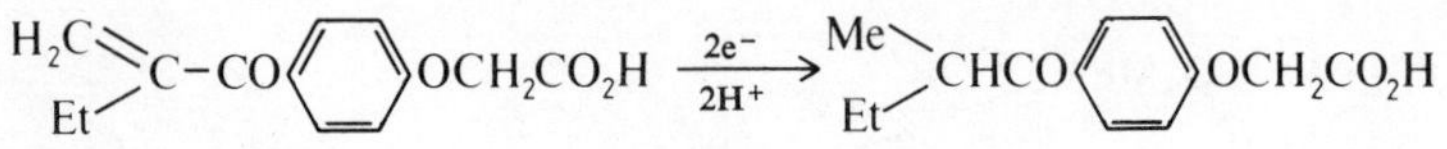

Scheme 10

preferentially before the carbonyl group (Scheme 10). A remarkable case of stereoselectivity and enantioselectivity has been reported for a cathodic pinacolization.[54] The presence of an existing optical centre within the reacting molecule determines the geometry of the new optical centre created during the reduction, and in addition it discriminates between like and unlike radicals in the dimerization step. Thus at pH 6, 1,9,10,10a-tetrahydro-3-(2H)-phenanthrone, when optically pure, gives only the *cis-threo-cis*-isomer instead of the expected three diastereoisomers when reduced

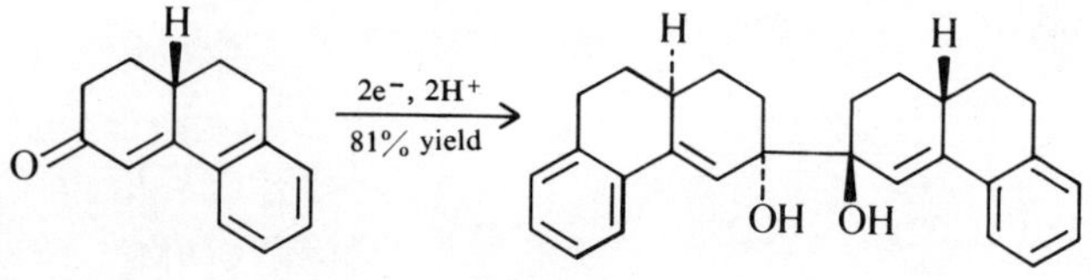

Scheme 11

at controlled potential (Scheme 11). The same principle is followed when a racemic mixture of the ketones is reduced. The only products are a racemic mixture of glycols instead of the eight possible pinacols.

Other Carbonyl Compounds.—Intramolecular cyclization reactions occur in the reductions of some diketones. In acetonitrile, 1,3-dibenzoylpropane is reduced to *cis*-1,2-diphenylcyclopentane-1,2-diol through cyclization of the dianion which results from disproportionation of the dibenzoylpropane radical anion.[55] A nearly quantitative yield can be obtained provided either that lithium ions are present to ion-pair with the organic anions or that acid is added continuously, so as to avoid strongly basic conditions.

1,2-Dibenzoylbenzene is reduced in monoglyme containing lithium salts to a radical anion which dimerizes or is reduced further, under more severe conditions, to an isobenzofuran.[56] In mineral acid the dimer dissociates to the reactant and the isobenzofuran (Scheme 12).

Reduction of benzophenone with TEA salts as supporting electrolytes in acetoni-

---

[53] L. Deshler and P. Zuman, *Analyt. Chim. Acta*, 1974, **73**, 337.
[54] E. Touboul and G. Dana, *Compt. rend.*, 1974, **278**, C, 1063.
[55] F. Ammar, C. P. Andrieux, and J. M. Saveant, *J. Electroanalyt. Chem.*, 1974, **53**, 407.
[56] J. A. Campbell, R. W. Koch, J. V. Hay, M. A. Ogliaruso, and J. F. Wolf, *J. Org. Chem.*, 1974, **39**, 146.

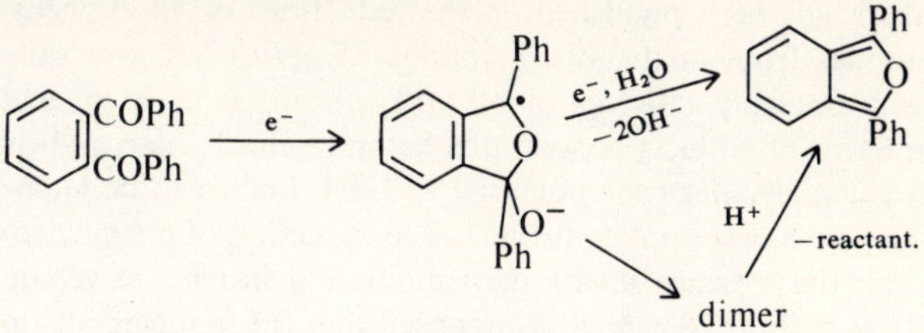

Scheme 12

trile containing acetic anhydride as the added electrophile in place of the more familiar proton donor leads to an acylation reaction (Scheme 13).[57]

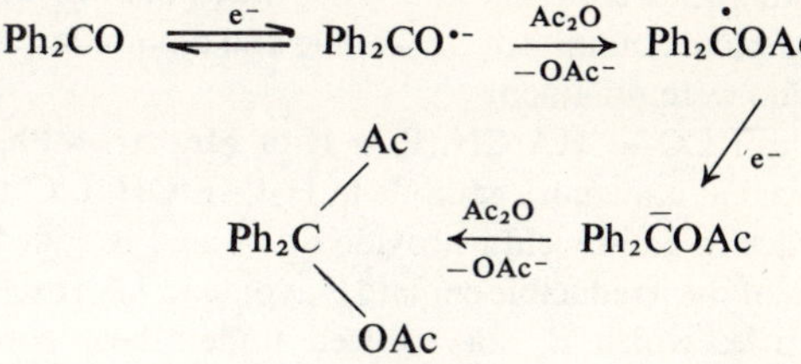

$$Ph_2CO \xrightleftharpoons{e^-} Ph_2CO^{\cdot -} \xrightarrow[-OAc^-]{Ac_2O} Ph_2\dot{C}OAc$$

$$Ph_2C\begin{matrix}Ac\\ \diagup\\ \diagdown\\ OAc\end{matrix} \xleftarrow[-OAc^-]{Ac_2O} Ph_2\bar{C}OAc \nwarrow e^-$$

Scheme 13

Several papers have described the familiar cathodic reductions of aromatic carbonyl compounds to alcohols or pinacols. Thus acetylferrocene hydrodimerizes to 2,3-diferrocenylbutane-2,3-diol, which dehydrates in two stages to 2,3-diferrocenylbuta-1,3-diene.[58]

In the absence of added proton donors, 2-benzoylthiophen[59] and 2-acetylthiophen[60] are reduced to resins at mercury in MeCN–TEAP. However, in the presence of acetic or benzoic acids, the corresponding pinacols are formed. When the added proton donor is water or phenol, then 2-benzoylthiophen yields thienyl-2-phenylmethanol at the potential of the second wave.[61]

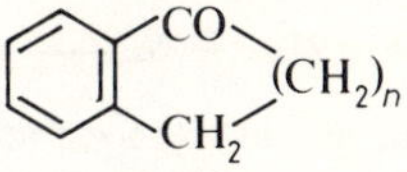

(1)

Reductions of three benzocyclenones (1) on mercury have been examined in aqueous methanol.[62] Indanone ($n = 1$) yields the secondary alcohol, tetralone ($n = 2$) produces the pinacol, and benzosuberone ($n = 3$) gives the pinacol under mild conditions and its alcohol at more negative potentials.

Dimerization rates of substituted benzaldehyde radical anions, electrogenerated at a platinum cathode in sulpholane–TBAP, have been measured by cyclic voltammetry.[63]

[57] J. J. Curphey, L. D. Trivedi, and T. Layloff, *J. Org. Chem.*, 1974, **39**, 3831.
[58] M. Lacan and Z. Ibrisagic, *Croat. Chem. Acta*, 1974, **46**, 107.
[59] P. Foulatier, J. P. Salaün, and C. Caullet, *Compt. rend.*, 1974, **279**, C, 779.
[60] P. Foulatier and C. Caullet, *Compt. rend.*, 1974, **279**, C, 25.
[61] P. Foulatier, J. P. Salaün, and C. Caullet, *Compt. rend.*, 1974, **279**, C, 679.
[62] V. Toure, M. Levy, and P. Zuman, *J. Electroanalyt. Chem.*, 1974, **56**, 285.
[63] N. R. Armstrong, R. K. Quinn, and N. E. Vanderborgh, *Analyt. Chem.*, 1974, **46**, 1759.

Further work has been published on the reductions of alkyl phenyl ketones at mercury cathodes from methanolic solutions of optically active salts.[64] Product distributions of optically active carbinols and optically inactive pinacols have been explained in terms of adducts between the ketone radical anion and the electrolyte cation. Unfortunately, electrode potential and pH, known to be important factors in ketone reductions, were not controlled. The reduction of phenylglyoxylic acid to mandelic acid in the presence of various alkaloids (*e.g.* brucine, strychnine) has led to optical yields as high as 20%.[65] It is supposed that the inducing alkaloid forms an adsorbed complex with the carbanion, causing it to retain its configuration at least partially whilst undergoing protonation.

The electroreduction of aliphatic carbonyl compounds has also received attention. Various aliphatic ketones were reduced at Ni–Pd cathodes in 5% aqueous NaOH solution containing equal amounts of the ketone and ammonia.[66] High yields of the corresponding amine were obtained.

The β-keto-nitriles RCO—CHArCN (R = H or Me; Ar = Ph, 1-naphthyl, or 2-naphthyl) are reduced at a mercury cathode in H_2O–EtOH–LiCl to the corresponding β-hydroxy-nitriles in good yields, provided that acid is added progressively, so that decomposition of the irreducible enolate is avoided.[67] An exception is naphthyl-2′-α-formylacetonitrile, which is not reduced under these conditions. Propionaldehyde in an aqueous alkaline phosphate buffer solution is converted, at a lead cathode, into 2-methylpentane-1,3-diol.[68]

Some higher alkyl methyl ketones have been hydrocoupled with acrylonitrile or ethyl acrylate at a mercury cathode in acidic aqueous solutions.[69] Good yields of the γ-lactones are obtained from acrylonitrile with ketones up to 2-heptanone, whereas 2-undecanone does not couple at all. 2-Octanone gives moderate yields. Patents have been issued for the corresponding coupling reactions between acrylonitrile and lower aliphatic aldehydes.[70] Less acidic conditions and graphite cathodes can be used with these systems.[71]

Ketyl radicals derived from acetone have been successfully hydrocoupled with chlorotrifluoroethylene:[72]

$$Me_2CO + CF_2=CFCl + 2H^+ + 2e^- \longrightarrow Me_2C(OH)CF_2CHFCl$$

Mercury cathodes were used in acidic aqueous solutions. Several simple electrocatalytic hydrogenations of ketones and aldehydes to alcohols have been reported. The effects of electrode potential and Pt–Ir alloy catalyst composition upon acetone reduction were studied.[73] Non-aqueous solutions of TBAB have been used to reduce

---

[64] L. Horner and D. Degner, *Electrochim. Acta*, 1974, **19**, 611.
[65] M. Jubault, E. Raoult, and D. Peltier, *Electrochim. Acta*, 1974, **19**, 865.
[66] I. V. Kirilysus, V. L. Mirzoyan, and D. V. Sokol'skii, *Electrochim. Acta*, 1974, **19**, 859.
[67] G. Le Guillanton, *Bull. Soc. chim. France*, 1973, 3458.
[68] A. P. Tomilov, B. L. Klyuev, V. D. Nechepurnoi, and N. Sh. Fuks, Russ. P. 419 502 (*Chem. Abs.*, 1974, **80**, 145 382).
[69] A. Froeling, *Rec. Trav. chim.*, 1974, **93**, 47.
[70] N. Seko, A. Yontanza, Y. Takahashi, F. Yamamoto, K. Harada, and S. Matsumoto, Japan P. 74 20 298 (*Chem. Abs.*, 1975, **82**, 139 427).
[71] A. P. Tomilov, B. L. Klyuev, V. D. Nechepurnoi, and N. Sh. Fuks, Russ. P. 427 927 (*Chem. Abs.*, 1974, **81**, 49 292).
[72] F. Liska, V. Dedek, and M. Nemec, *Coll. Czech. Chem. Comm.*, 1974, **39**, 689.
[73] A. D. Semenova, N. V. Kropotova, and G. D. Vovchenko, *Soviet Electrochem.*, 1974, **10**, 916, 1178.

2-ethylhex-2-enal, acetophenone, benzophenone, 2-butanone, and cyclohexanone.[74]

Platinum cathodes in methanol–NaOMe electrolytes have been used to reduce various α-oxo-esters to the hydroxy-esters (Scheme 14).[75] This reduction did not occur for R = CO_2Et, 4-MeOC_6H_4, or $4\text{-}O_2NC_6H_4$. Humulone has been converted into 4-deoxyhumulone in buffered aqueous methanol (pH 4.2).[76]

$$RCOCO_2Me \longrightarrow RCHOHCO_2Me$$

(R = Pr^n, Me, Ph, 4-ClC_6H_4, or 4-MeOCOC_6H_4)

Scheme 14

An interesting novel synthesis takes place when extreme cathodic conditions are applied to solutions of acetophenone, benzophenone, or benzaldehyde in MeCN containing TEAF electrolyte and 0.01% water.[77] Highly basic conditions near the

$$RCOPh + MeCN \longrightarrow PhCR(OH)CH_2CN \xrightarrow{-H_2O} PhCR\!=\!CHCN$$

$$\downarrow {\scriptstyle 35-45\%\ |\ MeCN}$$

$$PhCR(CH_2CN)_2$$

Scheme 15

mercury cathode promote coupling (Scheme 15). The intermediate cinnamonitrile can be reduced cathodically:

$$PhCR\!=\!CHCN + 2e^- + 2H^+ \xrightarrow{30-50\%} PhCHRCH_2CN$$

This last reaction accounts essentially for the total charge passed.

Such extremes of pH can be avoided by the use of an undivided cell with a platinized platinum anode over which hydrogen is bubbled.[78] Under these conditions acetophenone can be reduced to its hydro-dimer in MeCN–TEAP.

Quaternary Salts.—N-Alkyl-2,3,4,6-tetra-arylpyridinium perchlorate is reduced in yields of 50—60% to the 5,6-dihydro-product at a mercury cathode in MeCN–TEAP with phenol as proton donor.[79] 1-Methylnicotinamide in aqueous media forms a free radical reversibly.[80] This radical can form the 6,6'-dimer or, under severe cathodic conditions, can be reduced further to give the 1,6-dihydro-product. Nicotinamide mononucleotide behaves analogously.[80,81] Several 2-cyano-Δ^7-hexahydroindolium salts have been reduced at mercury cathodes in neutral aqueous buffer solutions to give either tertiary 2-(β-cyanoethyl)cyclohexylamines or 2-(β-cyanoethyl)cyclohexanone and the secondary amine (Scheme 16).[82] In DMF

---

[74] G. Filardo, M. Galluzzo, B. Giannici, and R. Ercoli, *J.C.S. Dalton*, 1974, 1787.
[75] B. Wladislaw, V. L. Pardini, and H. Vertier, *J.C.S. Perkin II*, 1974, 625.
[76] K. L. Schroeder, *Brauwissenschaft*, 1974, **27**, 274.
[77] E. M. Abbot, A. J. Bellamy, and J. Kerr, *Chem. and Ind.*, 1974, 828.
[78] J. M. Saveant and Su Khac Binh, *J. Electroanalyt. Chem.*, 1974, **50**, 417.
[79] M. Libert and C. Caullet, *Compt. rend.*, 1974, **278**, C, 495.
[80] C. O. Schmakel, K. S. V. Santhanam, and P. J. Elving, *J. Electrochem. Soc.*, 1974, **121**, 1033.
[81] H. Hanschmann, *Studia Biophys.*, 1974, **45**, 183.
[82] P. E. Iversen and J. Ø. Madsen, *Tetrahedron*, 1974, **30**, 3477.

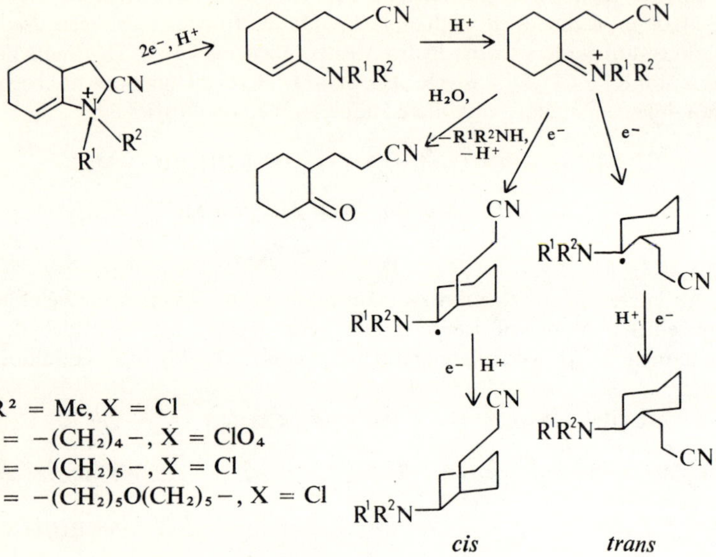

a $R^1 = R^2 = Me$, $X = Cl$
b $R^1R^2 = -(CH_2)_4-$, $X = ClO_4$
c $R^1R^2 = -(CH_2)_5-$, $X = Cl$
d $R^1R^2 = -(CH_2)_5O(CH_2)_5-$, $X = Cl$

Scheme 16

solution, 7,9-dimethylpurinium cations are reduced progressively to the 1,6-dihydro- and 1,2,3,6-tetrahydro-derivatives.[83]

Hydroxytriphenylarsonium perchlorate in acetonitrile is reduced to a free radical, which decomposes on platinum (Scheme 17).[84] If cathodes having a high hydrogen

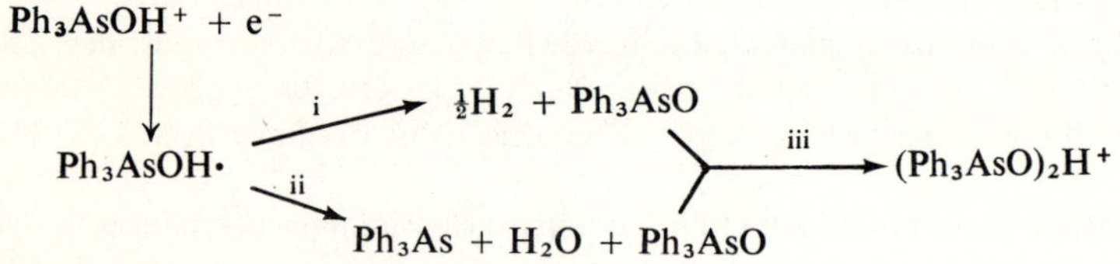

Reagents: i, Pt; ii, Hg or C; iii, Ph_3AsOH^+

Scheme 17

overvoltage are used, the radical disproportionates. At more negative potentials the radical is reduced further (Scheme 18). On cathodes of Hg, Pb, Cd, or Sn, alkyl-

$$Ph_3AsOH^\bullet \xrightarrow[-Ph_3As]{e^-} OH^- \xrightarrow[-H_2O]{i,} Ph_3AsO \xrightarrow{i} (Ph_3AsO)_2H^+$$

Reagents: i, Ph_3AsOH^+

Scheme 18

[83] Z. N. Timofeeva, L. S. Tikhonova, Kh. L. Muravich-Aleksandr, and A. V. El'tsov, *Russ. J. Gen. Chem.*, 1974, **44**, 1976.
[84] G. Schiavon, S. Zecchin, G. Cogoni, and G. Bontempelli, *J. Electroanalyt. Chem.*, 1974, **52**, 459.

triphenyl-arsonium or -phosphonium salts are reduced to arsines and phosphines.[85] The ratio of possible products depends upon the alkyl group, temperature, cathode material, and solvent (Scheme 19).

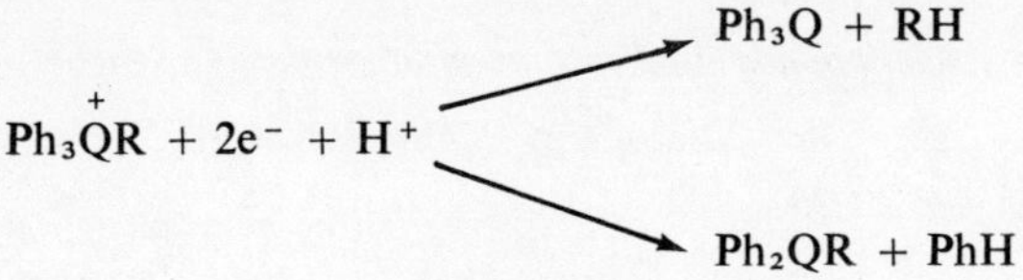

$$Ph_3\overset{+}{Q}R + 2e^- + H^+ \longrightarrow \begin{cases} Ph_3Q + RH \\ Ph_2QR + PhH \end{cases}$$

Scheme 19

Triphenylphosphine phenyl imide in aprotic solvents is reduced to aniline and triphenylphosphine, with some diphenylphosphinic acid, at mercury electrodes:[86]

$$Ph_3P=NPh + 2e^- + 2H^+ \longrightarrow PhNH_2 + PPh_3$$

Nitro-compounds.—Iversen and co-workers have involved the reduction products of various nitro-compounds in coupling reactions with suitable adducts.[87, 88] Nitro radical anions derived from $MeNO_2$, Bu^tNO_2, $PhNO_2$, $p\text{-}MeC_6H_4NO_2$, and $m\text{-}MeC_6H_4NO_2$ couple with acetic anhydride (Scheme 20).[87] A previously known

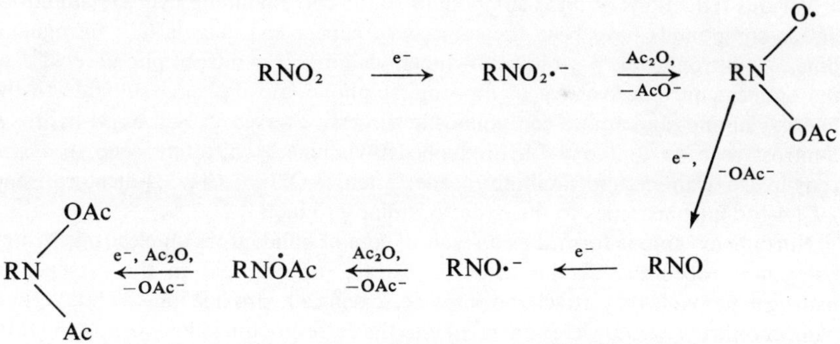

$$RNO_2 \xrightarrow{e^-} RNO_2^{\cdot -} \xrightarrow[-AcO^-]{Ac_2O} RN\begin{smallmatrix}O\cdot\\ \\OAc\end{smallmatrix}$$

$$RN\begin{smallmatrix}OAc\\ \\Ac\end{smallmatrix} \xleftarrow[-OAc^-]{e^-, Ac_2O} R\dot{N}OAc \xleftarrow[-OAc^-]{Ac_2O} RNO^{\cdot -} \xleftarrow{e^-} RNO$$

Scheme 20

reaction has been extended in scope; reduction of 2,2′-dinitrobiphenyl in aqueous acetate buffer solutions (pH ≈ 5) containing the cosolvents ethanol, methanol, or

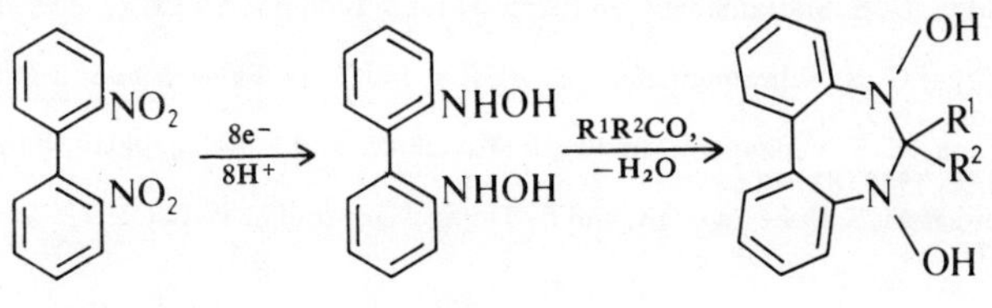

Scheme 21

---

[85] L. Horner, J. Roeder, and D. Gammel, *Phosphorus*, 1973, **3**, 175.
[86] Chong Min Pak and W. M. Gulick, *Taehan Hivahak Hoechi*, 1974, **18**, 341.
[87] L. H. Klemm, P. E. Iversen, and H. Lund, *Acta Chem. Scand.* (*B*), 1974, **28**, 593.

DMF, followed by the addition of various ketones or aldehydes, leads to substituted diazepines (Scheme 21).[88] Carbonyl compounds coupled with electrogenerated 2,2′-dihydroxyaminobiphenyl are given in Table 1.

Table 1 *Carbonyl compounds used in diazepine synthesis (Scheme 21)*

R^1	Me	Et	Pr^n	Bu^n	4-py	4Me-2-thiazolyl	Me	Et	Pr^n
R^2	H	H	H	H	H	H	Me	Me	Me

An intramolecular coupling by a reduction product of *o*-nitro-anilines occurs when the aniline is *N*-substituted by one alkylmethyl group and one alkyl group (Scheme 22).[89] Some diamine is formed in addition to the benzimidazole. The *para*-isomer gives the diamine exclusively.

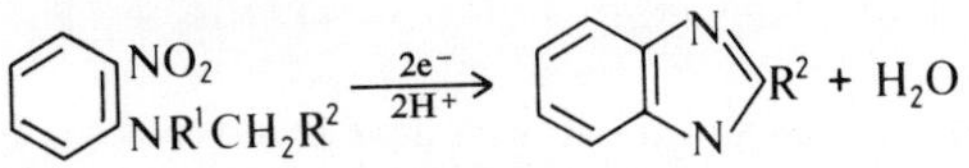

$$\underset{NR^1CH_2R^2}{\overset{NO_2}{\bigcirc}} \xrightarrow[2H^+]{2e^-} \underset{N}{\overset{N}{\bigcirc}} R^2 + H_2O$$

Scheme 22

The electrosynthesis of phenylalanine from *o*-nitrocinnamate esters described in earlier Reports has now been patented.[90]

Various reductions of nitro-compounds to the corresponding hydroxylamino- or amino-compounds have been the subjects of papers and patents (*viz.* nitroguanidine,[91] *p*-nitrophenol,[92] *p*-nitro-*NN*-dimethylaniline,[93] *p*-nitrosulphonic acid[94] to the amines; nitro-derivatives of diphenyl sulphone and diphenyl sulphide to the hydroxylamino- and amino-compounds in successive waves;[95] 3-alkoxy-4-hydroxy-β-nitrostyrene to 3-alkoxy-4-hydroxyphenethylamine;[96] nitroferrocene to ferrocenylhydroxylamine and aminoferrocene;[97] and 3-OH-, 3-Cl-, 3,4-dichloro-, and 2-Cl-4-Me-nitrobenzenes to the hydroxylamino-products[98]).

Nitrophenyl anions formed as a result of loss of halide from an electrogenerated halogenonitrobenzene dianion usually abstract protons, but in the presence of hydrogen peroxide they attack the peroxide, leaving a hydroxide ion.[99] Thus various halogenonitrobenzenes were converted into the corresponding nitrophenols in DMF solution.

[88] J. Becher and P. E. Iversen, *Acta Chem. Scand. (B)*, 1974, **28**, 539.
[89] A. Darchen and D. Peltier, *Bull. Soc. chim. France*, 1974, 673.
[90] E. V. Zaporozhets, I. A. Avrutskaya, M. Ya. Fioshin, K. K. Babievski, and V. M. Belnikov, Russ. P. 432 132 (*Chem. Abs.*, 1974, **81**, 91 930).
[91] H. V. Udupa, G. S. Subramanian, and K. S. Udupa, Indian P. 99 181 (*Chem. Abs.*, 1974, **81**, 57 659).
[92] H. V. Udupa, G. S. Subramanian, K. S. Udupa, and T. D. Belakrishnan, Indian P. 126 677 (*Chem. Abs.*, 1975, **82**, 36 695).
[93] H. V. Udupa, M. S. Venkatachalapathy, R. Kanakam, and S. Balagopalan, Indian P. 130 295 (*Chem. Abs.*, 1975, **82**, 36 696).
[94] H. V. Udupa, G. S. Subramanian, and S. Thangavelu, Indian P. 119 015 (*Chem. Abs.*, 1975, **82**, 36 697).
[95] W. Darlewski, *Roczniki. Chem.*, 1974, 48.
[96] M. Ono and M. Matsuoka, Japan. P. 74 13 777 (*Chem. Abs.*, 1975, **82**, 16 549).
[97] L. N. Nekrassov, *Faraday Discuss. Chem. Soc.*, 1973, **56**, 308.
[98] P. Jaeger and B. Zeeh, Ger. Offen., 2 262 851 (*Chem. Abs.*, 1974, **81**, 120 636).
[99] I. M. Sosonkin, T. K. Polynikova, and G. N. Strogov, *Doklady Akad. Nauk. S.S.S.R.*, 1974, **218**, 130.

Trifluoromethyl derivatives of nitrobenzenes are reduced in DMF, *via* the nitro radical anions, to the substituted azo- and azoxy-benzene products.[100]

Other Nitrogen-containing Compounds.—Hydrazo products obtained in the polarographic reduction of pyridylazo- and thiazolylazo-compounds have been shown to have a smaller propensity for disproportionation than do their homo-aromatic analogues.[101]

Coarse graphite cathodes in a diaphragm flow cell have been found to give improved chemical and current yields of phenylhydrazine from diazoaminobenzene.[102]

Radicals formed during the electroreduction of aryldiazonium salts have been trapped by nitrones to produce stable nitroxide radicals. A 0.01 mol l^{-1} solution of α-phenyl-*N*-t-butylnitrone, which is electroinactive over a wide potential range, in MeCN–0.1M-TBAP was used with a mercury pool cathode (Scheme 23).[103]

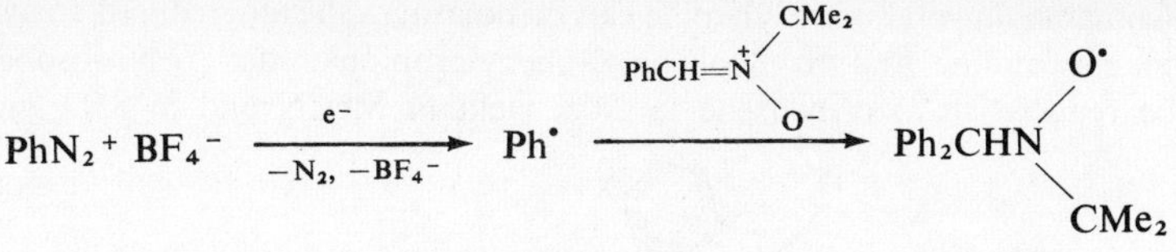

$$PhN_2^+ BF_4^- \xrightarrow[-N_2, -BF_4^-]{e^-} Ph^\bullet \xrightarrow{PhCH=\overset{+}{N}(CMe_2)/O^-} Ph_2CHN\begin{pmatrix}O^\bullet\\CMe_2\end{pmatrix}$$

Scheme 23

Lead cathodes have been found to give optimum yields of dimethylhydrazine when dimethylnitrosamine is reduced in aqueous H_2SO_4.[104]

Polarographic reduction of diazoacetone in 90% ethanolic solutions (pH < 5) leads to aminoacetone, which cyclizes to 2,5-dimethyldihydropyrazine.[105] This product is believed to yield the piperazine at more negative potentials. At 5 < pH < 7, mild reducing conditions cause formation of the hydrazone MeCOCH:NNH$_2$.

In DMF solution, 1-arylidene-2,2,2-trimethylhydrazinium iodides are reduced to the benzonitriles (Scheme 24), which are themselves reducible at more negative potentials.[106]

$$RC_6H_4CH=\overset{+}{N}NMe_3 \xrightarrow[-NMe_3]{e^-} RC_6H_4CH=\overset{\bullet}{N} \longrightarrow RC_6H_4CN + \tfrac{1}{2}H_2$$

Scheme 24

Polarographic reductions of naphthofurazan and naphthofuroxan derivatives have been conducted in aqueous and DMF media, with the formation of diaminotetralin (Scheme 25).[107]

[100] D. L. Dickerson and J. W. Rogers, *Analyt. Chim. Acta*, 1974, **71**, 433.
[101] T. M. Florence, D. A. Johnson, and G. E. Batley, *J. Electroanalyt. Chem.*, 1974, **50**, 113.
[102] J. Cramer and H. Alt, Ger. Offen., 2 305 574 (*Chem. Abs.*, 1974, **81**, 135 690).
[103] A. J. Bard, J. C. Gilbert, and R. C. Goodin, *J. Amer. Chem. Soc.*, 1974, **96**, 620.
[104] P. Thirunavukkarasu, P. Subbiah, K. S. Udupa, and H. V. K. Udupa, *Proc. Semin. Electrochem. 14th*, 1973, 97.
[105] G. Paliani, S. Sorrisco, and S. M. Murgia, *Z. Naturforsch.*, 1974, **29b**, 489.
[106] V. Kh. Ivanova, L. N. Orlova, V. I. Savin, and Yu. P. Kitaev, *Izvest. Akad. Nauk S.S.S.R., Ser. khim.*, 1974, 2474.
[107] Z. I. Fodiman, Z. V. Todres, and E. S. Levin, *Zhur. Vsesoyuz. Khim. Obshch. im D.I. Mendeleeva.*, 1974, **19**, 236.

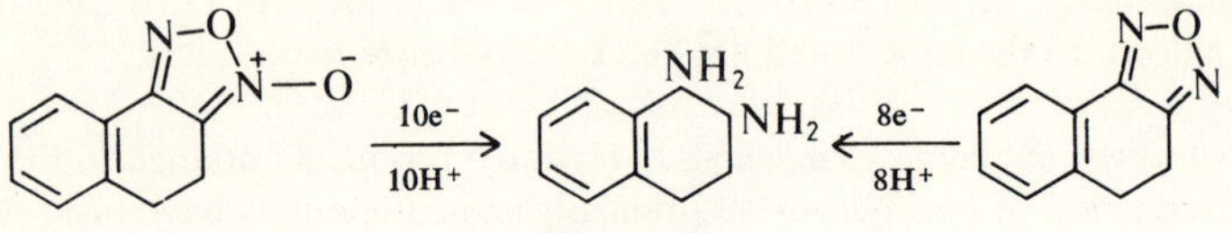

Scheme 25

β-Keto-α-amino-acid esters can be reduced to β-keto-acid esters in aqueous methanolic HCl (e.g. Scheme 26).[108]

$$\text{PhCOCMe(NH}_2\text{)CO}_2\text{Me} \xrightarrow[-NH_3]{2e^-,\ 2H^+} \text{PhCOCHMeCO}_2\text{Me}$$

Scheme 26

N-Cinnamoylphenylhydroxylamine can be polarographically reduced stepwise to N-cinnamoylaniline and to N-phenyl-2-phenylpropionamide.[109] Nitrosobenzene can be reduced to azoxybenzene in 73% yield in MeCN and in 85% yield in DMF.[110]

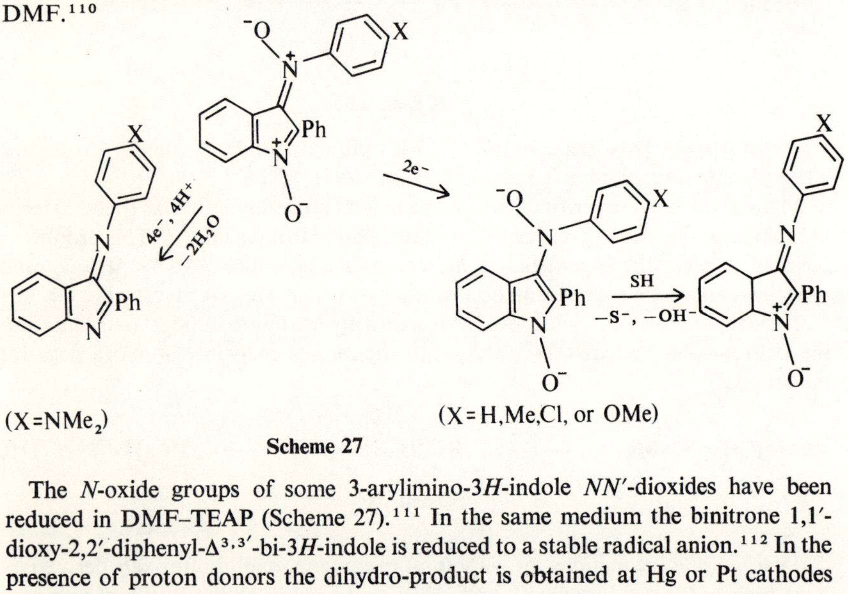

Scheme 27

The N-oxide groups of some 3-arylimino-3H-indole NN'-dioxides have been reduced in DMF–TEAP (Scheme 27).[111] In the same medium the binitrone 1,1'-dioxy-2,2'-diphenyl-$\Delta^{3,3'}$-bi-3H-indole is reduced to a stable radical anion.[112] In the presence of proton donors the dihydro-product is obtained at Hg or Pt cathodes (Scheme 28).

Reduction of the N-methyl derivatives of 2-pyrimidone and 4-methyluracil gives rise to the hydro-dimer 6,6'-bis(3,6-dihydro-2-pyrimidone), which is also obtained from the unsubstituted pyrimidone.[113]

---

[108] M. Miyoshi, M. Matsuoka, K. Matsumoto, and M. Suzuki, Japan. Kokai 74 47 317 (*Chem. Abs.*, 1974, **81**, 120 249).
[109] D. S. Sopilnyak, Yu. I. Usatenko, and S. N. Schcherbak, *Izvest. V.U.Z. Khim. i khim. Tekhnol.*, 1974, **17**, 352.
[110] M. R. Asirvatham and M. D. Hawley, *J. Electroanalyt. Chem.*, 1974, **57**, 179.
[111] R. Andruzzi, I. Carelli, A. Trazza, P. Bruni, and M. Colonna, *Tetrahedron*, 1974, **30**, 3741.
[112] R. Andruzzi, A. Trazza, and P. Bruni, *J. Electroanalyt. Chem.*, 1974, **51**, 341.
[113] B. Czochralska, M. Wrona, and D. Shugar, *Bioelectrochem. Bioenerg.*, 1974, **1**, 40.

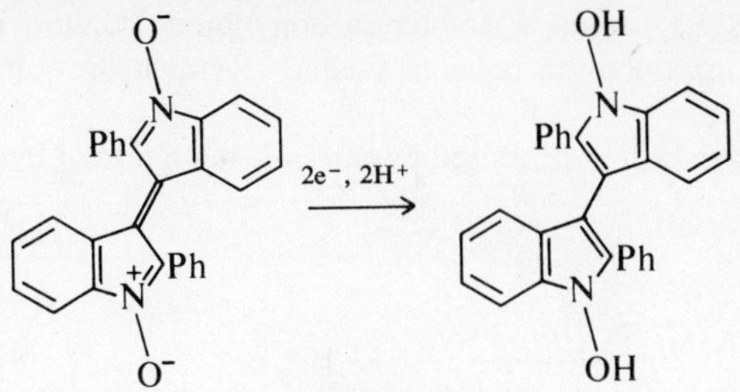

Scheme 28

Pyrazines in alkaline aqueous–organic media are reduced to 1,4-dihydropyrazines, which isomerize to the 1,2- or 1,6-dihydropyrazines.[114]

Nicotinamide and *N*-methylnicotinamide can be reduced under mild conditions to the 6,6′-dihydro-dimer and at more negative potentials to 1,6-dihydropyridine (*cf.* ref. 80).[115] In DMF, MeCN, or DMSO, 6-substituted purines are reduced on mercury to the hydro-dimers in the absence of a proton donor. In the presence of a proton donor, purine and 6-methylpurine each give the dihydro-purine at less negative potentials and the tetrahydro-purine at more negative potentials.[116] 6-Dimethylaminopurine, 6-methylaminopurine, 6-methoxypurine, and adenine each proceed to the tetrahydro-product in a single step when a proton donor is present.[117]

At an amalgamated copper cathode, benzotriazole is reduced in $2N$-H_2SO_4 to give 2,3-diaminophenazine.[118]

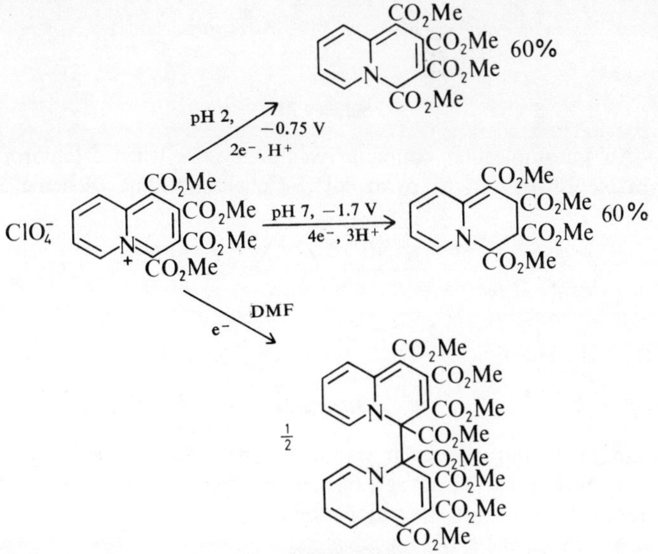

Scheme 29

[114] J. Armand, K. Chekir, and J. Pinson, *Canad. J. Chem.*, 1974, **52**, 3971.
[115] C. O. Schmakel, K. S. V. Santhanam, and P. J. Elving, *J. Electrochem. Soc.*, 1974, **121**, 345.
[116] K. S. V. Santhanam and P. J. Elving, *J. Amer. Chem. Soc.*, 1974, **96**, 1653.
[117] T. Yao and S. Musha, *Bull. Chem. Soc. Japan*, 1974, **47**, 2650.
[118] A. A. Rysakov, Yu. M. Loshkarev, V. F. Vargalyuk, and V. A. Omel'chenko, *Soviet Electrochem.*, 1974, **10**, 1785.

Reduction of 1,2,3,4-tetrakis(methoxycarbonyl)quinolizinium perchlorates at a mercury cathode produces, in aqueous media, several hydroquinolizines (Scheme 29).[119]

Amino-acids have been synthesized by cathodic reduction of hydantoins (Scheme 30).[120]

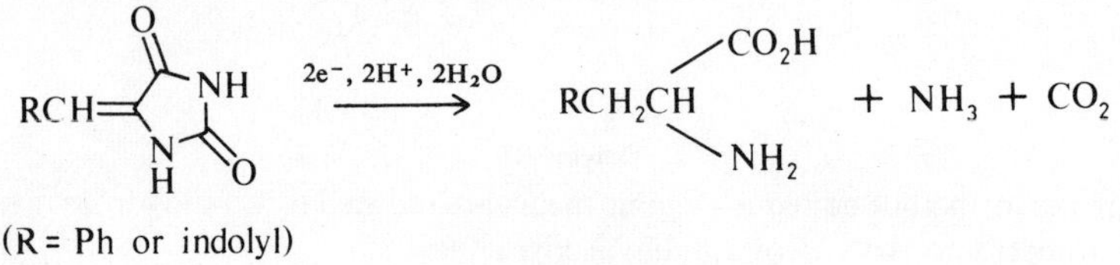

(R = Ph or indolyl)

Scheme 30

In DMF containing proton donors, porphyrins are reduced to phlorins (2e$^-$) and porphyrinogens (6e$^-$).[121]

Prazepam can be selectively reduced more effectively by an electrochemical method at a mercury pool than by catalytic hydrogenation (Scheme 31).[122]

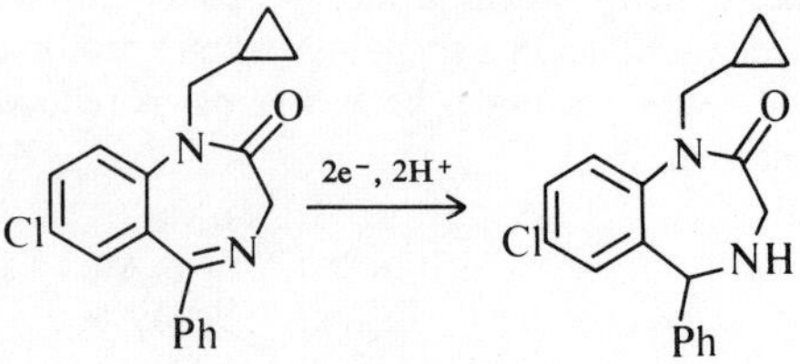

Scheme 31

Halides.—An intramolecular coupling reaction occurs when 5-(chlorophenyl)-1-phenylpyrazole is reduced to pyrazolo[1,5-f]phenanthridine (Scheme 32).[123] A

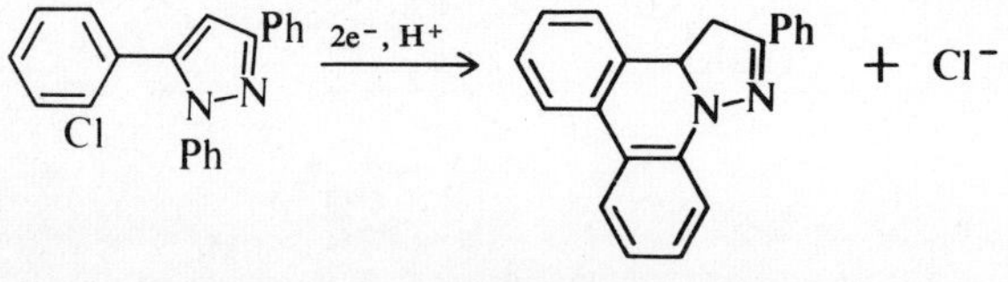

Scheme 32

further example of intramolecular trapping is provided by the reduction of the amide (2) in DMF–0.1M-TPAP at a mercury electrode (Scheme 33).[124] The open-chain amides (3) and (4) are the major products.

[119] S. Kato, Y. Tanaka, and J. Nakaya, *Denki Kagaku Oyobi Kogyo Butsuri Kagaku*, 1974, **42**, 223.
[120] M. Ya. Fioshin, E. P. Krysin, I. A. Avrutskaya, E. V. Zaporozhets, J. G. Tsar'kova, I. I. Gubenko, and B. M. Kotlyarevskaya, Russ. P. 433 144 (*Chem. Abs.*, 1974, **81**, 105 969).
[121] V. G. Mairanovskii, V. M. Mamaev, G. V. Ponomarev, R. I. Marinova, and R. P. Evstigneeva, *Russ. J. Gen. Chem.*, 1974, **44**, 2468.
[122] H. Oelschlaeger and F. I. Senguen, *Arch. Pharm.* (*Weinheim*), 1974, **307**, 909.
[123] W. J. Begley, J. Grimshaw, and J. Trocha-Grimshaw, *J.C.S. Perkin I*, 1974, 2633.
[124] J. Grimshaw and J. Trocha-Grimshaw, *Tetrahedron Letters*, 1974, 993.

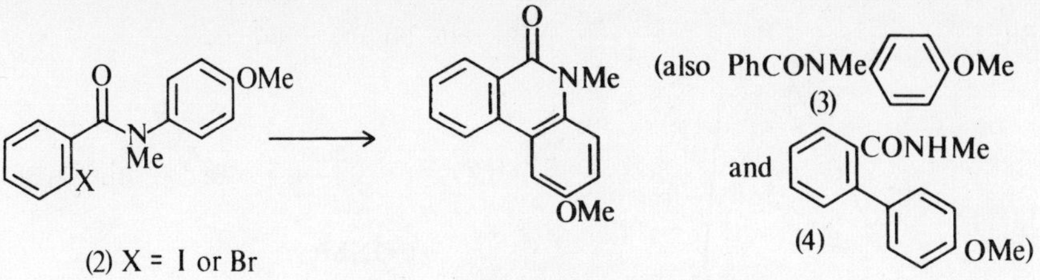

Scheme 33

(2) X = I or Br

(also PhCONMe-C₆H₄-OMe (3)

and 2-(PhCONHMe)-biphenyl-OMe (4))

Intermolecular coupling reactions during halide reductions are also of synthetic interest. Addition of alkyl groups to Schiff bases by controlled-potential electrolysis

$$RX + 2e^- \xrightarrow{-X^-} R^- \xrightarrow{i} PhCH_2NH-CRMeCO_2R'$$

(RX = benzyl chloride,
benzyl bromide,
ClCH$_2$CN, or
BrCH$_2$CO$_2$Et)

Reagents: i, PhCH$_2$N=CMeCO$_2$R' (R' = Et or benzyl)

Scheme 34

is possible when the latter compounds are electroinactive (Scheme 34).[125] DMF solutions were used, in a divided cell, with a mercury pool cathode. When butyl bromide is reduced in the presence of the electroinactive carbon dioxide, coupling products are not obtained at mercury cathodes in DMF–0.1M-TEAB. However, at potentials where CO$_2$ is also reduced, dibutyl oxalate and butyl valerate are formed

$$BuBr + e^- + Hg \xrightarrow{-Br^-} BuHg\cdot \longrightarrow \tfrac{1}{2}Hg + \tfrac{1}{2}Bu_2Hg$$

$$CO_2 + e^- \longrightarrow CO_2\cdot^-$$

$$(CO_2^-)_2 \xrightarrow{2BuBr} 2Br^- + (CO_2Bu)_2$$

$$CO_2\cdot^- \xrightarrow{BuBr}_{-Br^-} BuCO_2\cdot \xrightarrow{e^-} BuCO_2^- \xrightarrow{BuBr}_{-Br^-} BuCO_2Bu$$

Scheme 35

in addition to dibutylmercury (Scheme 35).[126] At graphite cathodes, both reactants are simultaneously electroreactive, and a product mixture of butane, butene, octane, butyl valerate, butyl NN-dimethoxamate, and dibutyl 2-methylmalonate is formed (Scheme 36).

The cathodic synthesis of high-purity sulphones has been patented. Cyclic and polymeric sulphones are obtained when SO$_2$ is reduced in the presence of suitable

---

[125] T. Iwasaki and K. Harada, *J.C.S. Chem. Comm.*, 1974, 338.
[126] J. H. Wagenknecht, *J. Electroanalyt. Chem.*, 1974, **52**, 489.

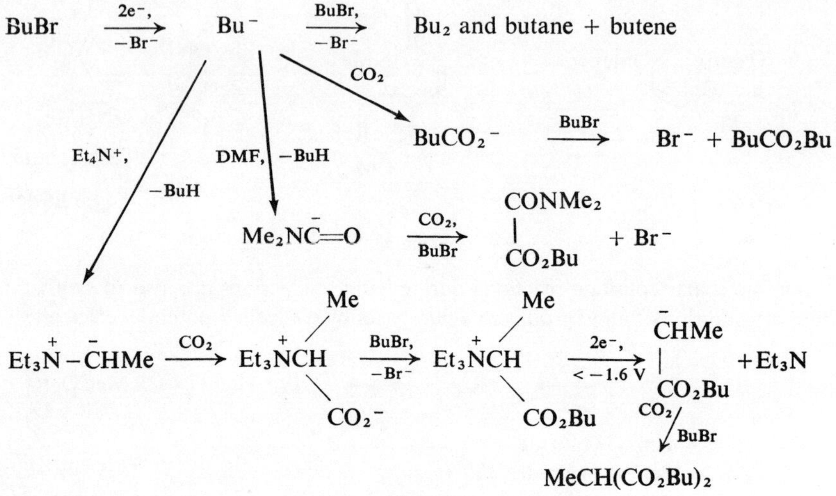

Scheme 36

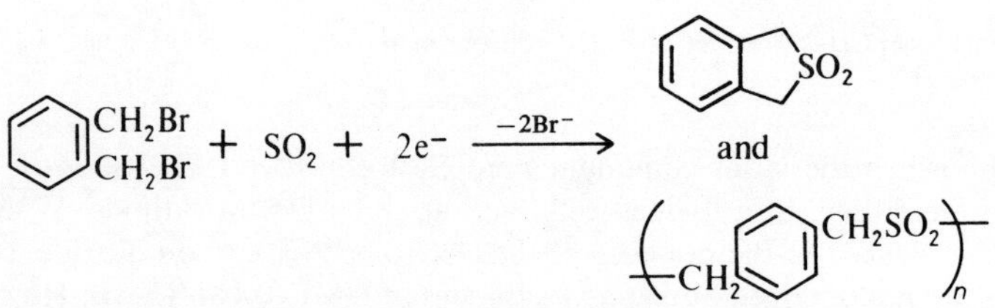

Scheme 37

dibromides in acetonitrile solutions at 80 °C in a divided cell (Scheme 37).[127] The rate of reaction of SO_2 radical anions with benzyl bromide has been investigated by e.s.r. spectroscopy.[128]

A substitution reaction catalysed by polarization of a cathode in the reaction mixture has been reported.[129] The proposed mechanism for conversion of *p*-bromobenzophenone into *p*-phenylthiobenzophenone is given in Scheme 38.

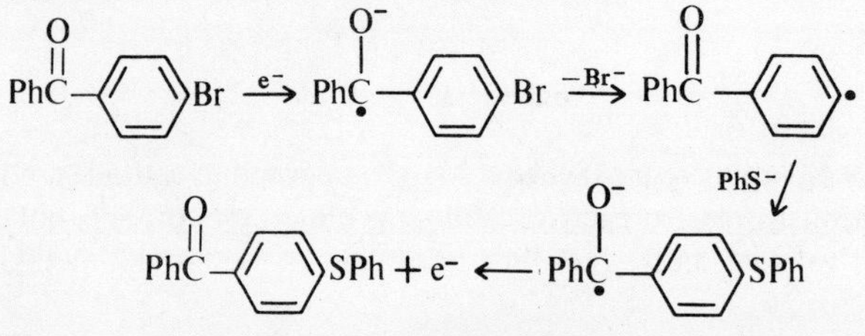

Scheme 38

[127] D. Knittel and B. Kastening, Ger. Offen., 2 328 196 (*Chem. Abs.*, 1975, **82**, 125 086).
[128] B. Kastening, B. Gostisa-Mihelcic, and K. Divisek, *Faraday Discuss. Chem. Soc.*, 1973, **56**, 341.
[129] J. Pinson and J. M. Saveant, *J.C.S. Chem. Comm.*, 1974, 933.

A further example of the formation of radical anions from aryl halides is the reduction of the various monohalide isomers of fluorenone in DMF–TEAP. The relative stabilities of the radical anions with respect to halide fission were consistent with the free electron densities at the position of substitution.[130] Selective dehalogenation has been achieved in the cathodic reduction of 5-(4-chlorophenyl)-3-(4-chlorostyryl)-1-phenyl-Δ^2-pyrazoline in DMF (Scheme 39).[131] The corresponding

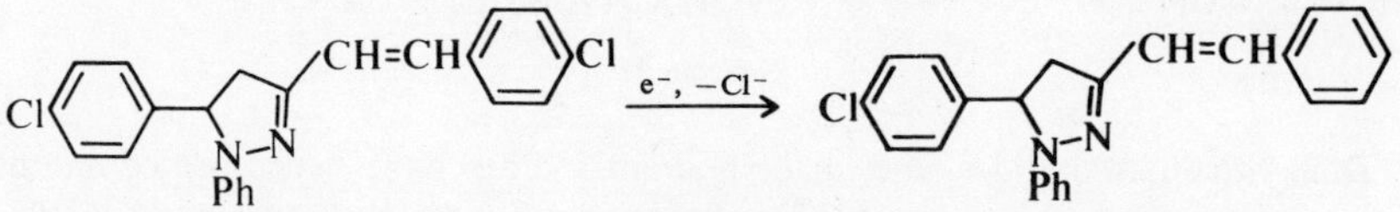

Scheme 39

dibromo-compound behaved similarly. The compounds 3-(4-chlorophenyl)-5-(4-chlorostyryl)pyrazole and its dibromo-analogue followed the same pattern on reduction. Free-radical intermediates were shown to abstract hydrogen atoms from the solvent. In these examples, loss of halide ion from the primary radical anion is determined by the distribution of free electron density.

In the reduction of 4,6-dichloro-2-diethanolamino-s-triazine, one chloride is first removed and then ring reduction takes place.[132] Corresponding behaviour is observed when 2-chloro-4-(ethylamino)-6-isopropylamino-s-triazine is reduced.[133] Similarly, the reduction of 6-chloro-7,9-dimethylpurinium cations in acidic solutions occurs progressively (Scheme 40).[83]

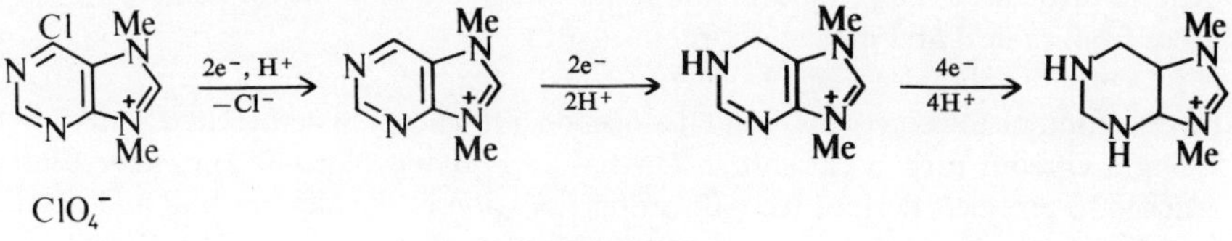

Scheme 40

The sequence of reduction of electrophores in the isometric iodoformylpyrroles has been examined in 50% aqueous ethanol, as a function of pH. In most cases iodide loss preceded reduction of the aldehyde group.[134]

Interest continues in the synthesis of metal alkyls from alkyl halides in cathodic reactions. Dibutyltin dibromide and tributyltin bromide are formed in the electrolysis of butyl bromide at tin electrodes.[135] The corresponding methyl compounds have been prepared similarly. Dibutyltin and tributyltin chloride have been formed from butyl chloride in acetonitrile containing a quaternary ammonium salt.[136]

---

[130] J. Grimshaw and J. Trocha-Grimshaw, *J. Electroanalyt. Chem.*, 1974, **56**, 443.
[131] J. Grimshaw and J. Trocha-Grimshaw, *J.C.S. Perkin I*, 1974, 1383.
[132] E. Yu. Khmelnitskaya, *Soviet Electrochem.*, 1974, **10**, 165.
[133] G. S. Supin, I. A. Melnikov, T. N. Bykhovskaya, and N. N. Melnikov, *Russ. J. Gen. Chem.*, 1974, **44**, 1165.
[134] M. Person, R. Mora, and M. Farnier, *Compt. rend.*, 1974, **278**, C, 1125.
[135] H. Matschiner, H. Schilling, R. Voigtländer, and K. Trautner, East Ger. P. 107 289 (*Chem. Abs.*, 1975, **82**, 98 145).
[136] H. E. Ulery, U.S.P. 3 823 077 (*Chem. Abs.*, 1974, **81**, 98 675).

Organobismuth compounds have been formed by attack on a bismuth cathode by radicals generated at its surface in aqueous solutions (Scheme 41).[137] Reduction of MeI on bismuth, followed by bromination, affords Me_2BiBr.

$$3CH_2=CHCN + 3H_2O + 3e^- \xrightarrow{Bi} 3OH^- + Bi(CH_2CH_2CN)_3$$

$$2ICH_2CH_2CN + e^- \xrightarrow{Bi} Bi(CH_2CH_2CN)_2I + I^-$$

Scheme 41

Dialkylthallium iodides and dialkylindium iodides have been formed in the reductions of alkyl iodides at cathodes of the respective metals in aqueous K_2HPO_4 or in MeCN–NaClO$_4$ solution.[138] Formation of tetraethyl-lead from ethyl iodide on lead cathodes in DMF containing quaternary ammonium electrolytes has been shown to be essentially quantitative.[139]

The intermediacy of adsorbed ethyl and butyl radicals in the reduction of the alkyl iodides on mercury has been demonstrated by means of rapid linear-sweep voltammetry.[140]

Several reductions of simple monohalides have been reported. Dibenzoylmethyl bromide in DMSO gives roughly equal amounts of dibenzoylmethane enolate and dibenzoylmethane.[141] The latter must have formed by abstraction of hydrogen atoms by the intermediate radical, because the enolate anion is known to be stable against protonation in the reaction medium.

Reduction of 9-bromoanthracene gave no evidence for the formation of a Dewar-benzene structure. The products anthracene (75%) and dianthranylmercury (10%) arose from radical and carbanion intermediates.[142]

1-Deuterionaphthalene and 1-deuterio-4-methylnaphthalene have been prepared by reduction of the corresponding 1-halogenonaphthalenes in deuteriated water.[143] Using a vitreous carbon cathode in DMF–TEAP, monochloro-alkanes have been reduced to products derived from disproportionation of the alkyl radical and from its further reduction.[144]

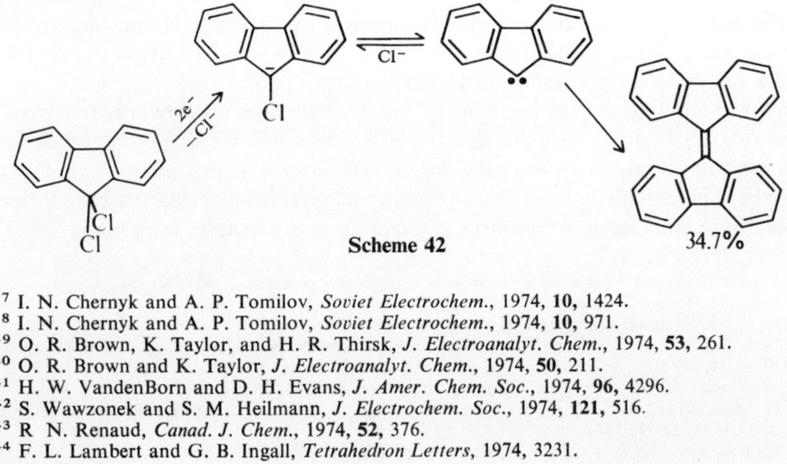

Scheme 42

[137] I. N. Chernyk and A. P. Tomilov, *Soviet Electrochem.*, 1974, **10**, 1424.
[138] I. N. Chernyk and A. P. Tomilov, *Soviet Electrochem.*, 1974, **10**, 971.
[139] O. R. Brown, K. Taylor, and H. R. Thirsk, *J. Electroanalyt. Chem.*, 1974, **53**, 261.
[140] O. R. Brown and K. Taylor, *J. Electroanalyt. Chem.*, 1974, **50**, 211.
[141] H. W. VandenBorn and D. H. Evans, *J. Amer. Chem. Soc.*, 1974, **96**, 4296.
[142] S. Wawzonek and S. M. Heilmann, *J. Electrochem. Soc.*, 1974, **121**, 516.
[143] R N. Renaud, *Canad. J. Chem.*, 1974, **52**, 376.
[144] F. L. Lambert and G. B. Ingall, *Tetrahedron Letters*, 1974, 3231.

The carbene intermediate fluorenylidene has been postulated in the reduction of 9,9-dichlorofluorene at mercury in DMF–TBAB (Scheme 42).[145] The yield of difluorenylidene is non-quantitative because the carbene can be further reduced at the cathode. In the presence of water the primary carbanion becomes protonated, and carbene formation is pre-empted.

2-Dichloromethyl-3-arylthiazolin-4-ones have been prepared quantitatively by cathodic reduction of the corresponding trichloromethyl compounds (Scheme 43).[146] Similarly, dichlorovinyl phosphates have been produced from 2,2,2-trichloro-1-hydroxyethyl phosphates (Scheme 44).[147]

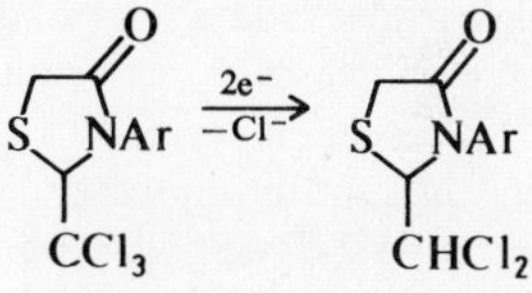

$$\underset{CCl_3}{\overset{O}{\underset{S}{\bigcirc}}NAr} \xrightarrow[-Cl^-]{2e^-} \underset{CHCl_2}{\overset{O}{\underset{S}{\bigcirc}}NAr}$$

Ar = Ph, 3-MeC$_6$H$_4$, 4-ClC$_6$H$_4$, or 4-BrC$_6$H$_4$

Scheme 43

$$Cl_3CCH(OH)PO(OR)_2 \xrightarrow[-Cl^-]{2e^-} Cl_2C=CHPO(OR)_2$$

(R = Me, Et, or Bu)

Scheme 44

Reduction at a mercury cathode of compounds of the type R(CH$_2$)$_n$CHBrCCl$_2$Br, where R and n are specified in Table 2, led to elimination of two bromide ions, with formation of the alkenes.[148]

Table 2 *Compounds of the type* R(CH$_2$)$_n$CHBrCCl$_2$Br *that have been reduced at a mercury cathode*

n	1	3	3	3	4
R	Cl	Cl	CO$_2$Me	OH	CO$_2$H

In the reduction of *meso-* and *dl*-1,2-dibromo-1,2-diphenylethane at mercury cathodes in DMF–TEAP, only *trans*-stilbene was produced, whether or not acids were added.[149] *meso*-2,3-Dibromobutane is reduced quantitatively to *trans*-2-butene under mild conditions at a stirred mercury pool cathode in DMF, but the *dl*-isomers produce *cis*-2-butene, indicating bromide elimination from the *trans*-

---

[145] R. C. Duty, G. Biolchini, and W. Matthews, *Analyt. Chem.*, 1974, **46**, 167.
[146] H. Lettau, H. Matschiner, R. Matuschke, and J. Hauschild, East Ger. P. 107 047 (*Chem. Abs.*, 1975, **82**, 72 976).
[147] H. Matschiner, C. Richter, H. Schilling, K. Trautner, and G. Erfurt, East. Ger. P. 104 980 (*Chem. Abs.*, 1975, **82**, 43 596).
[148] G. S. Supin, G. I. Kaplan, V. S. Mikhailov, and S. D. Volodkovich, *Russ. J. Gen. Chem.*, 1974, **44**, 1625.
[149] A. Inesi and L. Rampasso, *J. Electroanalyt. Chem.*, 1974, **54**, 289.

periplanar conformation when it is accessible.[150] Other vicinal dibromides behave similarly.

In aqueous ethanolic LiClO$_4$ solution the N-phenyl-imide of 2,3-dichlorobicyclo-[2,2,1]hept-5-ene-2,3-dioic acid is reduced in a single four-electron polarographic wave, both chlorines being replaced by hydrogen.[151] *trans*-2,3-Dichloro-1,4-dioxan and *trans*-2,3-dichlorotetrahydropyran are each reduced in a two-electron process at a mercury cathode in DMF–TEAP.[152] The *cis*-isomers were not reducible.

At a mercury cathode in DMF–TEAF[153] or TEAB[154] at −20 °C, 1-bromo-4-chlorobicyclo[2,2,0]hexane eliminated two halide ions to give $\Delta^{1,4}$-bicyclo[2,2,0]-hexene, identified as its Diels–Alder adduct with cyclopentadiene. This result contrasts with that obtained with Na–NH$_3$ as reducing agent; then the dimer, $\Delta^{3,6}$-tetracyclo[6,2,2,0^{1,8},0^{3,6}]dodecene, is formed.

The reduction of *meso*- and *dl*-2,4-dibromopentane in DMSO yielded roughly equal amounts of *cis*- and *trans*-1,2-dimethylcyclopropane.[155] The (+)-(2S,4S) optical isomer was reduced to the (−)-(1R,2R)-*trans*-product in high optical yields. This result was taken to indicate that the monocarbanion MeCHCH$_2$CHBrMe is an intermediate which cyclizes *via* a semi-w transition state.

Relative amounts of cyclopropane and propane formed in the reduction of 1,3-dibromopropane on platinum in DMF–TEAB have been studied as a function of potential.[156] Mild negative potentials favour propane and severe conditions cyclopropane. Similarly, 1,4-dibromobutane at extreme negative potentials yields mainly cyclobutane. Benzyl methyl ether was present as a hydrogen source to scavenge radicals. Propane is believed to be formed *via* a radical intermediate. At a mercury cathode in DMF, αω-dibromides above 1,3-dibromopropane are reported to give dialkylmercury products in high yield (Scheme 45).[157]

$$2Br(CH_2)_nBr + 6e^- \longrightarrow 4Br^- + {}^-(CH_2)_nHg(CH_2)_n{}^-$$

Scheme 45

In DMF–TEAB at −20 °C, 1,4-dibromobicyclo[2,2,2]octane is reduced on platinum in the presence of chlorine to give 1,4-dichlorobicyclo[2,2,2]octane as a minor product (12%).[158] It was deduced that the parent [2,2,2]propellane had been produced in the electrode reaction (Scheme 46). The major product was 1,4-dimethylenecyclohexane.

The ethyl esters of difluoroacetic acid and trifluoroacetic acid each undergo one-electron reductions at mercury cathodes in DMF–TEAP with F$^-$ fission.[159] The presence of a proton donor doubles the wave height. Benzotrifluoride is reduced

---

[150] J. Casanova and H. R. Rogers, *J. Org. Chem.*, 1974, **39**, 2408.
[151] R. F. Mahishko, V. A. Semenov, V. V. Simonov, and A. I. Naimishu, *Russ. J. Gen. Chem.*, 1974, **44**, 1372.
[152] B. A. Arbuzov, E. A. Berdnikov, L. K. Yuldasheva, and E. N. Klimovitskii, *Izvest. Akad. Nauk. S.S.S.R. Ser. khim.*, 1974, 732.
[153] J. Casanova and H. R. Rogers, *J. Org. Chem.*, 1974, **39**, 3803.
[154] K. B. Wiberg, W. F. Bailey, and M. F. Jason, *J. Org. Chem.*, 1974, **39**, 3803.
[155] A. J. Fry and W. E. Britton, *J. Org. Chem.*, 1973, **38**, 4016.
[156] K. B. Wiberg and G. A. Epling, *Tetrahedron Letters*, 1974, 1119.
[157] J. Casanova and H. R. Rogers, *J. Amer. Chem. Soc.*, 1974, **96**, 1942.
[158] K. B. Wiberg, G. A. Epling, and M. Mason, *J. Amer. Chem. Soc.*, 1974, **96**, 912.
[159] A. Inesi and L. Rampasso, *J. Electroanalyt. Chem.*, 1974, **49**, 85.

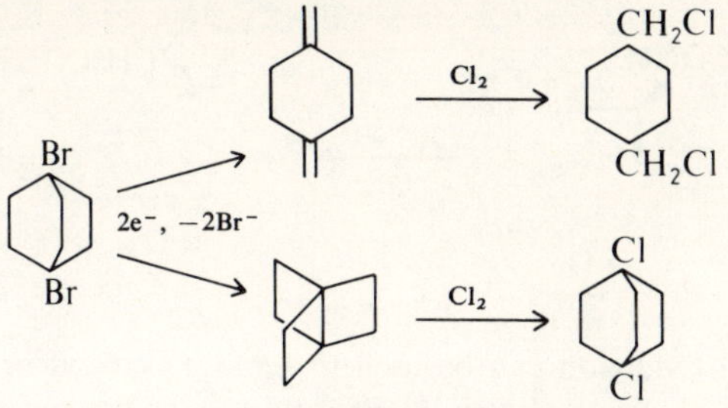

Scheme 46

stepwise to toluene in DMF.[160] The chlorofluoromethyl-benzenes lose chloride preferentially and then fluoride. In neutral or alkaline part-aqueous media, freons dehalogenate to give fluoro-olefins.[161]

Carboxylic Acids and Derivatives.—Mechanistic aspects of the reduction of benzoic acid at cathodes of materials having high overvoltages have been discussed.[162] Benzoic acid is reduced to benzyl alcohol in an electrolyte of 20—30% sulphuric acid at 50 °C with a current yield of 53% when a lead cathode is used in a divided cell.[163] Yields of salicylaldehyde as high as 70—78% have been obtained from the reduction of salicylic acid.[164] Engineering aspects of the reduction of phthalic acid to dihydrophthalic acid have been discussed.[165] The oxalic acid–glyoxylic acid conversion has also been optimized.[166]

Reduction of D-ribono-γ-lactone to D-ribose can be carried out at amalgamated solid cathodes between 5 and 15 °C and 3 < pH < 4 in yields of approximately 70%.[167]

A mercury pool cathode has been used to reduce cephalosporanic acid esters to the corresponding 3-methylene-cephams (Scheme 47).[168] 3-Hydroxymethyl-3-cepham-4-carboxylic acid lactone behaves similarly (Scheme 48).

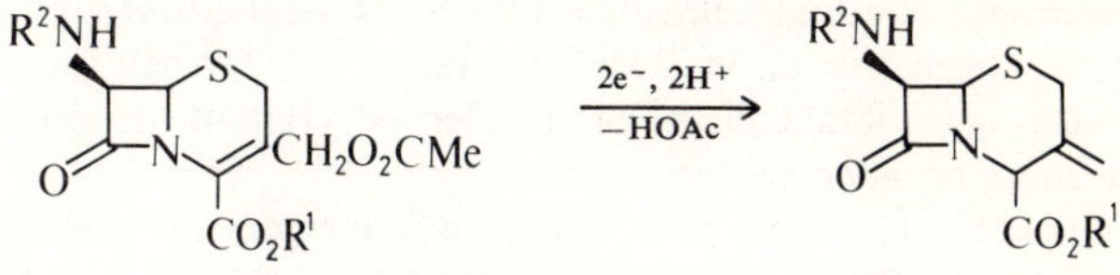

Scheme 47

[160] H. Lund and N. J. Jensen, *Acta Chem. Scand.* (B), 1974, **28**, 263.
[161] K. M. Smirnov, L. G. Feoktistov, M. M. Goldin, A. P. Tomilov, and S. L. Varshavskii, Russ. P. 230 131 (*Chem. Abs.*, 1975, **82**, 42 963).
[162] R. G. Barradas, O. Kutowy, and D. W. Shoesmith, *Electrochim. Acta*, 1974, **19**, 49.
[163] G. Schwarzlose, Ger. Offen. 2 237 612 (*Chem. Abs.*, 1974, **80**, 127 576).
[164] L. G. Kurkina, E. P. Sturostenko, and N. K. Sturostenko, *Tr. Mosk. Kkim.-Tekhnol. Inst.*, 1973, **75**, 126.
[165] H. Nohe, *Chem.-Ing.-Tech.*, 1974, **46**, 594.
[166] D. J. Pickett and K. S. Yap, *J. Appl. Electrochem.*, 1974, **4**, 17.
[167] (a) A. Korczynski, L. Piszczek, and J. Swiderski, *Zeszyty Nauk Politech. Slask Chem.*, 1973, 177; (b) E. V. Gromova, I. A. Avrutskaya, and M. Ya. Fioshin, *Tr. Mosk. Khim-Tekhnol. Inst.*, 1973, **75**, 211.
[168] M. Ochiai, O. Aki, A. Morimoto, T. Okada, K. Shinozaki, and Y. Asahi, *J.C.S. Perkin I*, 1974, 258.

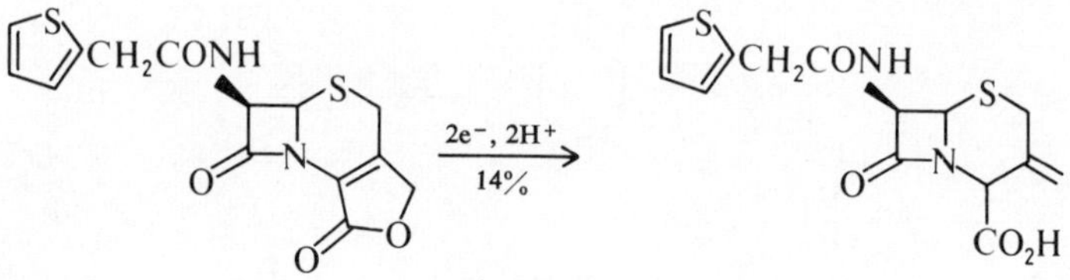

Scheme 48

Electrochemical reduction can be used to remove benzoyl or benzylidene protecting groups from carbohydrates.[169] Benzoyl groups are removed more readily than benzylidene groups. For example, at a mercury cathode in DMF, the 2-benzoate of methyl-4,6-o-benzylidene-β-D-galactopyranoside, at -2.35 V, gives 85% methyl-4,6-o-benzylidene-β-D-galactopyranoside, which at -2.9 V yields 80% of methyl-β-D-galactopyranoside.

Reduction can also be employed in removing protecting groups of the type $X_m CH_{3-m} CH_2 O(CO)_n$, where $n = 0$ or 1 and $m = 1, 2,$ or 3, from amino-acids, peptides, proteins, hormones, and antibiotics.[170] Other protecting groups (tosyl, benzyloxy, *etc.*) are not affected by the reduction at a mercury cathode.

An example of removal of benzoyl protecting groups from amino-acids is in the synthesis of D-($-$)-α-aminophenylacetic acid.[171] It is obtained in 77—81% yield and in a 98% optically pure state by reduction of its N-benzoyl derivatives in MeOH–TMAC at $4 < \text{pH} < 6$.

In liquid ammonia, propionamide is reduced to propionaldehyde or propyl alcohol, depending upon the reaction conditions. Propionaldehyde is favoured by low temperatures and by the avoidance of basic conditions.[172]

Reduction of phthalimides on mercury cathodes in acidic H_2O–MeCN solutions yields quantitative amounts of the corresponding hydroxyphthalimidines.[173] Further reduction only takes place in the potential region of background electrolyte decomposition.

Sulphur Compounds.—Cathodic cleavage of the S—O bond in aryl sulphenates of the type PhSOR has been studied in DMF solutions.[174, 175] When the ring is NO_2-substituted, cleavage still take place, and further reduction of the nitro-thiophenolate occurs at more negative potentials.

Arylsulphonyl chlorides have been reduced at a mercury cathode in acetonitrile to give the sulphinic acid in high yield:[176]

$$ArSO_2Cl + 2e^- \longrightarrow Cl^- + ArSO_2^-$$

However, in DMF or DMSO, traces of water catalyse additional reactions, and the

---

[169] V. G. Mairanovskii, N. F. Loginova, A. M. Ponomarev, and A. Ya. Veinberg, *Soviet Electrochem.*, 1974, **10**, 164.
[170] E. Kasafirek and L. Novak, Czech. P. 154 978 (*Chem. Abs.*, 1975, **82**, 171 456).
[171] G. Bison and H. Schuebel, Ger. Offen. 2 300 959 (*Chem. Abs.*, 1974, **81**, 105 975).
[172] O. R. Brown and P. D. Stokes, *J. Electroanalyt. Chem.*, 1974, **57**, 425.
[173] O. R. Brown, S. Fletcher, and J. A. Harrison, *J. Electroanalyt. Chem.*, 1974, **57**, 351.
[174] Z. V. Todres, K. P. Butin, and S. P. Avagyan, *Izvest. Akad. Nauk S.S.S.R., Ser. khim.*, 1974, 75.
[175] Z. V. Todres and S. P. Avagyan, *Internat. J. Sulfur Chem.*, 1973, **8**, 373.
[176] G. Jeminet, J. Simonet, and J. G. Gourey, *Bull. Soc. chim. France*, 1974, 1102.

overall process becomes:

$$3ArSO_2Cl + 6e^- \longrightarrow ArS^- + 2ArSO_3^- + 3Cl^-$$

In acidic conditions the protonated sulphonyl chlorides are reduced to thiosulphinates, which can be further reduced to disulphides and thence to thiols:

$$ArSO_2Cl + 3e^- + 2H^+ \longrightarrow H_2O + Cl^- + \tfrac{1}{2}ArSO_2SAr$$

$$ArSO_2SAr + 4e^- + 4H^+ \xrightarrow{-2H_2O} ArSSAr \xrightarrow{2e^-} 2ArS^-$$

Electrolysis in aprotic media containing alkyl halides leads to the production of organic sulphides and sulphones:

$$ArS^- + RX \longrightarrow X^- + RSAr$$

$$ArSO_2^- + RX \longrightarrow X^- + RSO_2Ar$$

Arylsulphonyl cyanides have been reduced at a mercury cathode in DMF.[177] Under mild conditions, organomercury products are obtained:

$$2ArSO_2CN + Hg + 2e^- \longrightarrow 2CN^- + (ArSO_2)_2Hg$$

Progressively negative potentials reduce the arylsulphonyl radical to the anion, then to the dianion, and eventually to a trianion:

$$ArSO_2\cdot \xrightarrow{e^-} ArSO_2^- \xrightarrow{e^-} ArSO_2^{2-} \xrightarrow{e^-} ArSO_2^{3-}$$

Several *para*-substituted phenyl methyl sulphoxides have been reduced in various solvents to the corresponding sulphides.[178]

Diphenyl diselenide is reduced to the selenol at a mercury cathode:[179, 180]

$$PhSeSePh + 2e^- \longrightarrow 2PhSe^-$$

Diphenyl disulphide similarly gives the thiophenolate ion on reduction in DMF or DMSO.[181] The reduction of the bis(nitrophenyl) disulphides (*ortho*- and *para*-isomers) in DMF proceeds in similar fashion, with further reduction of the nitrothiophenolate ion taking place at more negative potentials.[182]

A large range of organic disulphides have been reduced with fission of the S—S bond to give anions, which have then been coupled with a wide range of electrophiles present in the DMF–LiCl medium.[183] In this way a variety of alkylation and acylation reactions were performed in high yields.

The cleavage of several cyclic sulphones at a mercury cathode has been studied in DMF or methanol solutions of quaternary ammonium salts:[184]

$$-CH_2SO_2- \xrightarrow{2e^-, H^+} -CH_3 + -SO_2^-$$

The position of cleavage and the structure of the resulting sulphinate anion have been

---

[177] D. Van der Meer, A. Spaans, and H. Thijsse, *Rec. Trav. chim.*, 1974, **93**, 7.
[178] G. Grandi, L. Benedetti, R. Andreoli, and G. Battiotuzzi Gavioli, *J. Electroanalyt. Chem.*, 1974, **54**, 221.
[179] G. Palianai and M. L. Cataliotti, *Z. Naturforsch.*, 1974, **29b**, 376.
[180] F. Fagioli, F. Pulidori, C. Bighi, and A. De Battisti, *Gazzetta*, 1974, **104**, 639.
[181] B. Persson and B. Nygard, *J. Electroanalyt. Chem.*, 1974, **56**, 373.
[182] S. P. Avagyan, *Russ. J. Gen. Chem.*, 1974, **44**, 1764.
[183] P. E. Iversen and H. Lund, *Acta Chem. Scand. (B)*, 1974, **28**, 827.
[184] B. Lamm and J. Simonet, *Acta Chem. Scand. (B)*, 1974, **28**, 147.

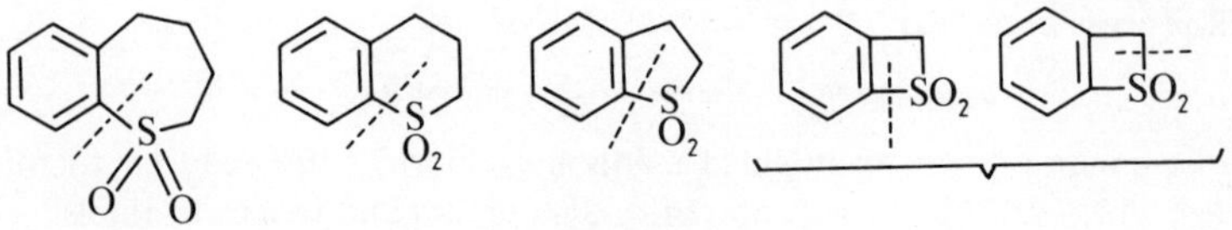

Scheme 49

determined, and are shown in the cases of certain sulphones (Scheme 49). The mechanism of the reduction of diphenyl sulphone in DMSO has been shown to be *ECE* under mild conditions, in which the initial radical anion is protonated and further reduced before cleavage occurs.[185] Alternatively, at more negative potentials, a dianion is formed before protonation and cleavage take place. Rate constants have been given for the protonation reactions.

Various vinyl sulphones have also been examined in aqueous methanolic media.[186] The anion formed by the transfer of two electrons and one proton can decompose in either of two ways:

$$R^1SO_2CH_2CH_2R^2 \xleftarrow[H^+]{(B)} R^1SO_2CH_2\bar{C}HR^2 \xrightarrow{(A)} R^1SO_2^- + CH_2=CHR^2$$

Alkyl vinyl sulphones follow route A, but route B is competitive in the cases of phenyl styryl sulphone, alkyl styryl sulphones, and phenyl vinyl sulphone.

Trityl and carbobenzoxy-groups used to protect sulphur-containing amino-acids can be removed by controlled-potential reduction at a mercury cathode in dry DMF–TBAP without disturbing similar groups used to protect —OH or —NH$_2$ functions.[187]

Reduction of 2-thioethoxy-1,3-dithiolium ions at a platinum cathode in MeCN–TEAP leads to the orthothio-oxalate (Scheme 50).[188] Reductions of other cyclic trithiosulphonium ions (Scheme 51) have also been examined.[189] 5-Methyl-1,2 dithiole-3-thione in MeCN is reduced stepwise at a mercury cathode (Scheme 52).[190]

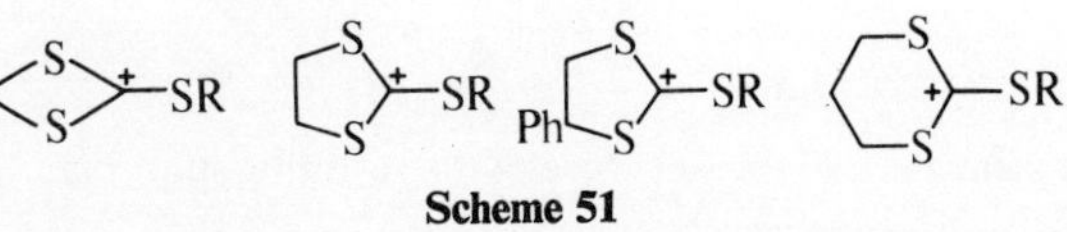

Scheme 50

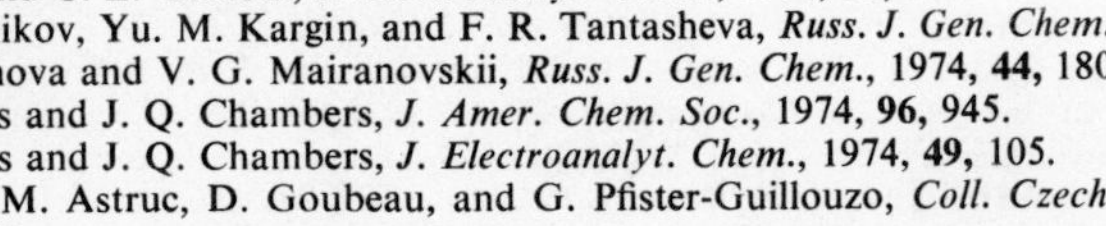

Scheme 51

---

[185] J. A. Cox and C. L. Ozment, *J. Electroanalyt. Chem.*, 1974, **51**, 75.
[186] E. A. Berdnikov, Yu. M. Kargin, and F. R. Tantasheva, *Russ. J. Gen. Chem.*, 1974, **44**, 1341.
[187] N. F. Loginova and V. G. Mairanovskii, *Russ. J. Gen. Chem.*, 1974, **44**, 1805.
[188] P. R. Moses and J. Q. Chambers, *J. Amer. Chem. Soc.*, 1974, **96**, 945.
[189] P. R. Moses and J. Q. Chambers, *J. Electroanalyt. Chem.*, 1974, **49**, 105.
[190] A. Astruc, M. Astruc, D. Goubeau, and G. Pfister-Guillouzo, *Coll. Czech. Chem. Comm.* 1974, **39**, 861.

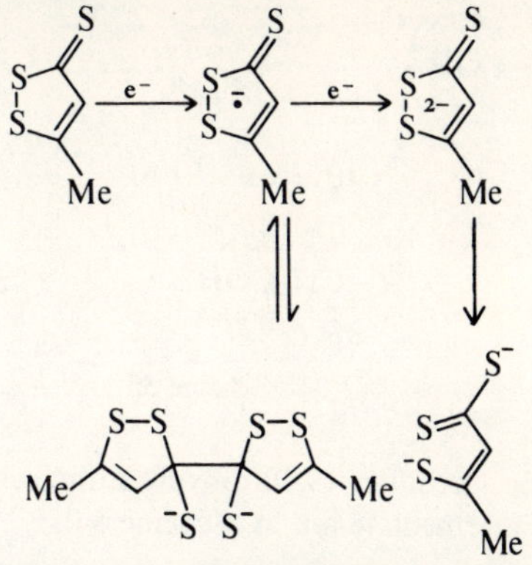

Scheme 52

Reduction of 1,2-dimethylindolizine-3-thial at a mercury electrode in MeCN–TPAP takes place in two stages (Scheme 53).[191] The four-co-ordinate quadrivalent sulphur dilactones (5; R = H, OMe, or NO_2) are reduced to divalent sulphur dicarboxylates (Scheme 54).[192] The nitro-substituted dicarboxylate is reduced further at more negative potentials.

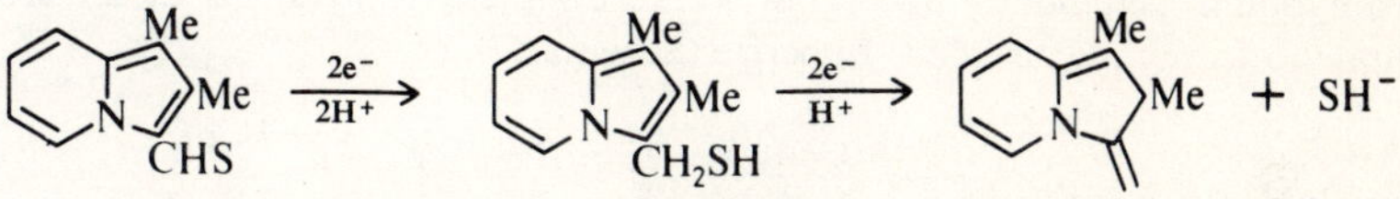

Scheme 53

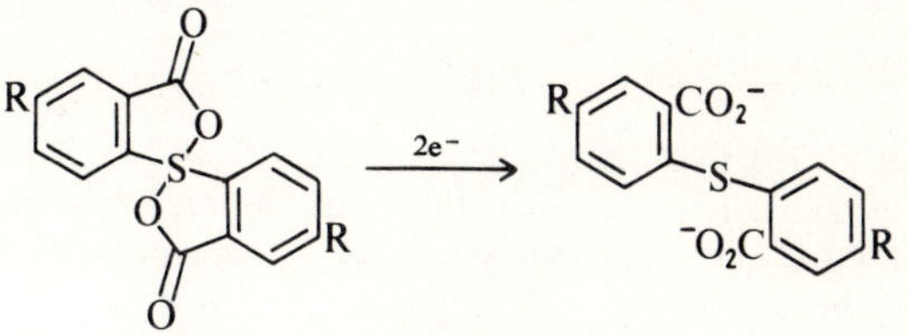

Scheme 54

Oxygen Compounds.—A mechanism previously proposed for the reduction of unsaturated alcohols at mercury in DMF has now been put forward by different authors.[193] The primary radical anion can either eliminate hydroxide or become protonated before being further reduced (Scheme 55). In the case of α-hydroxy-9-

---

[191] J. N. Cape and C. A. Vincent, *J. Electroanalyt. Chem.*, 1974, **56**, 427.
[192] Chi-Sheng Liao, J. Q. Chambers, I. Kapovits, and J. Rabai, *J.C.S. Chem. Comm.*, 1974, 149.
[193] H. Lund, H. Doupeux, M. A. Michel, G. Mousset, and J. Simonet, *Electrochim. Acta*, 1974, **19**, 629.

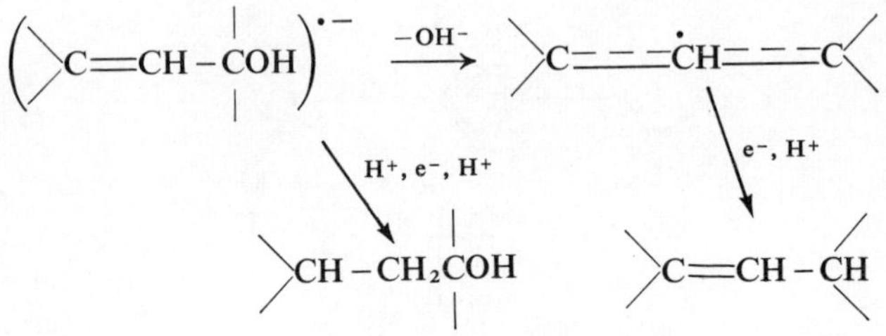

ethylanthracene, a minor product (9,10-dihydroanthracene) arises from loss of acetaldehyde from the intermediate anion (Scheme 56).

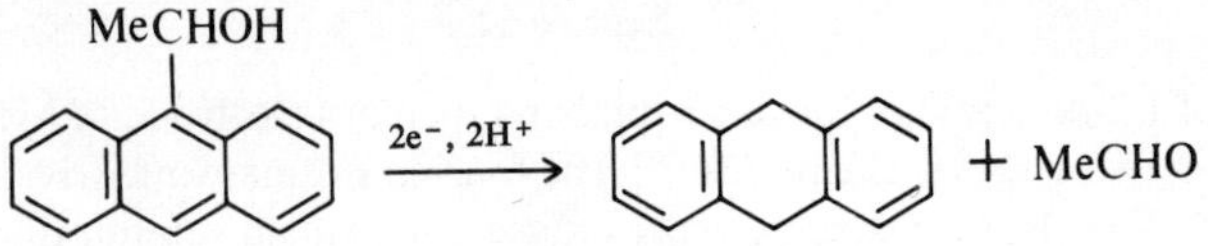

Scheme 56

In aqueous alcoholic H_2SO_4 solutions, 5,6,7,8-tetrahydro-2,4-diphenyl-1-benzopyrylium perchlorate undergoes a reverisible electron transfer at a mercury cathode to form radicals which dimerize (Scheme 57).[194]

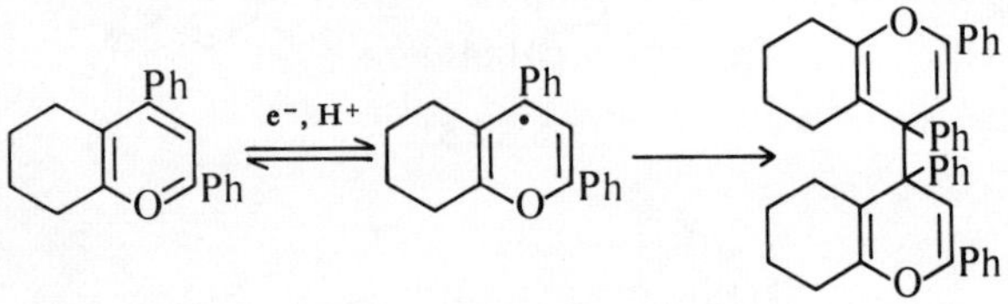

Scheme 57

The four substituted pyrones (6; R^1 = Me or Ph, R^2 = H or Me) have been reduced polarographically in aqueous ethanolic solution.[195] 6-Phenyl-4-hydroxy-2-pyrone in 0.4M-TBAI gave as products 6-phenyl-5,6-dihydro-4-hydroxy-2-pyrone at

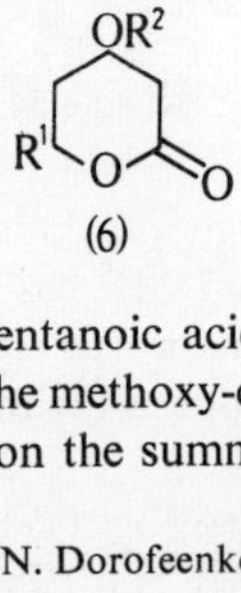

(6)

−2.05 V and 5-phenyl-3-hydroxypentanoic acid at −2.2 V, with mixtures being formed at intermediate potentials. The methoxy-compounds undergo hydrogenation of the C-5–C-6 bond at potentials on the summit of the polarographic wave, but

---

[194] M. M. Evstifeev, F. Kh. Aminova, G. N. Dorofeenko, and E. P. Olekhanovich, *Russ. J. Gen. Chem.*, 1974, **44**, 2225.
[195] G. Le Guillanton, *Bull. Soc. chim. France*, 1974, 627.

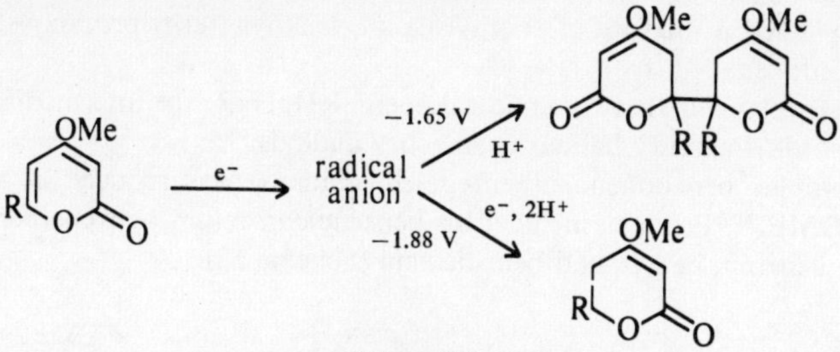

Scheme 58

yield a dimer under less severe conditions (Scheme 58). At potentials of its second wave (−2.05 V), 6-phenyl-4-methoxy-2-pyrone gives a mixture of 6-phenyl-5,6-dihydro-4-methoxy-2-pyrone and 5-phenyl-3-hydroxypentanoic acid. The esters of the 4-hydroxy-compounds have also been reduced.[196] The acetates, benzoates, and octanoates were examined. At potentials of the first polarographic wave, the reaction

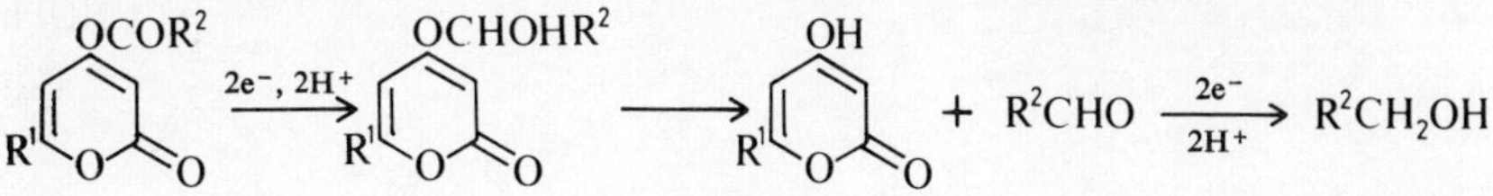

Scheme 59

is indicated by Scheme 59. The product from the 6-phenyl derivative is reduced further at more negative potentials, to give 6-phenyl-5,6-dihydro-4-hydroxy-2-pyrone and 5-phenyl-3-hydroxypentanoic acid.

Organometallic Compounds.—The aromatic rings of aryltrimethylsilanes and aryltrimethylgermanes have been reduced in methylamine–LiCl (Scheme 60).[197] However, the analogous tin compounds behave differently when treated similarly (Scheme 61).

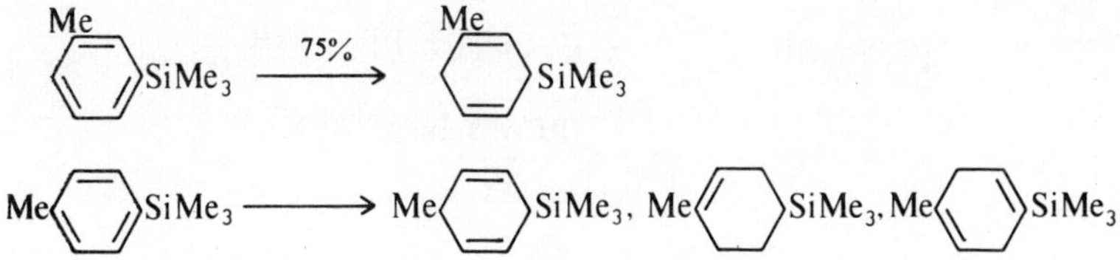

Scheme 60

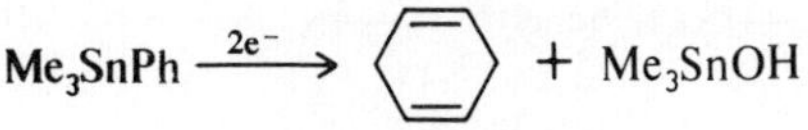

Scheme 61

[196] G. Le Guillanton, *Bull. Soc. chim. France*, 1974, 1699.
[197] C. Eaborn, R. A. Jackson, and R. Pearce, *J.C.S. Perkin I*, 1974, 2055.

Electrolysis of a mixture of triethylsilane and 3-pentanone between platinum electrodes yielded a mixture of triethylsilanol, triethyl(1-ethylpropoxy)silane, and hexaethyldisiloxene.[198]

It has been claimed that 'organic calomels' RHgHgR are intermediates in the reduction of alkylmercury halides to mercury dialkyls.[199]

The π-complex benzophenone(benzene)chromium(I) was reduced at a mercury cathode in DMF.[200] Products included bis(benzene)chromium, benzene(benzhydrol)-chromium, benzoin, benzil, and benzilic acid (Scheme 62).

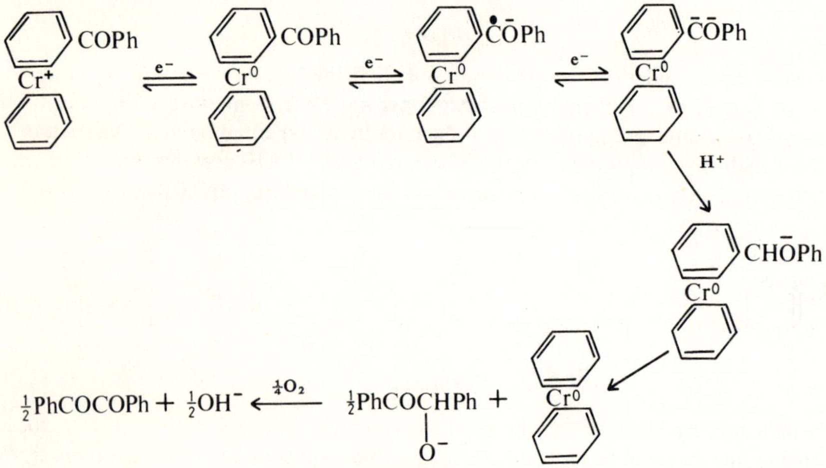

Scheme 62

Reduction of Co^{II} complexes with planar quadridentate ligands at a mercury cathode in DMF containing alkyl halides, quaternary ammonium ions, or sulpho-

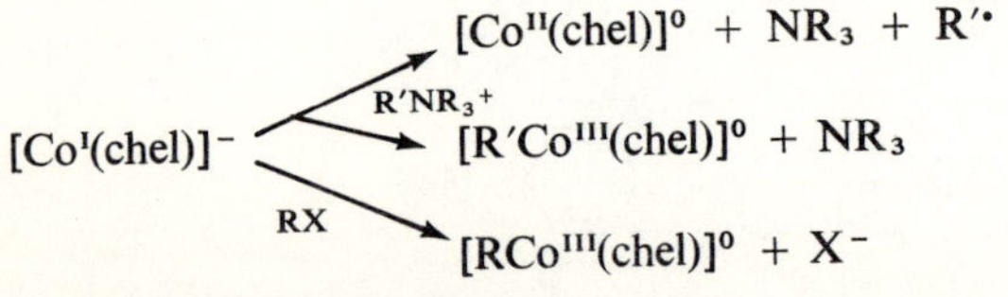

Scheme 63

nium ions gives rise to transalkylation reactions of the Co^I primary product (Scheme 63).[201, 202] In the presence of suitable acceptor groups, the alkylation reaction can be reversed:

$$A^- + [RCo^{III}(chel)]^0 \longrightarrow RA + [Co^I(chel)]^-$$

---

[198] G. T. Fedorova, N. P. Lharitonov, and B. P. Nechaev, *Russ. J. Gen. Chem.*, 1974, **44**, 121.
[199] A. B. Ershler, V. V. Strelets, K. P. Butin, and A. N. Kushin, *J. Electroanalyt. Chem.*, 1974, **54**, 75.
[200] S. Valcher and G. Casalbore, *J. Electroanalyt. Chem.*, 1974, **50**, 359.
[201] G. Costa, A. Puxedda, and E. Reisenhofer, *J.C.S. Dalton*, 1973, 2034.
[202] G. Costa, A. Puxedda, and E. Reisenhofer, *Bioelectrochem. Bioenerg.*, 1974, **1**, 29.

Anodic oxidation of the Co^{III} species facilitates such reversal:

$$[RCo^{III}(chel)]^0 \xrightarrow{-e^-} [RCo^{IV}(chel)]^+ \xrightarrow{X^-} [Co^{II}(chel)]^0 + RX$$

Vitamin B_{12} has been converted into vitamin B_{12a} by a sequence of controlled-potential cathodic reduction and aerial oxidation.[203] In the reduced (Co^{II}) state the cyanide ligands are relatively labile, and are precipitated by adding a silver salt. Re-oxidation of the aquo-complex yields aquocobalamin.

Syntheses of tetra-aza[16]annulene complexes of cobalt(III), copper(III), and nickel(III) in which the chelate exists as a rigid dianion have been performed cathodically in methanol or acetonitrile:[204]

$$[M^{II}(TAAB)]^{2+} \xrightarrow{e^-} [M^{I}(TAAB)]^+ \xrightarrow{slow} [M^{III}(TAAB)^{2-}]^+$$

Similarly $[Co^{II}(TAAB)^{2-}]^0 MeCN$ and $[Ni^{II}(TAAB)^{2-}]^0$ were formed.

Changes in ligand structure were also noted in the reduction of iron(III) porphyrins in DMF to the Fe^{II} and Fe^{I} states.[205]

Iron tris(acetylacetone) is reduced at a mercury cathode in methanol containing triphenylphosphine and an acyclic 1,3-diolefin (isoprene or buta-1,3-diene) at $-15\,°C$ to give a 13% yield of the complex $[Fe^0(Ph_3P)L_2]$.[206]

Similarly, the reduction of cobalt(II) chloride in the presence of tributylphosphine and buta-1,3-diene at a mercury cathode in methanolic LiCl yielded 68% of the product (buta-1,3-diene)(π-butenyl)(tributylphosphine)cobalt(I):[207]

$$2C_4H_6 + MeOH + CoCl_2 + Bu_3P \xrightarrow{e^-} MeO^- + (C_4H_6)Co^I(C_4H_7)(PBu_3)$$

When triphenylphosphine was used in place of tributylphosphine, a mixture of $[(PPh_3)_3CoCl]$ and $[(\pi-C_4H_6)(\pi-C_8H_{13})Co^I(PPh_3)]$ was obtained.

Metmyoglobin has been reduced to myoglobin at a mercury pool cathode in aqueous solution.[208]

Simple Inorganic Materials.—Reduction of carbon disulphide in DMF–0.2M-TEAB,

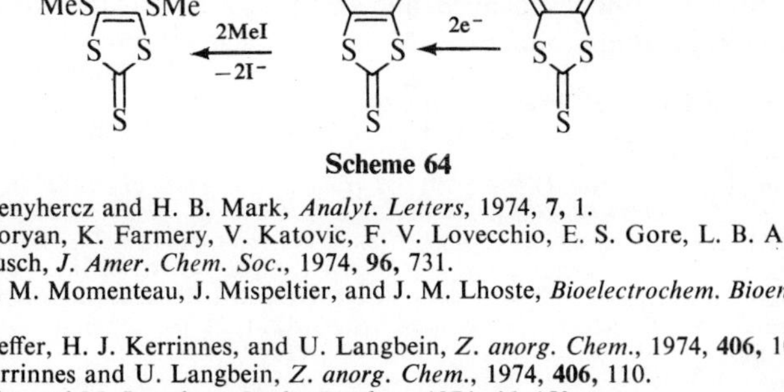

Scheme 64

---

[203] T. M. Kenyhercz and H. B. Mark, *Analyt. Letters*, 1974, **7**, 1.
[204] N. Takvoryan, K. Farmery, V. Katovic, F. V. Lovecchio, E. S. Gore, L. B. Anderson, and D. H. Busch, *J. Amer. Chem. Soc.*, 1974, **96**, 731.
[205] D. Lexa, M. Momenteau, J. Mispeltier, and J. M. Lhoste, *Bioelectrochem. Bioenerg.*, 1974, **1**, 108.
[206] W. Schaeffer, H. J. Kerrinnes, and U. Langbein, *Z. anorg. Chem.*, 1974, **406**, 101.
[207] H. J. Kerrinnes and U. Langbein, *Z. anorg. Chem.*, 1974, **406**, 110.
[208] F. Scheller and M. Jaenchen, *Studia Biophys.*, 1974, **46**, 153.

followed by treatment with methyl iodide, gave 4,5-bis(methylthio)-1,3-dithiole-2-thione in yields as high as 41% (Scheme 64).[209] The radical anion formed in the cathodic reduction of carbon dioxide is protonated in aqueous solutions, but in aprotic media it attacks a second CO_2 molecule.[210] The synthesis of oxalic acid by reduction of CO_2 on a Cr–Ni steel cathode in $(Et_4N)_2CO_3$–propylene carbonate is claimed to give yields of 60%.[211] Cobalt and nickel phthalocyanines are claimed to be effective electrocatalysts for the reduction of CO_2 to oxalic or glycolic acids.[212]

2 Oxidations

Hydrocarbons and Unsaturated Ethers.—Several branched alkanes undergo fragmentation and acetamidation by carbonium ion mechanisms when oxidized at a platinum anode in MeCN–TEAF at -45 °C in a divided cell.[213] For example, t-butylcyclohexane yielded 63% t-butylacetamide and 47% cyclohexylacetamide.

Fluorosulphonic acid has played a role in several studies of oxidation of alkanes. The presence of the acid in organic solvents when cyclohexane, ethane, or methane were oxidized anodically shifted the potential of oxidation to less positive values.[214] In TFA, cyclohexane formed cyclohexyl trifluoroacetate. Butane, pentane, hexane, and dodecane in fluorosulphonic acid were converted into carbocations under anodic conditions.[215] Oxidation of saturated organic materials by the electrolysis of their solutions in a molar solution of fluorosulphonate ions in fluorosulphonic acid led to their attack by the fluorosulphonate radical, which probably reacts by first abstracting a hydrogen atom.[216] Thus propionic acid was converted into its 3-fluorosulphonate derivative, which yielded acrylic acid on standing. Acetic acid yielded $CH_2(FSO_3)_2$. Methane gave the same product at 0 °C and current density 0.1 A cm^{-2}, although a finer gas stream at 45 °C caused formation of CH_3OSO_2F when 50 mA cm^{-2} was passed.

The conversion of hexene into caproic acid in yields exceeding 80% has been claimed to occur on palladized carbon in $HClO_4$.[217] Current efficiencies as high as 40% and high product purity have been achieved for the oxidation of benzene to benzoquinone, with subsequent reduction to hydroquinone on the cathode, when a pilot plant was operated for 4000 h with aqueous H_2SO_4 as electrolyte.[218] Benzene, naphthalene, and toluene have been successfully oxidized on PbO_2-coated titanium mesh anodes in sulphuric acid solution, to give benzoquinone, naphthoquinone, and benzoic acid, respectively.[219]

The formation of aromatic aldehydes and ketones by the oxidation of suspensions of alkylbenzenes at platinum anodes in 7N-sulphuric acid solutions has been studied.

[209] S. Wawzonek and S. M. Heilmann, *J. Org. Chem.*, 1974, **39**, 511.
[210] A. W. B. Aylmer-Kelly, A. Bewick, P. R. Cantrill, and A. M. Tuxford, *Faraday Discuss. Chem. Soc.*, 1973, **56**, 96.
[211] E. Heitz and U. Kaiser, Ger. Offen. 2 301 032 (*Chem. Abs.*, 1974, **81**, 130 200).
[212] S. Meshitsaka, M. Ichikawa, and K. Tamera, *J.C.S. Chem. Comm.*, 1974, 158.
[213] T. M. Siegel, L. L. Miller, and J. Y. Becker, *J.C.S. Chem. Comm.*, 1974, 341.
[214] H. P. Fritz and T. Würminghausen, *J. Electroanalyt. Chem.*, 1974, **54**, 181.
[215] F. Bobilliart, A. Thiebault, and M. Herlem, *Compt. rend.*, 1974, **278**, C, 1485.
[216] J. P. Coleman and D. Pletcher, *Tetrahedron Letters*, 1974, 147.
[217] H. Kinza, East Ger. P. 100 703 (*Chem. Abs.*, 1974, **81**, 52 151).
[218] M. Fremery, H. Hoever, and G. Schwarzlose, *Chem.-Ing.-Tech.*, 1974, **46**, 635.
[219] J. P. Millington, B. P. 1 377 681 (*Chem. Abs.*, 1975, **82**, 125 099).

Yields are improved by the addition of manganous sulphate.[220] Thus p-tolualdehyde can be obtained from p-xylene in yields as high as 95%, whereas p-cymene can be converted into 60% p-methylacetophenone and 30% cinnamaldehyde.[221] In the latter study, 4.2N-KOH was investigated as an alternative electrolyte.

The effects of changing pH, temperature, and potential upon the product distribution in the oxidation of ethylene on gold anodes have been examined.[222] The addition of anodically generated nitro-radicals to isoprene to form methyldinitrobutane has been carried out at a platinum mesh anode (c.d. 1 A dm^{-2}) at 0 °C in a divided cell.[223] The anolyte was 0.1N-HNO_3–0.1N-$NaNO_3$–1M-isoprene.

Several studies of the oxidative dimerization of olefins were published in 1974. α-Methylstyrene in wet MeCN–$NaClO_4$ yielded a mixture of cis- and trans-2,5-dimethyl-2,5-diphenyltetrahydrofuran, believed to occur by a symmetrical coupling of α-methylstyrene radical cations followed by the reaction of the resulting dication with water.[224]

Styrene, in wet MeCN containing Bu_4NHSO_4 or TEAT, undergoes the analogous reaction to form 2,5-diphenyltetrahydrofuran. Minor products obtained are shown

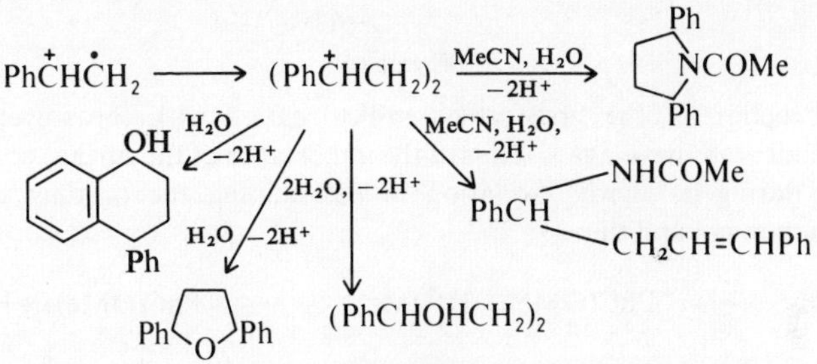

Scheme 65

in Scheme 65.[225] In an aqueous methylene chloride suspension containing Bu_4NHSO_4, the acetamidated products are, of course, not formed, but additional products 1,4-diphenyl-1-hydroxybutane and 1,4-diphenyl-4-hydroxybut-1-ene are obtained in an undivided cell.

In an undivided cell containing $NaClO_4$ and 2,6-lutidine in methanol, several vinyl ethers have been found to undergo dimerizations at a graphite anode to form acetals of dicarbonyl compounds, or the dicarbonyl compounds themselves, in yields of 30—60% (Scheme 66).[226] A mechanism involving dimerization of cation radicals is believed to pertain.

Acetylenes also undergo some oxidative dimerization at a rotating carbon anode in methanol–0.2M-$NaClO_4$, but monomeric products predominate as a result of the

---

[220] A. V. Solomin, E. N. Kopachevskii, and E. I. Kryuchkova, *Zhur. priklad. Khim.*, 1974, **47**, 939.
[221] A. V. Solomin, E. I. Kryuchkova, and V. A. Glybovskaya, *Soviet Electrochem.*, 1974, **10**, 954.
[222] C. Cwiklinski and J. Perichon, *Electrochim. Acta*, 1974, **19**, 297.
[223] A. K. Erimbetov, M. Ya. Fioshin, and M. Zh. Zhurinov, *Soviet Electrochem.*, 1974, **10**, 1240.
[224] H. Sternerup, *Acta Chem. Scand.* (B), 1974, **28**, 579.
[225] E. Steckhan and H. Schäfer, *Angew. Chem.*, 1974, **86**, 480.
[226] D. Koch, H. Schäfer, and E. Steckhan, *Chem. Ber.*, 1974, **107**, 3640.

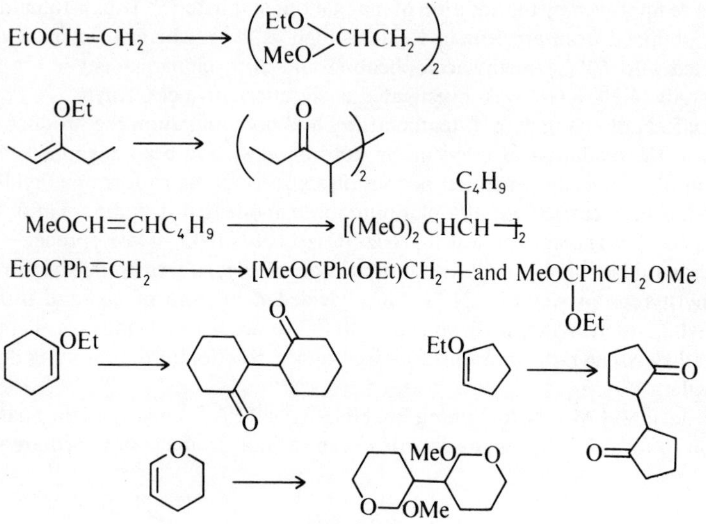

$$EtOCH=CH_2 \longrightarrow \left(\begin{matrix}EtO\\MeO\end{matrix}\!\!\!>\!\!CHCH_2\!\!-\!\!\right)_2$$

[Scheme 66 with additional reactions]

$$MeOCH=CHC_4H_9 \longrightarrow [(MeO)_2CHCH(C_4H_9)\!-\!]_2$$

$$EtOCPh=CH_2 \longrightarrow [MeOCPh(OEt)CH_2\!-\!]\text{ and } MeOCPhCH_2OMe$$
(with OEt group)

Scheme 66

preferred reaction of the first-formed radical cation with the solvent (Scheme 67).[227] Wendt *et al.* have again stressed the importance of the surface concentration of styrene during its anodic oxidation in determining the product distribution between monomers and dimers.[228]

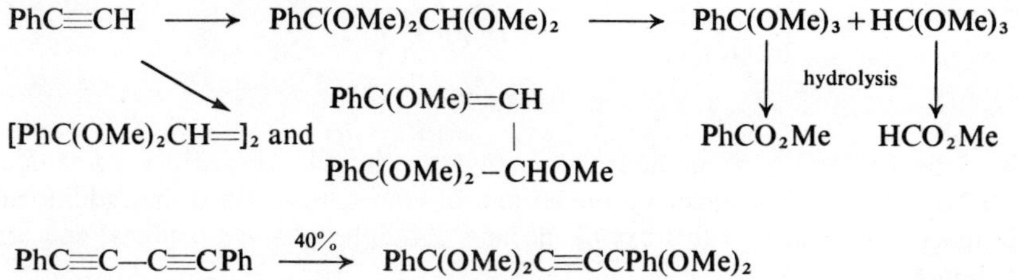

$$PhC\equiv CH \longrightarrow PhC(OMe)_2CH(OMe)_2 \longrightarrow PhC(OMe)_3 + HC(OMe)_3$$

$$[PhC(OMe)_2CH=]_2 \text{ and } PhC(OMe)=CH \mid PhC(OMe)_2-CHOMe$$

hydrolysis ↓ ↓

$$PhCO_2Me \quad HCO_2Me$$

$$PhC\equiv C\!-\!C\equiv CPh \xrightarrow{40\%} PhC(OMe)_2C\equiv CCPh(OMe)_2$$

Scheme 67

Various anode materials (Pt, C, PbO_2) have been used in the oxidation of buta-1,3-diene in methanol. Products included 2 monomers, 3 dimers, and 4 trimers.[229] Although a mixture of the monomers could be favoured by careful choice of conditions at a PbO_2 anode, it was found to be impossible to adjust conditions in order to select a particular product.

In strongly basic methanolic solutions the oxidation of olefins at a platinum electrode can produce organic carbonates.[230] For example, bicyclo[2,2,1]heptene (7), after passage of 0.3 A cm^{-2} for 3 F mol^{-1}, gave products (8)—(10) (R = CO_2Me). Minor products were (8)—(10) (R = H or R = Me). Electrolysis of the

---

[227] M. Katz and H. Wendt, *J. Electroanalyt. Chem.*, 1974, **53**, 456.
[228] V. Plzak, H. Schneider, and H. Wendt, *Ber. Bunsengesellschaft phys. Chem.*, 1974, **78**, 1373.
[229] M. Katz, Oe. Saygin, and H. Wendt, *Electrochim. Acta*, 1974, **19**, 193.
[230] R. Brettle and J. R. Sutton, *J.C.S. Chem. Comm.*, 1974, 449.

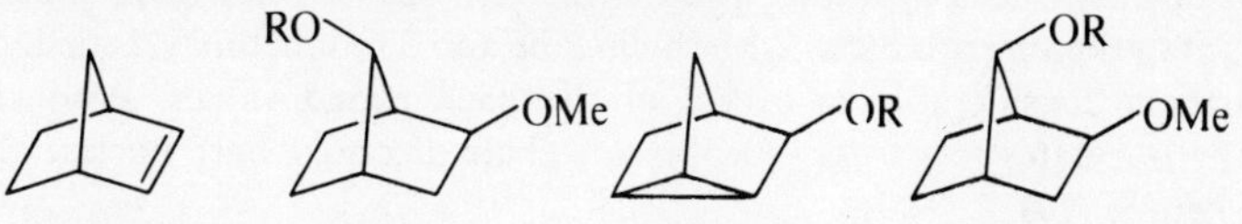

(7) (8) (9) (10)

background electrolyte is believed to yield methylsodium carbonate. Similar results are obtained with other olefins, *e.g.* cyclohexane, 2,3-dimethylbut-2-ene, and 3,3-dimethylbut-1-ene (Scheme 68). However, when less basic conditions were employed (methylsodium carbonate) and a higher charge was passed at a lower c.d., the

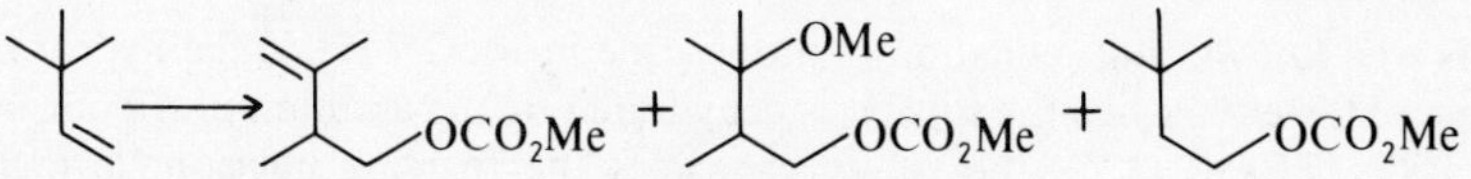

Scheme 68

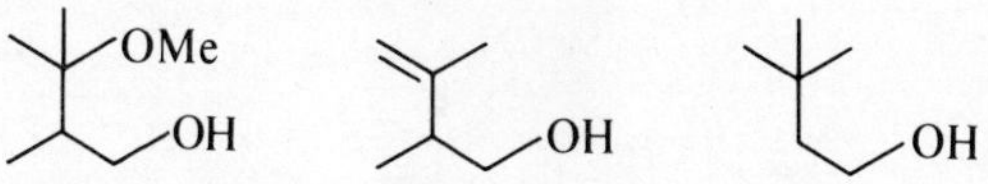

(11) (12) (13)

products formed were (11)—(13). These same alcohols can be formed from the corresponding carbonates by the reaction with MeONa–MeOH over 48 hours.

Cyclopropanes are oxidized in the presence of methylsodium carbonate (Scheme 69).

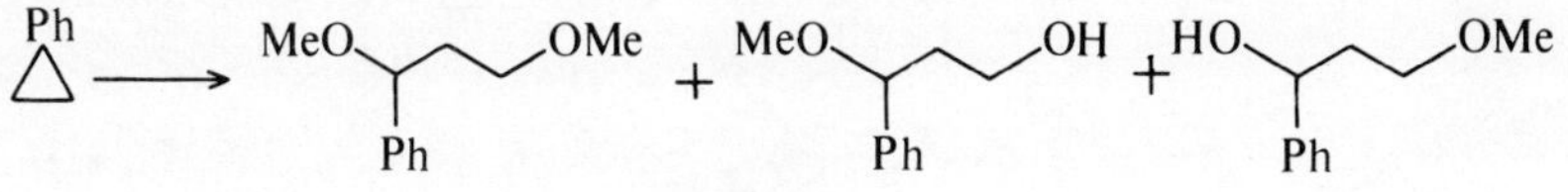

Scheme 69

Cyanations of naphthalene and anisole at a platinum anode have been carried out in an emulsified aqueous organic system.[231] Using the principle of ion-pair extraction, whereby the TBA ion extracts cyanide ion almost quantitatively from water to methylene chloride, monocyano-products have been isolated in chemical yields of up to 70%. This result compares favourably with previous work in MeOH or MeCN solvent systems. Cell voltages have been reduced by an order of magnitude, a factor which easily compensates for the inferior current efficiency. Conduction occurs in the aqueous phase (3M-NaCN), whereas the reaction is believed to be confined to the organic phase. Corresponding acyloxylations were less successful. Anodic substitution reactions of aromatic materials are well known. Cumene has been converted into 2-phenylpropanol by oxidation in MeCN containing $NaClO_4$ or TBAF and 2% water.[224] Anodic acetoxylation at Pt, C, or PbO_2 in NaOAc–HOAc has been compared with chemical acetoxylation using cobalt(III) acetate in order to investigate the role of adsorption.[232] 2,2-Dimethylindane and neopentylbenzene gave

[231] L. Eberson and B. Helgee, *Chemica Scripta*, 1974, **5**, 47.
[232] H. Sternerup, *Acta Chem. Scand.* (*B*), 1974, **28**, 969.

mainly side-chain acetoxylation, whereas the chemical method gave much larger amounts of the ketone products. 2-Methylindane and 5,6-dimethoxy-2-methylindane gave currents considerably in excess of the background values. Products were side-chain 1-acetates (*cis*:*trans*, 50:50), side-chain alcohols, both nuclear alcohols, and the ketone (14).

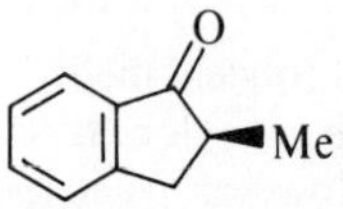

(14)

It is well known that oxidation of anthracene in MeCN yields the hydroxylated product. However, when the solvent is dehydrated by the addition of trifluoroacetic anhydride (TFAn) (4%), the solvent becomes the reacting nucleophile (Scheme 70).[233] It has been shown that thorough drying of solvents, *e.g.* with suspended

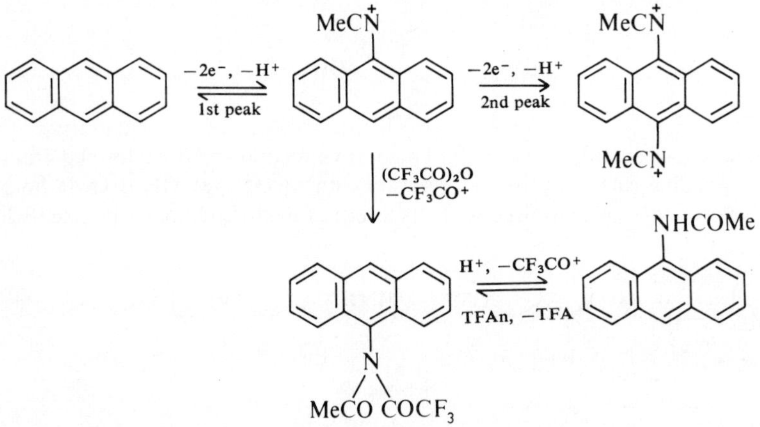

Scheme 70

alumina in acetonitrile, methylene chloride, or nitrobenzene, increases the stabilities of cation radicals derived from anthracenes. In the case of 9,10-dianisylanthracene even the dication is sufficiently stable for the equilibrium constant to be measured for the disproportionation reaction of the radical cation.[234] Dications of the anthracenes react with TFA (Scheme 71).

The biphenylene ring system also supports relatively stable radical cations and dications.[235] Biphenylene and its 2,3,6,7-tetramethoxy-derivative have been compared with biphenyl and its 3,3′,4,4′-tetramethoxy-derivative. In methylene chloride containing TFA, biphenylene radical cations are stable and can be isolated as their trifluoroacetates. In very dry MeCN at -60 °C, even the dication of tetramethoxybiphenylene is stable.

[233] O. Hammerlich and V. D. Parker, *J.C.S. Chem. Comm.*, 1974, 245.
[234] O. Hammerlich and V. D. Parker, *J. Amer. Chem. Soc.*, 1974, **96**, 4289.
[235] A. Ronlan and V. D. Parker, *J.C.S. Chem. Comm.*, 1974, 33.

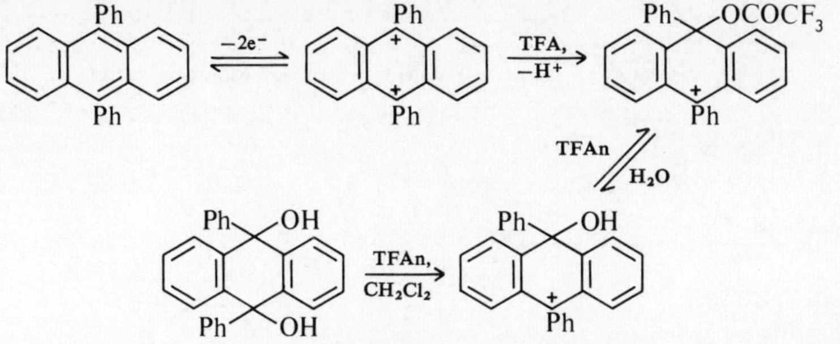

Scheme 71

The anodic pyridination of 9,10-diphenylanthracene in MeCN–TEAP containing pyridine has been shown to proceed by an *ECeC* mechanism.[236] Triethylamine formed at the cathode set up a base-exchange equilibrium with the pyridine adduct 9,10-diphenyl-9,10-dipyridinium-9,10-dihydroanthracene diperchlorate. Pyridine and 4-picoline form the substitution compound *N*-(6-benzo[*a*]pyrenyl)pyridinium perchlorate when benzo[*a*]pyrene is oxidized in MeCN–TEAP.[237] When 1-methylimidazole was used as the base, coupling took place at the 1-position of the hydrocarbon.

The stability of cation radicals and dications derived from methoxy-substituted biphenyls has been investigated by cyclic voltammetry in MeCN, CH_2Cl_2, and CCl_3CO_2H.[238] In the series $MeO(C_6H_4)_nOMe$, optimum stability is achieved for $n = 2$ or 3. Coplanarity of both phenyl rings is important (*e.g.* Scheme 72).

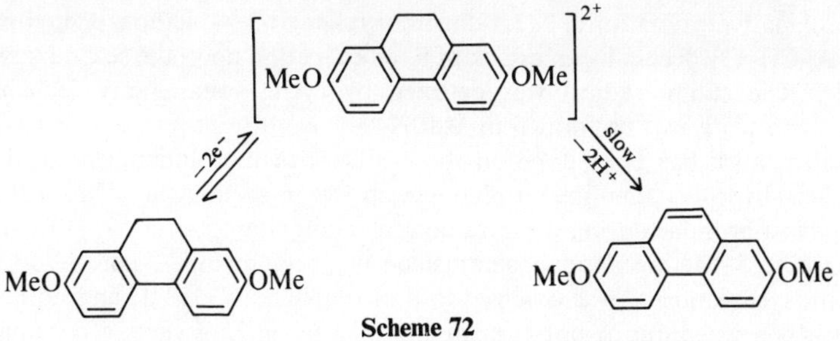

Scheme 72

Anodic coupling between benzenoid rings of various bis(3,4-dimethoxyphenyl)-alkanes in CH_2Cl_2–TFA solvent mixture occurs in higher yields than when the chemical variant, using tris(acetonylacetonate)manganese, is employed (Scheme 73).[239]

Intramolecular coupling reactions of various methyl- and methoxy-substituted bibenzyls have been performed at platinum anodes in MeCN–$LiClO_4$ at 0 °C.[240]

---

[236] D. T. Shang and H. N. Blount, *J. Electroanalyt. Chem.*, 1974, **54**, 305.
[237] G. M. Blackburn and J. P. Will, *J.C.S. Chem. Comm.*, 1974, 67.
[238] A. Ronlan, J. Coleman, O. Hammerlich, and V. D. Parker, *J. Amer. Chem. Soc.*, 1974, **96**, 845.
[239] A. Ronlan and V. D. Parker, *J. Org. Chem.*, 1974, **39**, 1014.
[240] J. R. Falck, L. L. Miller, and F. R. Stermitz, *J. Amer. Chem. Soc.*, 1974, **96**, 2981.

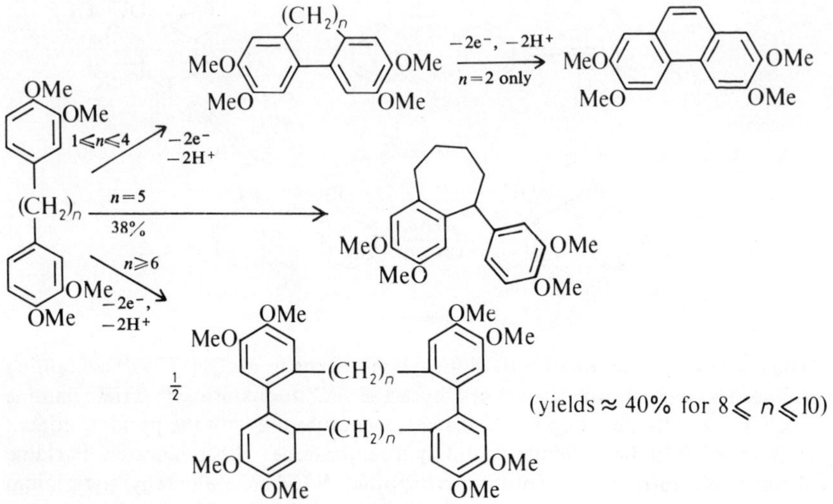

Scheme 73

The mechanism, which includes coupling of the dication diradicals, involves an interesting subsequent migration of the ethylene bridge. Phenanthrone products can be used to prepare steroids and terpenoids (Scheme 74).[241] An exception was 2-methyl-4,5-dimethoxybibenzyl, which, instead of cyclizing, gave 2-methyl-4,5-dimethoxy-α-hydroxybibenzyl, benzylacetonitrile, 6-methylveratryl alcohol, and 6-methylveratraldehyde.

The oxidative cyclization of 1-(3,4-dimethoxyphenyl)-3-(4-methoxyphenyl)propane in TFA–CH_2Cl_2 occurs by an attack of a radical cation upon the second benzenoid ring.[242] The cation radical was detected by cyclic voltammetry (Scheme 75). Preparative work was performed in MeCN.

Further work has been done on the synthesis of morphinandienones from 1-benzyltetrahydroisoquinolines at platinum anodes in acetonitrile.[243] 6-Bromolaudanosine lost bromine during the intramolecular coupling to form 16% of O-methylflavinantine. 5′-Methoxylaudanosine yielded 38% of O-methyl-C-norandrocymbine; 8′-methoxylaudanosine was oxidized to protostephanone (35%); and norlaudanosine was converted into N-nor-O-methylflavinantine in 22% yield, according to the now familiar pattern (Scheme 76). However, when R = CHO, oxidation causes fission. After reduction of the products by $LiAlH_4$, 3,4-dimethoxybenzyl alcohol and O-methylcorypalline were isolated.

Traces of triphenyloxonium cation have been produced in the oxidation of a mixture of diphenyl ether and benzene at a platinum anode in a divided cell containing TEAF in aprotic organic solvents.[244]

Rotating-disc and ring-disc studies of the oxidation of p-methoxystyrene have shown that the radical resulting from the transfer of one electron and one proton

---

[241] J. R. Falck, L. L. Miller, and F. R. Stermitz, U.S.P. 491 610 (*Chem. Abs.*, 1975, **82**, 117 889).
[242] V. D. Parker, U. Palmquist, and A. Ronlan, *Acta Chem. Scand.* (*B*), 1974, **28**, 1241.
[243] J. R. Falck, L. L. Miller, and F. R. Stermitz, *Tetrahedron*, 1974, **30**, 931.
[244] H. Behret and G. Sandstede, *J. Electroanalyt. Chem.*, 1974, **56**, 455.

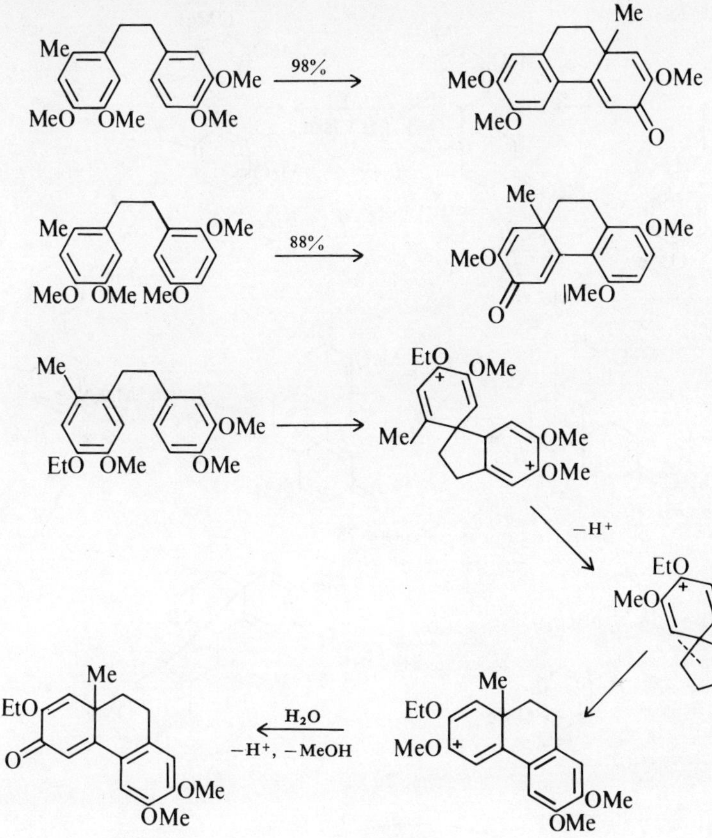

Scheme 74

can take up oxygen in a chain process (Scheme 77).[245] In the presence of pyridine, 4-methoxystyrene, 4-ethoxystyrene, and 3,4-dimethoxypropenylbenzene all lose two electrons and add two pyridine molecules in an *ECEC* sequence at platinum anodes in acetonitrile.[246] Quinoline can be used in place of pyridine.[247]

Alkyl Iodides.—Further studies have been conducted on the formation of alkylacetamides by anodic oxidation of alkyl iodides in acetonitrile. Electrogenerated iodonium ions as oxidant perform the same reactions with alkyl iodides as does anodic polarization, *e.g.* (*S*)-2-octyl iodide yields (*R*)-2-octylacetamide, (*S*)-2-octylacetamide, and 3-octylamine.[248] It is concluded that the anodic variant operates through the intermediacy of iodonium ions generated from iodine *in situ*. It has been deduced that carbonium ions involved in these processes are not free, and

[245] S. Müller and R. Landsberg, *Z. phys. Chem.* (*Leipzig*), 1974, **255**, 831.
[246] S. Müller and R. Landsberg, *Z. Chem.*, 1974, **14**, 442.
[247] R. Landsberg and S. Müller, *Electrochim. Acta*, 1974, **19**, 681.
[248] L. L. Miller and B. F. Watkins, *Tetrahedron Letters*, 1974, 4495.

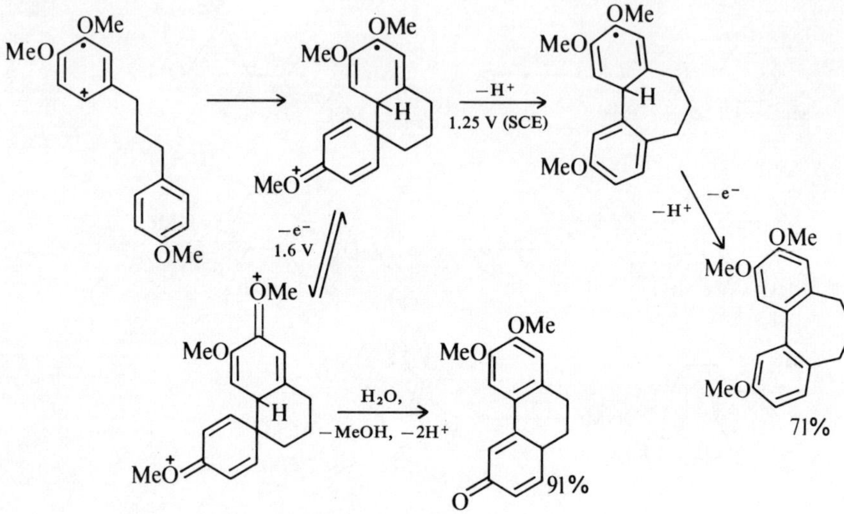

Scheme 75

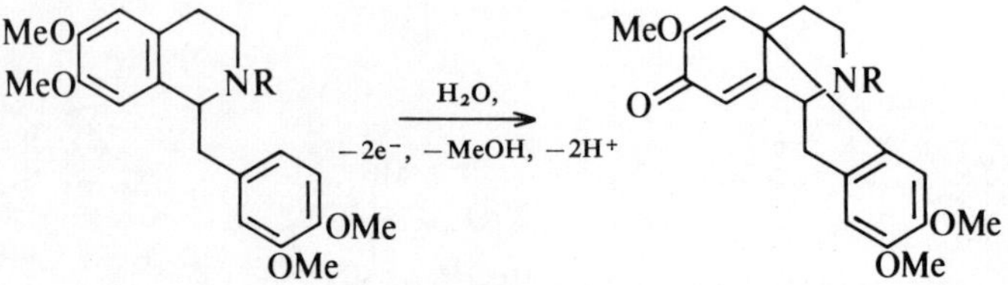

Scheme 76

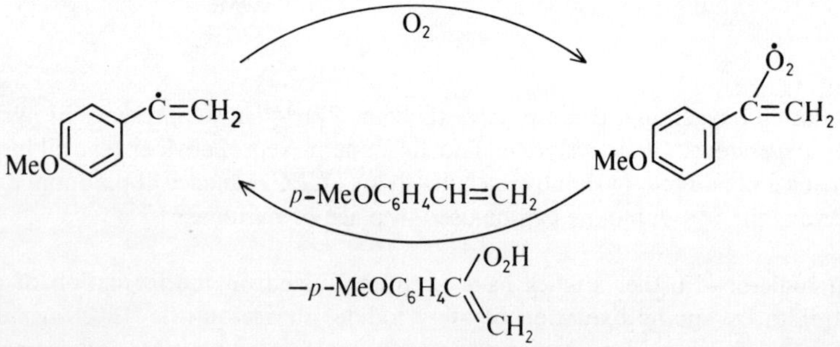

Scheme 77

that a concerted mechanism which includes a neighbouring-group effect and solvent assistance occurs.[249] This is indicated by a comparison of the product distributions with those obtained in reactions with known carbonium ion mechanisms (*e.g.* tosylate solvolysis). Further evidence is obtained from the fact that racemization is not complete.[250]

[249] A. Laurent, E. Laurent, and R. Tardevil, *Tetrahedron*, 1974, **30**, 3423.
[250] A. Laurent, E. Laurent, and R. Tardevil, *Tetrahedron*, 1974, **30**, 3431.

Phenols.—The lignin model compound 3,5-dimethoxy-4-hydroxy-α-methylbenzyl alcohol is oxidized at a platinum anode in wet MeCN–0.1M-TBAP in 2 two-electron stages (Scheme 78).[251] 3,5-Dimethoxy-4-hydroxyacetophenone, by contrast, retains its side-chain, to yield 3,4-dihydroxy-5-methoxyacetophenone.[252]

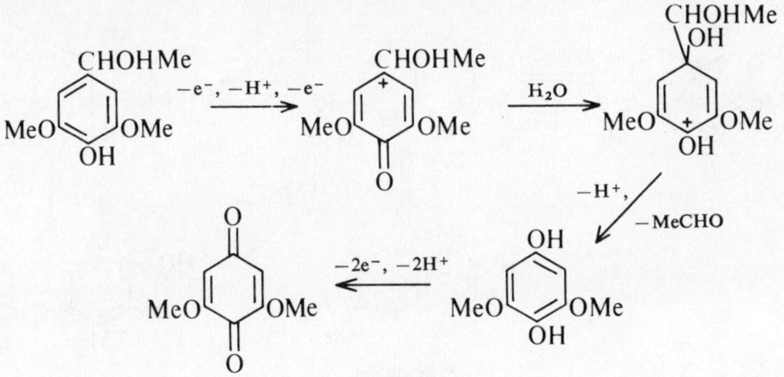

Scheme 78

Intramolecular anodic coupling reactions of diarylpropanes have been reported.[253] In an undivided cell containing MeCN–NaClO$_4$–Na$_2$CO$_3$ with a platinum anode, oxidation of the phenol ring occurs as shown in Scheme 79. The spiro-dienone

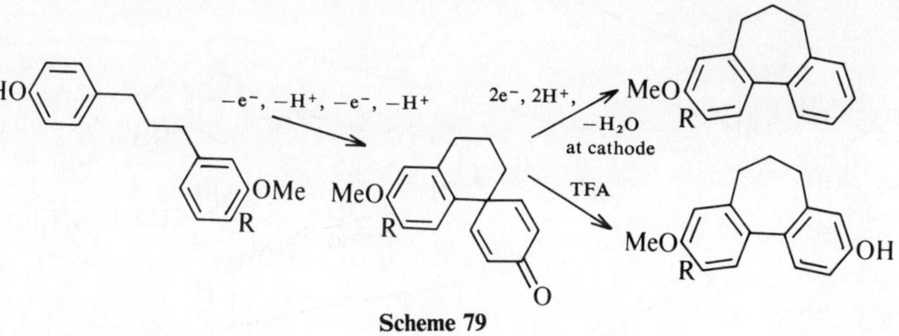

Scheme 79

(R = H or MeO) is formed in high yield, probably *via* a phenoxonium ion. Compound (15) behaves analogously: *o*- and *p*-hydroquinones are readily oxidized.[254] Consider the α-tocopherol model (Scheme 80). Pyrocatechol, dopamine, and adrenaline are oxidized in the same pattern as 3,4-dihydroxy-phenylalanine (Scheme 81).[255] At pH < 4, cyclization is precluded by protonation of the nitrogen atom. Nevertheless, at all pH values, a four-electron oxidation is observed, following an *ECe* mechanism.

---

[251] C. Steelink and W. E. Britton, *Tetrahedron Letters*, 1974, 2869.
[252] W. E. Britton and C. Steelink, *Tetrahedron Letters*, 1974, 2873.
[253] U. Palmquist, A. Ronlan, and V. D. Parker, *Acta Chem. Scand.* (*B*), 1974, **28**, 267.
[254] U. Svanholm, K. Bechgaard, and V. D. Parker, *J. Amer. Chem. Soc.*, 1974, **96**, 2409.
[255] A. Brun and R. Rossett, *J. Electroanalyt. Chem.*, 1974, **49**, 287.

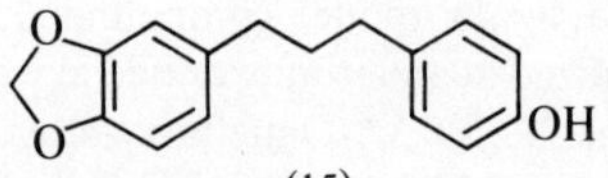

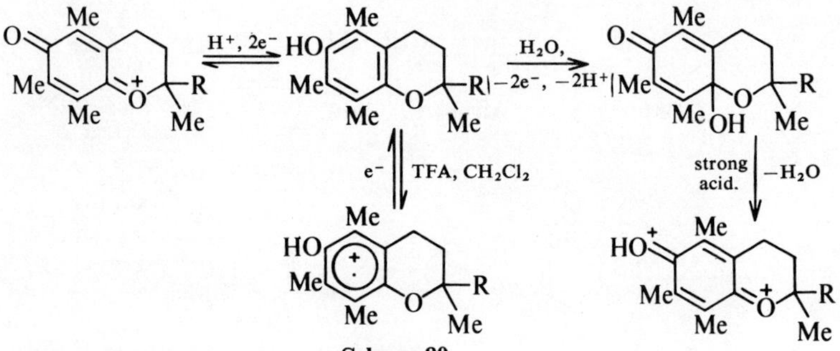

Scheme 80

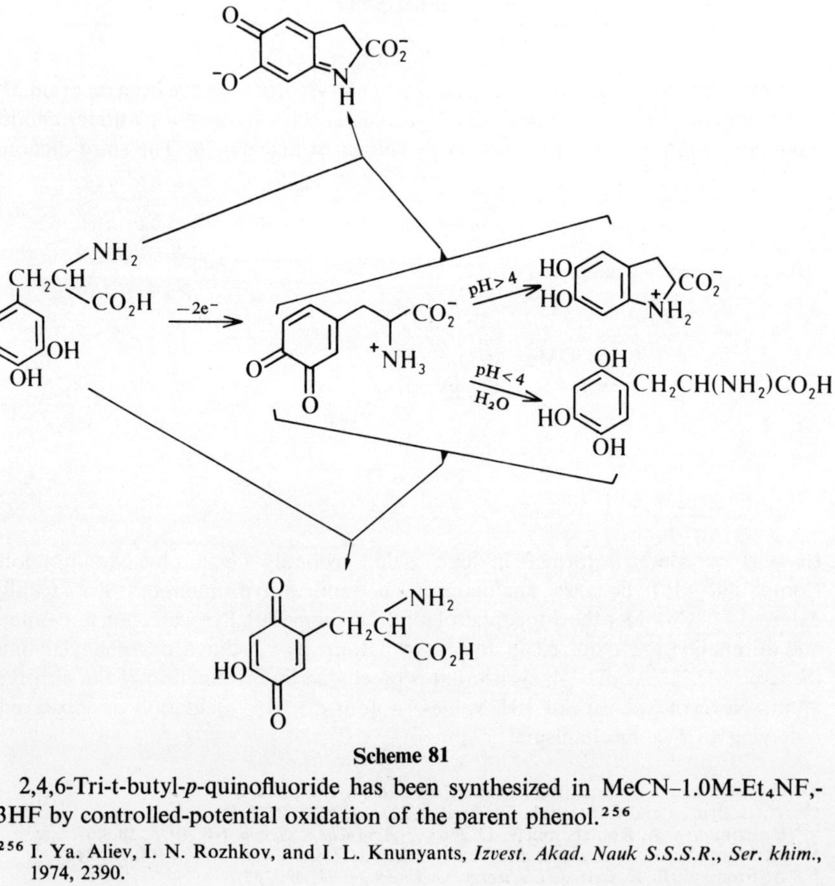

Scheme 81

2,4,6-Tri-t-butyl-*p*-quinofluoride has been synthesized in MeCN–1.0M-Et$_4$NF,-3HF by controlled-potential oxidation of the parent phenol.[256]

[256] I. Ya. Aliev, I. N. Rozhkov, and I. L. Knunyants, *Izvest. Akad. Nauk S.S.S.R., Ser. khim.*, 1974, 2390.

Esters.—Enol esters have been oxidized at carbon anodes in HOAc–TEAT.[257, 258] The carbonium ion intermediates give rise to four types of product: conjugated enones, α-acetoxycarbonyl compounds, *gem*-diacetoxy-compounds, and triacetoxy-compounds. For example, the second synthetic step in the conversion of 1-menthone into 1-citronellol can be carried out anodically in 97% yield (Scheme 82). The other patterns are exemplified in Scheme 83.[259]

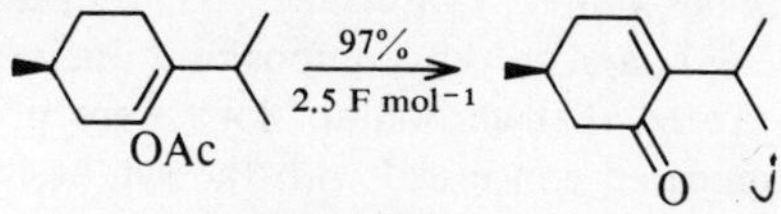

Scheme 82

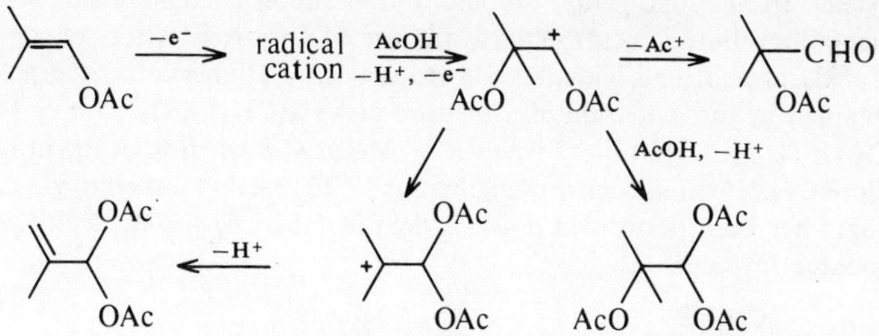

Scheme 83

Selective acetamidation of aliphatic esters in the ω-position has been achieved at platinum anodes in MeCN–LiClO$_4$.[260] Oxidation of the electrolyte occurs, leading to carbonium ion intermediates. The passage of 3 F mol^{-1} gives rise to yields as high as 70%, *e.g.*

$$MeCH_2CH_2CO_2Et \longrightarrow MeCONHCH_2CH_2CH_2CO_2Et$$

Carboxylic Acids.—Various symmetrical products have been prepared by means of the Kolbe reaction. Symmetrical γ-diketones can be produced in yields as high as 35% by oxidation of 0.1 mol l^{-1} solutions of β-keto-carboxylates in 40% dioxan–water mixtures at a platinum anode in an undivided cell:[261]

$$2R^1COCR^2R^3CO_2^- \longrightarrow 2e^- + 2CO_2 + (R^1COCR^2R^3—)_2$$

Muconic acid and its esters can be formed in 1:1 DMF–MeOH mixtures containing unsaturated hydrocarbon catalysts, by oxidation of maleic acid half-esters.[262, 263]

Polarization measurements of the Brown–Walker reaction have been made to investigate the effects of changing process variables. The shape of the curves for this

---

[257] T. Shono, Y. Matsumura, K. Hibino, and S. Miyawaki, *Tetrahedron Letters*, 1974, 1295.
[258] T. Shono, Japan. Kokai 74 11 851 (*Chem. Abs.*, 1974, **81**, 25 205).
[259] T. Shono, Y. Matsumura, and Y. Nakagawa, *J. Amer. Chem. Soc.*, 1974, **96**, 3532.
[260] L. L. Miller and V. Ramachandran, *J. Org. Chem.*, 1974, **39**, 369.
[261] D. Lelandais and M. Chkir, *Tetrahedron Letters*, 1974, 3113.
[262] L. A. Mirkind, L. V. Aniskova, and M. Ya. Fioshin, Russ. P. 427 922 (*Chem. Abs.*, 1974, **81**, 49 284).
[263] L. V. Aniskova, L. A. Mirkind, M. Ya. Fioshin, and A. G. Korienko, Russ. P. 436 815 (*Chem. Abs.*, 1974, **81**, 135 461).

reaction on Pt in methanol is similar to that for acetate oxidation. As the chain length is increased, the potential required to discharge the half-ester progressively falls.[264] In the oxidation of monomethyl glutarate, the maximum current for oxidation of the solvent methanol depends strongly upon the anode material.[265] A yield of dimethyl octanedioate in excess of 80% can be obtained under optimum conditions.[266] Mono-esters of perfluorodicarboxylic acids gave the Brown–Walker reaction on platinum anodes in MeCN–MeOH.[267] The influence on yields of changes of temperature, chain length, and solvent composition was explored.[268] The nature of the by-products indicated that carbonium ions were probably intermediates. Further patents have appeared concerned with the synthesis of sebacic acid ester from adipic acid half-esters in methanol.[269, 270]

Several crossed Kolbe reactions have been studied. The optimization of crossed-dimer yields from monomethyl adipate and perfluorinated aliphatic acids (*e.g.* TFA, heptafluorobutyric acid) requires the use of an excess of the weaker unsubstituted acids, and solvents with alkaline properties.[271] Higher carboxylic acids have been obtained by the oxidation of a mixture of $CH_3(CH_2)_m CO_2H$ ($m = 14$ or 16) and $ROCO(CH_2)_n CO_2H$ ($n = 7$ when $R = Me$, $n = 8$ when $R = Et$) in refluxing alcoholic KOH.[272] Similar chain lengthening of (*S*)-methyl 2-methoxy-3-carboxypropanoate has been performed in 21% yield in DMF by mixed electrolysis with dodecanoate:[273]

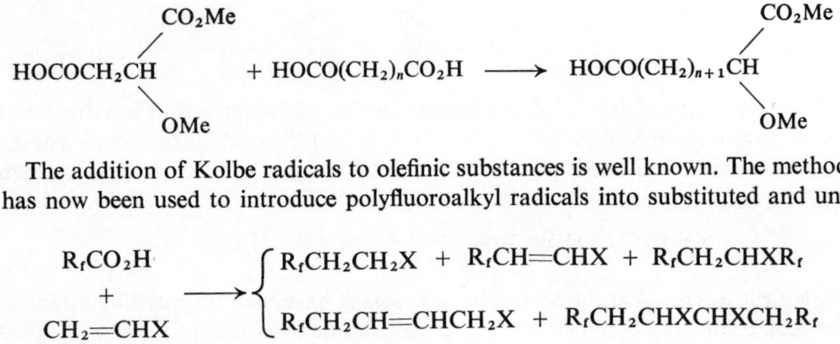

$$\underset{\underset{OMe}{\diagdown}}{\overset{\overset{CO_2Me}{\diagup}}{HOCOCH_2CH}} + HOCO(CH_2)_n CO_2H \longrightarrow \underset{\underset{OMe}{\diagdown}}{\overset{\overset{CO_2Me}{\diagup}}{HOCO(CH_2)_{n+1}CH}}$$

The addition of Kolbe radicals to olefinic substances is well known. The method has now been used to introduce polyfluoroalkyl radicals into substituted and un-

$$\begin{matrix} R_f CO_2H \\ + \\ CH_2=CHX \end{matrix} \longrightarrow \begin{cases} R_f CH_2CH_2X + R_f CH=CHX + R_f CH_2 CHXR_f \\ R_f CH_2CH=CHCH_2X + R_f CH_2 CHXCHXCH_2 R_f \end{cases}$$

(16)

Scheme 84

---

[264] G. A. Tarkhanov, E. P. Kovsman, G. N. Friedlin, and Yu. B. Vasilev, *Soviet Electrochem.*, 1974, **10**, 905.
[265] G. A. Tarkhanov, E. P. Kovsman, G. N. Friedlin, and Yu. B. Vasilev, *Soviet Electrochem.*, 1974, **10**, 600.
[266] E. P. Kovsman, S. S. Gluzman, G. A. Tarkhanov, N. I. Faingold, G. N. Friedlin, and T. F. Samoshina, *Soviet Electrochem.*, 1974, **10**, 521.
[267] V. V. Berenblit, E. S. Panitkova, V. P. Sass, and S. V. Sokolov, *Zhur. org. Khim.*, 1974, **10**, 2507.
[268] V. V. Berenblit, E. S. Panitkova, D. S. Rondarev, V. P. Sass, and S. V. Sokolov, *Zhur. priklad. Khim.*, 1974, **47**, 2427.
[269] I. Toshiro, K. Rinishi, and K. Chikayuki, Ger. Offen. 2 404 560 (*Chem. Abs.*, 1975, **82**, 49 312).
[270] F. Wenisch, H. Nohe, and H. Suter, Ger. Offen. 2 248 562 (*Chem. Abs.*, 1974, **81**, 32 626).
[271] O. N. Chechina and A. I. Levin, *Soviet Electrochem.*, 1974, **10**, 1112.
[272] G. A. Tember, Z. I. Getmanskaya, and P. R. Nichikova, *Zhur. priklad. Khim.*, 1974, **47**, 477.
[273] G. Odham, B. Petterson, and E. Stenhagen, *Acta Chem. Scand.* (*B*), 1974, **28**, 36.

substituted olefins in MeCN–H$_2$O in a divided cells with a platinum anode (Scheme 84).[274] The last product (16) predominated in most cases.

In the oxidation of the half-ester of phthalic acid in the presence of buta-1,3-diene at a platinum anode in DMF–MeOH, products formed include dimethyl phthalate and (17).[275] A further instance of an acetoxy radical adding to an olefinic bond is an intramolecular example (Scheme 85).[276]

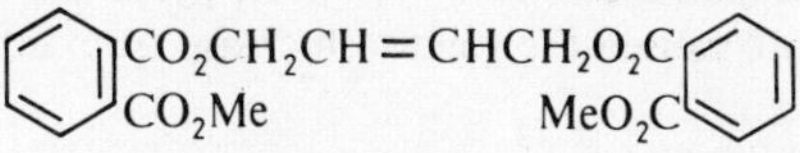

$$\text{[benzene]}\begin{array}{l}CO_2CH_2CH=CHCH_2O_2C\\CO_2Me \quad\quad\quad\quad\quad MeO_2C\end{array}\text{[benzene]}$$
(17)

$$PhCH=CHCH_2CH_2CO_2Na \xrightarrow[-2e^-, -2H^+]{MeOH} \underset{O}{\overset{O}{\bigcirc}}\!\!\!\!\!\!\underset{}{\overset{}{\diagup}}\!\!\!\!\underset{CH}{\overset{Ph}{\diagdown}}\!\!\!\!\underset{OMe}{}$$

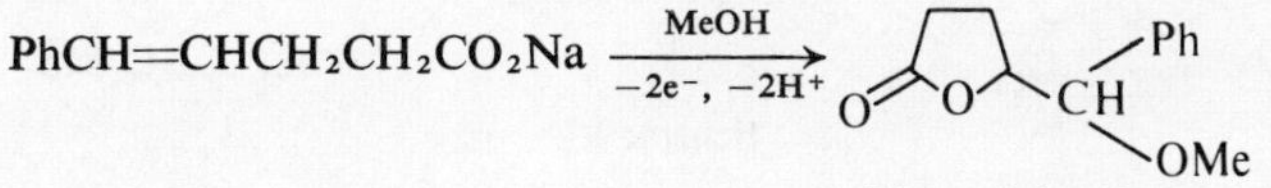

(90% yield of *threo* and *erythro* mixture)

Scheme 85

In the simple anodic dimerization of carboxylates, adsorbed aromatic, olefinic, and alicyclic hydrocarbons have been shown to act as catalytic promoters at platinum anodes.[277]

The well-known tendency of graphite anodes to favour products based upon carbonium ion intermediates was noted when α-alkoxy-acids were oxidized in dry MeCN.[278] Esters are formed in yields of 47—81% (Scheme 86).

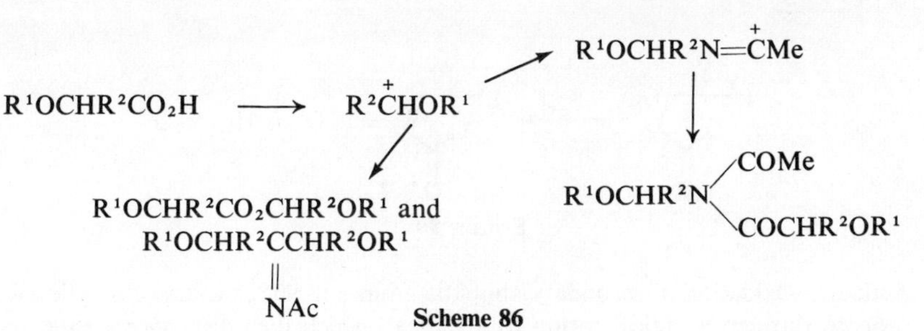

$$R^1OCHR^2CO_2H \longrightarrow R^2\overset{+}{C}HOR^1 \nearrow R^1OCHR^2N\overset{+}{=}CMe$$

$$R^1OCHR^2CO_2CHR^2OR^1 \text{ and}$$
$$R^1OCHR^2CCHR^2OR^1$$
$$\parallel$$
$$NAc$$

$$R^1OCHR^2N\!\!\!\begin{array}{c}\diagup COMe\\ \diagdown COCHR^2OR^1\end{array}$$

Scheme 86

The effects of nuclear substitution and electrochemical parameters upon the product distribution in the oxidation of various phenylacetate ions have been studied.[279] In addition to Kolbe dimer, (PhCH$_2$)$_2$, the products derived from carbonium ions,

---

[274] C. J. Brooks, P. J. Coe, D. M. Owen, A. E. Pedler, and J. C. Tatlow, *J.C.S. Chem. Comm.*, 1974, 323.
[275] L. V. Aniskova, L. A. Mirkind, and M. Ya. Fioshin, *Tr. Moskov. Khim.-Tekhnol. Inst.*, 1973, **75**, 212.
[276] F. M. Banda and R. Brettle, *J.C.S. Perkin I*, 1974, 1907.
[277] L. A. Mirkind, M. Ya. Fioshin, and L. V. Aniskova, *Doklady Akad. Nauk S.S.S.R.*, 1974, **219**, 1127.
[278] H. G. Thomas and E. Katzer, *Tetrahedron Letters*, 1974, 887.
[279] J. P. Coleman, R. Lines, J. H. P. Utley, and B. C. L. Weedon, *J.C.S. Perkin II*, 1974, 1064.

viz. PhCH$_2$OMe, PhCH$_2$OH, and PhCHO, are obtained in methanol. The aldehyde is favoured in more basic solutions, whereas when pH = pK_a the dimer predominates. However, the addition of NaClO$_4$ causes the yield of ether to increase at the expense of that of the dimer.

The synthesis of olefins from substituted succinic acids is a familiar variant of the Kolbe reaction. In this way, 1,4-bis(methoxycarbonyl)bicyclo[2,2,2]octa-2,5-diene has been prepared (Scheme 87).[280] In pyridine–water–Et$_3$N, the addition of 4-t-butylcatechol caused the yield to increase from 30 to 65%.

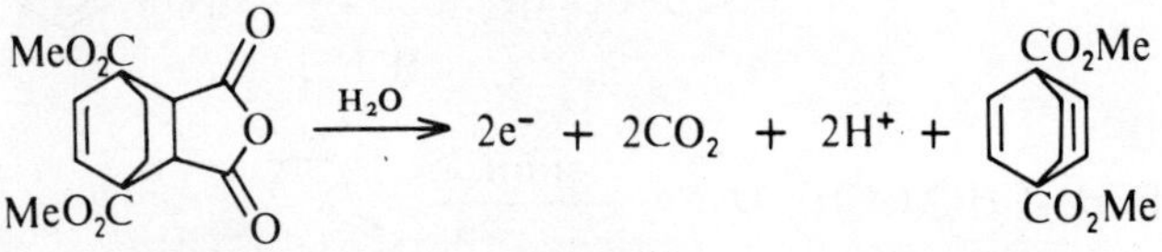

$$\text{MeO}_2\text{C-anhydride} \xrightarrow{\text{H}_2\text{O}} 2e^- + 2\text{CO}_2 + 2\text{H}^+ + \text{diene(CO}_2\text{Me)}_2$$

Scheme 87

γ-Substituted paraconic acids have been oxidized under basic conditions.[281] At platinum anodes in MeOH–NaOMe the carboxylate function is replaced either by hydrogen or by a methoxy-group (Scheme 88). When a carbon rod anode is used in pyridine–water–Et$_3$N, butenolides are formed (Scheme 89).

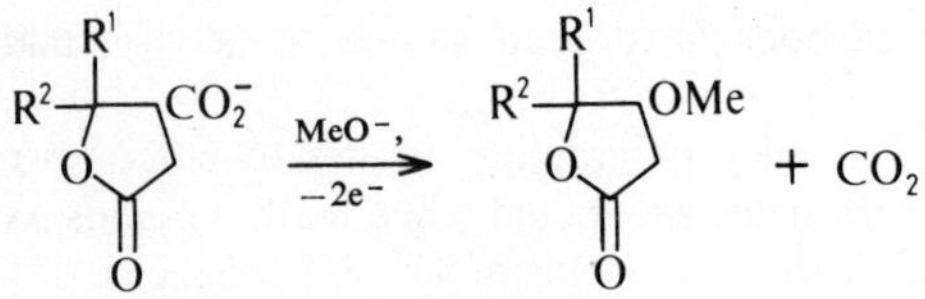

$$R^2\text{-}\underset{O}{\overset{R^1}{|}}\text{-CO}_2^- \xrightarrow[-2e^-]{\text{MeO}^-} R^2\text{-}\underset{O}{\overset{R^1}{|}}\text{-OMe} + \text{CO}_2$$

Scheme 88

$$R^2\text{-}\underset{O}{\overset{R^1}{|}}\text{-CO}_2^- \xrightarrow{49\%} \text{butenolide} + \text{CO}_2 + \text{H}^+ + 2e^-$$

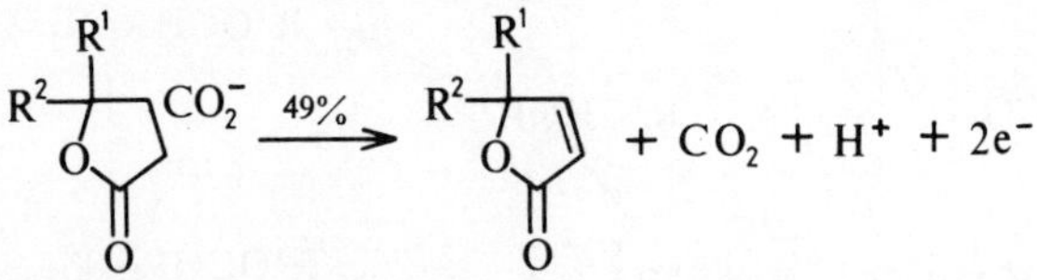

Scheme 89

Amines.—Oxidation of secondary aliphatic amines in 90% methanol is believed to proceed through a radical cation to a radical, which then disproportionates (*e.g.* Scheme 90).[282] Final products included BuNH$_2$, MeNH$_2$, and ammonia, with traces of Bu$_2$NMe and PhCH$_2$NMe$_2$.

$$\text{Bu}_2\text{NH} \longrightarrow e^- + \text{Bu}_2\text{NH}^{\cdot+} \xrightarrow{-\text{H}^+} \text{Bu}_2\dot{\text{N}} \xrightarrow{\times 2} \begin{array}{c} \text{Bu}_2\text{NH} \\ + \\ \text{BuN}=\text{CHCH}_2\text{CH}_2\text{Me} \end{array}$$

Scheme 90

[280] C. B. Warren and J. J. Bloomfield, *J. Org. Chem.*, 1973, **38**, 4011.
[281] S. Torii, T. Okamoto, and H. Tanaka, *J. Org. Chem.*, 1974, **39**, 2486.
[282] S. Wawzonek and S. M. Heilmann, *J. Electrochem. Soc.*, 1974, **121**, 378.

Phenethylamines are also oxidized to radical cations.[283] Benzaldehydes are formed when the side-chain is substituted by the hydroxy-group or when water is present in the solvent (Scheme 91).

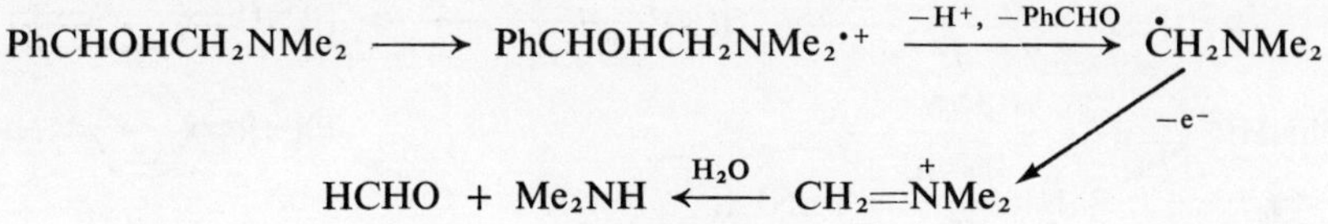

$$PhCHOHCH_2NMe_2 \longrightarrow PhCHOHCH_2NMe_2^{\cdot+} \xrightarrow{-H^+, -PhCHO} \overset{\cdot}{C}H_2NMe_2$$

$$HCHO + Me_2NH \xleftarrow{H_2O} CH_2=\overset{+}{N}Me_2 \xleftarrow{-e^-}$$

Scheme 91

Radical cations derived from NN-dimethylbenzylamine in methanol containing TBAF or KOH can be deprotonated in either of two positions, depending upon the base present.[284] Strong bases (hydroxide or alkoxide) lead to a statistical distribution of products, whereas if the base is an amine, attack of the methyl group predominates (90%) (Scheme 92). At higher potentials (1.5 V) the mechanism involves hydroxymethyl radicals obtained from direct oxidation of the solvent.

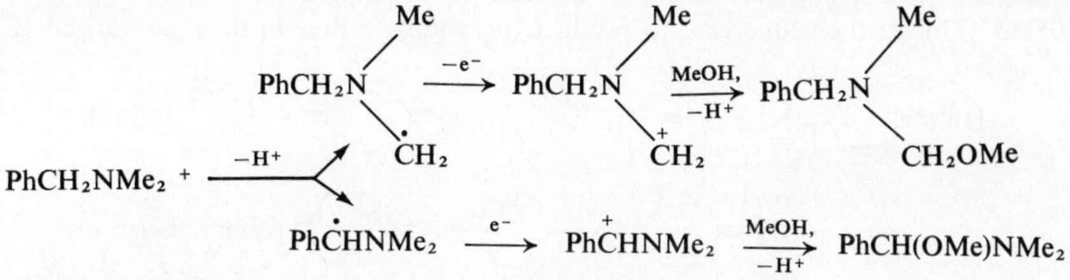

Scheme 92

The mechanism of anodic oxidation of NN-dimethylaniline to $NNN'N'$-tetramethylbenzidine in strongly acidic aqueous media has been shown to proceed through the coupling of neutral radicals derived from the cation radicals by deprotonation.[285] In $AlCl_3$–NaCl melts of various acidities the radical cations of substituted dimethylanilines and the dimeric dications were found to be relatively very stable (Scheme 93).[286] Although N-alkyl-anilines in strongly acidic media also yield

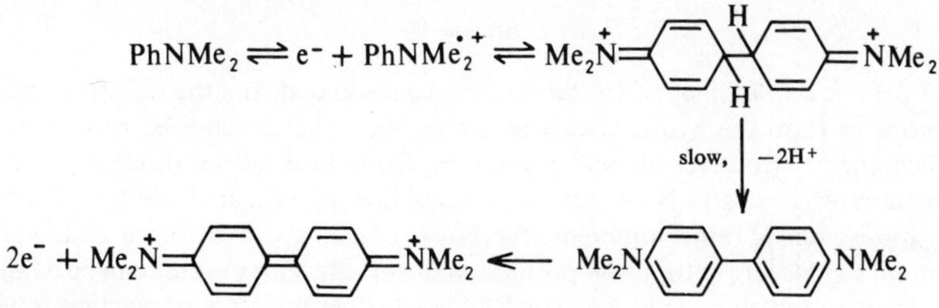

Scheme 93

[283] L. C. Portis, J. T. Klug, and C. K. Mann, *J. Org. Chem.*, 1974, **39**, 3488.
[284] J. E. Barry, M. Finkelstein, E. A. Mayeda, and S. D. Ross, *J. Org. Chem.*, 1974, **39**, 2695.
[285] G. Neubert and K. B. Prater, *J. Electrochem. Soc.*, 1974, **121**, 745.
[286] H. Lloyd Jones and R. A. Osteryoung, *J. Electroanalyt. Chem.*, 1974, **49**, 281.

benzidines, diphenylamines tend to be formed in non-aqueous media.[287] Product distribution is influenced by current density, reactant concentration, basicity, and

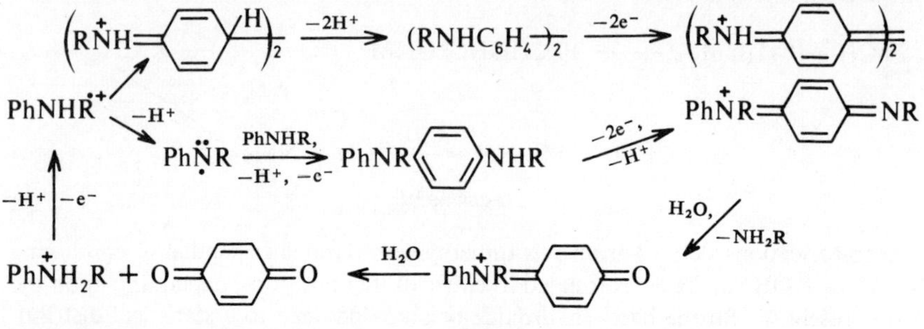

Scheme 94

size of the alkyl group (Scheme 94). Aniline itself is known to follow a similar pattern. However, under acidic conditions the quinone imine can dimerize to Willstatter's blue imine, which is oxidized to Willstatter's red imine (Scheme 95).[288] This further dimerizes, to produce nigraniline, which in turn is oxidized to

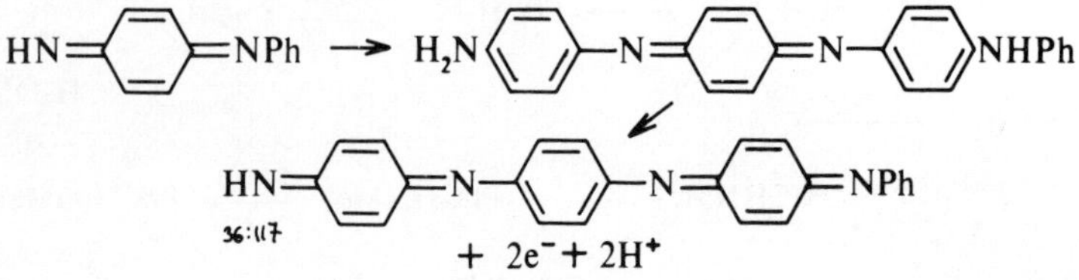

Scheme 95

aniline black. In neutral media the quinone imine reacts with aniline to yield *N*-phenyl-2,5-dianilino-*p*-benzoquinonedi-imine (Scheme 96).

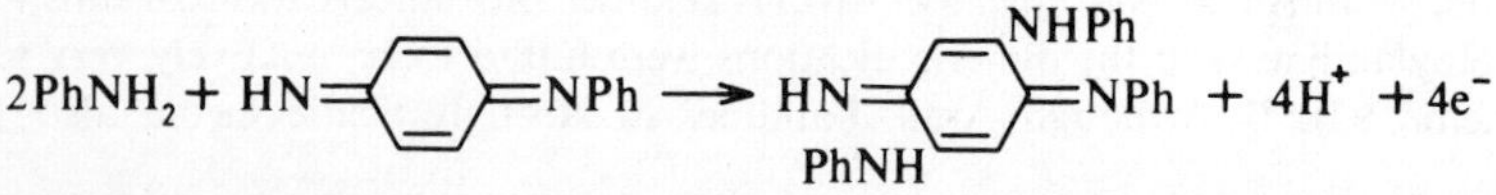

$$2PhNH_2 + HN{=}\!\!\bigcirc\!\!{=}NPh \longrightarrow HN{=}\!\!\bigcirc\!\!{=}NPh + 4H^+ + 4e^-$$

Scheme 96

At pH > 8, *o*-toluidine radical cations are deprotonated, and the resulting radicals dimerize to form the hydrazobenzene derivative, which is oxidized further to the azobenzene.[289] However, in acidic solutions, the radical cations dimerize, with the formation of C—C and N—C bonds. *m*-Toluidine behaves analogously. However, the *para*-isomer is rather different; the N—C coupling cannot occur at the *para*-position, so it occurs at the *ortho*-position, to form (18). This product can be oxidized anodically to the quinone imine, which adds a further molecule of reactant to yield (19).

[287] R. L. Hand and R. F. Nelson, *J. Amer. Chem. Soc.*, 1974, **96**, 850.
[288] L. Dunsch, *Z. Chem.*, 1974, **14**, 101.
[289] P. G. Desideri, L. Lepri, and D. Heimler, *J. Electroanalyt. Chem.*, 1974, **52**, 93.

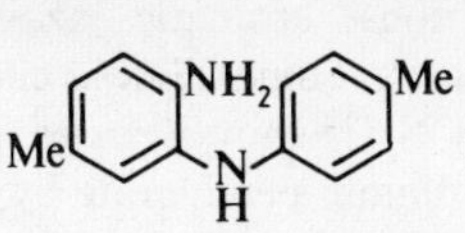

(18) 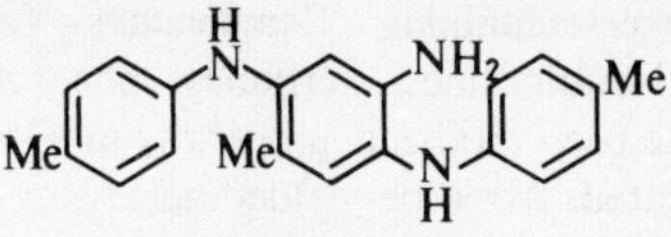 (19)

The conversion of substituted diphenylamines and triphenylamines into carbazoles at platinum anodes in MeCN–TEAP takes place if the intermediate cation radical is fairly stable (*e.g.* Scheme 97).[290]

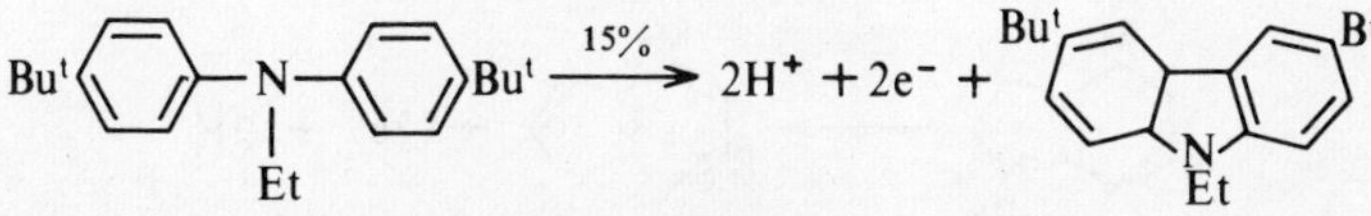

Scheme 97

The formation of *N*-phosphonomethylglycine by the oxidation of tertiary amines at a graphite anode in aqueous salt solution has been patented (Scheme 98).[291]

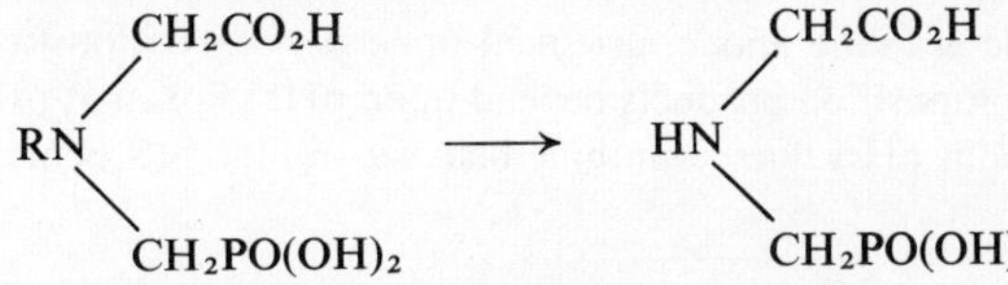

$R = CH_2CO_2H, CH_2Ph, CH_2CH=CH_2,$ or $CH_2PO(OH)_2$

Scheme 98

Nitrilotrimethylenetriphosphoric acid, formed in the Mannich reaction from ammonia, formaldehyde, and orthophosphorous acid, has been oxidized in 20% HCl solution at a graphite anode to give a 79% yield of iminodimethylenediphosphonic acid:[292]

$$N(CH_2PO_3H_2)_3 + H_2O \longrightarrow 2e^- + 2H^+ + HN(CH_2PO_3H_2)_2 + HCOPO_3H_2$$

α-Amino- and α-hydrazino-phosphinic acids have been oxidized in acidic aqueous solutions according to Scheme 99.[293]

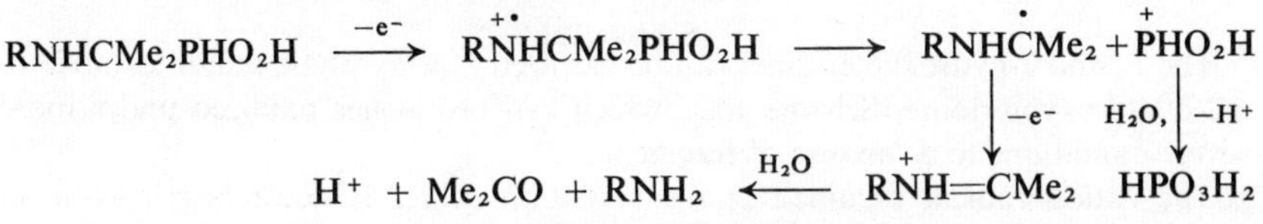

Scheme 99

[290] R. Reynolds, L. L. Line, and R. F. Nelson, *J. Amer. Chem. Soc.*, 1974, **96**, 1087.
[291] H. W. Frazier, L. R. Smith, and J. H. Wagenknecht, Ger. Offen., 2 363 634 (*Chem. Abs.*, 1974, **81**, 105 695).
[292] J. H. Wagenknecht, *Synth. React. Inorg. Metal-org. Chem.*, 1974, **4**, 567.
[293] G. H. Alt and J. H. Wagenknecht, *Synth. React. Inorg. Metal-org. Chem.*, 1974, **4**, 255.

Other Nitrogen-containing Compounds.—Various aromatic aza-hydrocarbons (quinolines, isoquinolines, acridines, phenazines, quinoxalines, cinnolines, and pyridine) have been oxidized anodically in MeCN.[294] In most cases the oxidations were complicated. Acridine yielded acridylacridinium perchlorate by attack of the radical cation on a second acridine molecule.

The dehydrogenation of 1,4-dihydro-3,5-dicarboxylic acid diethyl ester–2,6-dimethylpyridine has been carried out at a graphite anode in aqueous acetone solution.[295] The yield exceeded 65% and the current efficiency 80%. 6-Amino-7-azaindoline derivatives also undergo anodic dehydrogenation (Scheme 100).[296]

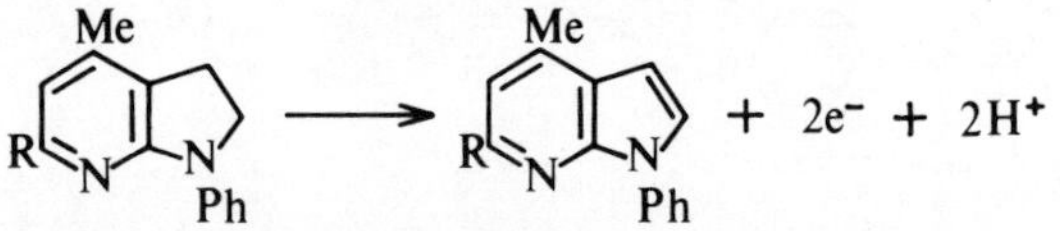

(R = NH_2, NHBu, NBu_2, NHPh, NEtPh, NMe_2, $NHSO_2C_6H_4Me$-p, or heterocyclic amino-)

Scheme 100

At a pyrolytic graphite anode, uric acid undergoes oxidation to its 4,5-diol. Its subsequent decomposition products depend upon pH: alloxan at pH 1, allantoin at pH 7, urea at both pH values. Xanthine behaves similarly (Scheme 101).[297]

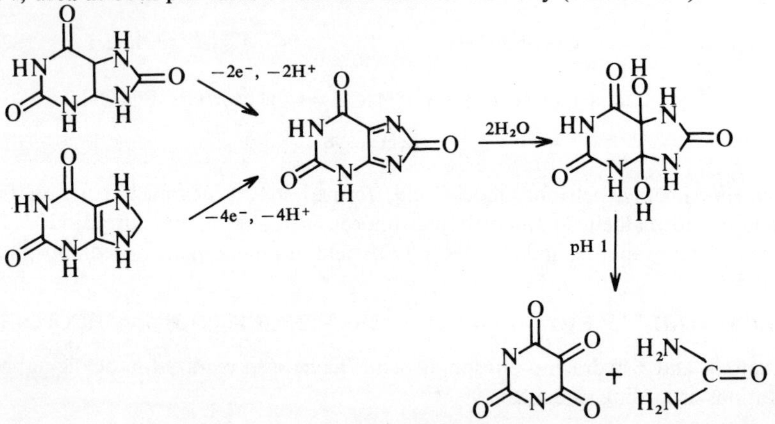

Scheme 101

The 6- and 7-hydroxypteridines can be oxidized at a pyrolytic graphite anode to 6,7-dihydroxypteridine (Scheme 102), which can be further oxidized under more severe conditions to a mixture of fragments.[298]

The cation radical obtained in the oxidation of 1,2,3,5-tetraphenylpyrrole in

---

[294] L. Marcoux and R. N. Adams, *J. Electroanalyt. Chem.*, 1974, **49**, 111.
[295] H. Berge, P. Jeroschewski, and K. Tellert, East Ger. P. 106 376 (*Chem. Abs.*, 1975, **82**, 23 757).
[296] I. N. Palint, Y. I. Vainshtein, L. N. Yakhontov, D. M. Krasnokutskaya, and M. M. Goldin, *Tr. Vses. Nauchno-Issled. Inst. Khim. Reakt. Osobo. Chist. Khim. Veshchestv.*, 1973, **35**, 68.
[297] G. Dryhurst, *Bioelectrochem. Bioenerg.*, 1974, **1**, 49.
[298] D. L. McAllister and G. Dryhurst, *J. Electroanalyt. Chem.*, 1974, **55**, 69.

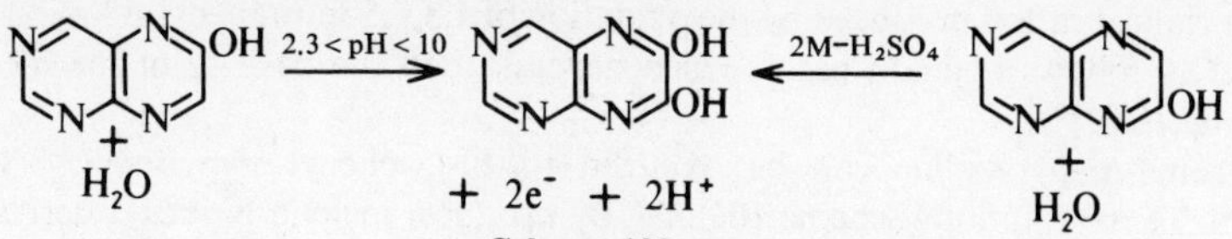

Scheme 102

MeCN dimerizes in neutral solution to 1,2,3,5,1',2',3',5'-octaphenyl-4,4'-bipyrrole, but in the presence of the nucleophilic electrolyte Et$_4$NCN the addition of cyanide

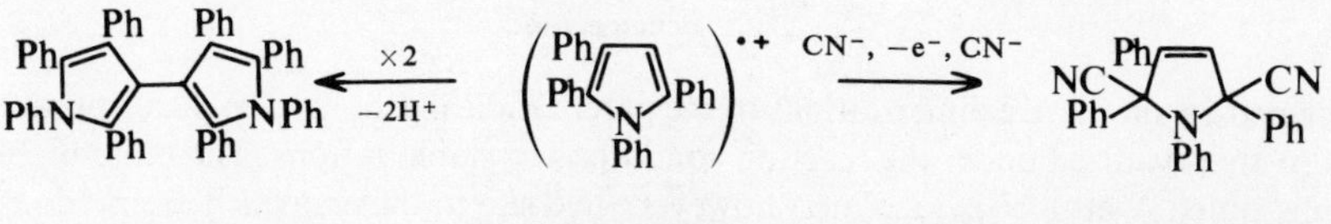

Scheme 103

occurs, to yield 1,2,3,5-tetraphenyl-2,5-dicyano-Δ^3-pyrroline (Scheme 103).[299] Fully aryl-substituted pyrroles, in the absence of nucleophiles, give fairly stable dications.[300] In nitromethane solutions, however, the anion arising from dissociation of the solvent behaves as a nucleophile, with the resulting expansion of the ring. In

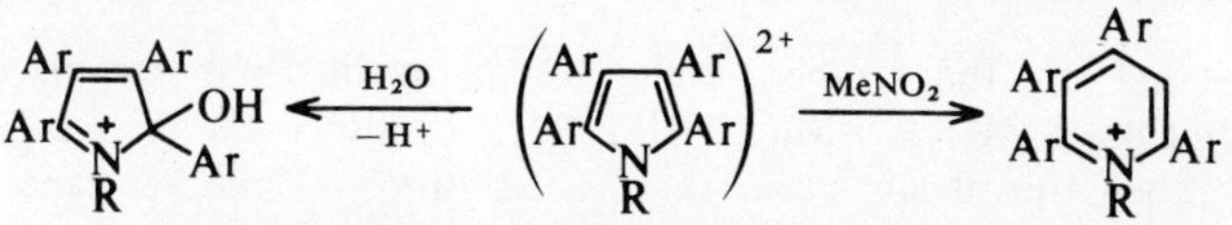

(Ar = Ph, MeC$_6$H$_4$, or MeOC$_6$H$_4$; R = Me, CD$_3$, or Et)

Scheme 104

the presence of water, addition occurs (Scheme 104). This result was found earlier for similar pyrroles that were not substituted at nitrogen.

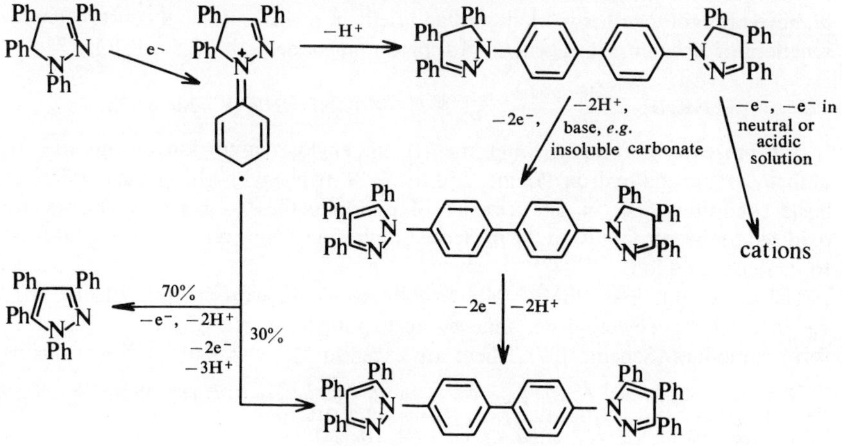

Scheme 105

[299] S. Longchamp, C. Caullet, and M. Libert, *Bull. Soc. chim. France*, 1974, 353.
[300] M. Libert and C. Caullet, *Bull. Soc. chim. France*, 1974, 800.

The radical cation produced by the oxidation of 1,3,4,5-tetraphenyl-Δ^2-pyrazoline in MeCN follows reaction paths which depend upon the basicity of the medium (Scheme 105).[301]

1-Phenyl-Δ^2-pyrazoline can be oxidized to bis(1-phenyl-Δ^2-pyrazolin-3-yl) in MeCN–TEAP solution (Scheme 106).[302] By-products include biphenyl derivatives

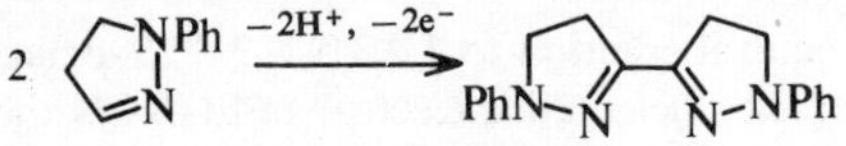

Scheme 106

formed by alternative dimerizations of the parent radical cations. The major product is further oxidized under the reaction conditions to stable cations. The more highly substituted N-aryl-Δ^2-pyrazolines, however, give rise to dimers which are predominantly of the biphenyl type (20).[303] Thus various 1,3,5-triaryl-Δ^2-pyrazolines have

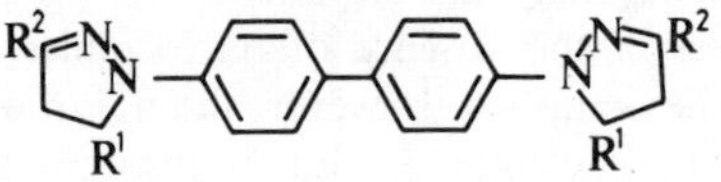

(20)

been converted into the 4,4'-bis(3,5-diaryl-Δ^2-pyrazolin-1-yl)biphenyls (20; R^1 = Ph, R^2 = p-ClC$_6$H$_4$; R^1 = p-Me$_2$NC$_6$H$_4$, R^2 = Ph; R^1 = Ph, R^2 = p-NO$_2$C$_6$H$_4$). However, when the phenyl group in the 1-position of the reactant was *para*-substituted by the nitro-group, the reaction was a simple dehydrogenation to the corresponding 1-p-nitrophenylpyrazole. The oxidations of several 1,3,5-triaryl-Δ^2-pyrazolines have been studied kinetically at the rotating disc electrode.[304] In the presence of pyridine these compounds are oxidized in MeCN–0.1M-TEAP to the corresponding pyrazoles and the 4,4'-bis(pyrazol-1-yl)biphenyls.[305]

Several studies have been made of the oxidation of organic hydrazines. Cauquis *et al.* have brought together work described briefly in earlier notes and concerned with reactions of diazenium ions generated at platinum anodes in MeCN–LiClO$_4$:[306, 307]

$$PhNRNH_2 \xrightarrow{\text{acidic media}} 2e^- + H^+ + Ph\overset{+}{N}R=NH \ (R = Me \text{ or } Ph)$$

In the presence of residual water the 1,1-diphenyldiazenium ion decomposes to diphenylamine and hydroxylamine, and to NNN'-triphenyl-p-phenylenediamine. In basic conditions (*e.g.* in the presence of pyridine) these organic hydrazines are oxidized to the tetrazenes. In the presence of pyridine, triphenylhydrazine is oxidized to its diazenium ion.

The diazenium ions Ph$\overset{+}{N}$R=NH usually condense with suitable olefins (*e.g.* styrene, 1,1-diphenylethylene, indene, acenaphthylene, phenylnorborn-2-ene) to form cinnolines (Scheme 107). There are exceptions to this pattern. Some olefins

---

[301] P. Corbon, G. Barbey, A. Dupre, and C. Caullet, *Bull. Soc. chim. France*, 1974, 768.
[302] F. Pragst and I. Schwertfeger, *J. prakt. Chem.*, 1974, **316**, 795.
[303] F. Pragst and B. Siefke, *J. prakt. Chem.*, 1974, **316**, 267.
[304] F. Pragst and W. Jugelt, *J. prakt. Chem.*, 1974, **316**, 981.
[305] F. Pragst, *Z. Chem.*, 1974, **14**, 236.
[306] G. Cauquis, B. Chabaud, and M. Genies, *Bull. Soc. chim. France*, 1973, 3482.
[307] G. Cauquis, B. Chabaud, and M. Genies, *Bull. Soc. chim. France*, 1973, 3487.

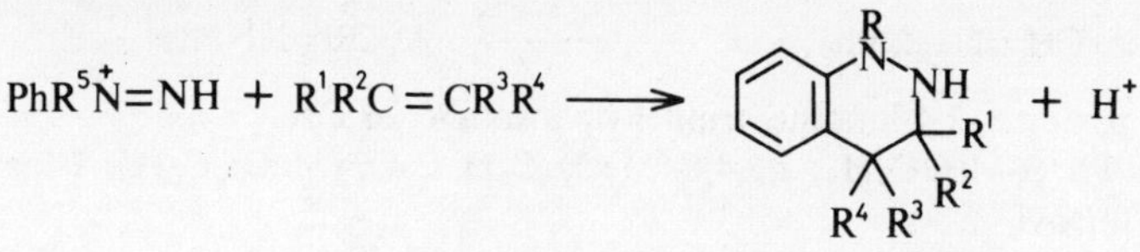

$$PhR^5\overset{+}{N}=NH + R^1R^2C=CR^3R^4 \longrightarrow \text{[bicyclic product]} + H^+$$

Scheme 107

(*e.g. cis*-stilbene) are unreactive. Others (*e.g. trans*-stilbene) gave products which simultaneously dehydrogenate. Cyclohexene gives rise to a quite different reaction,

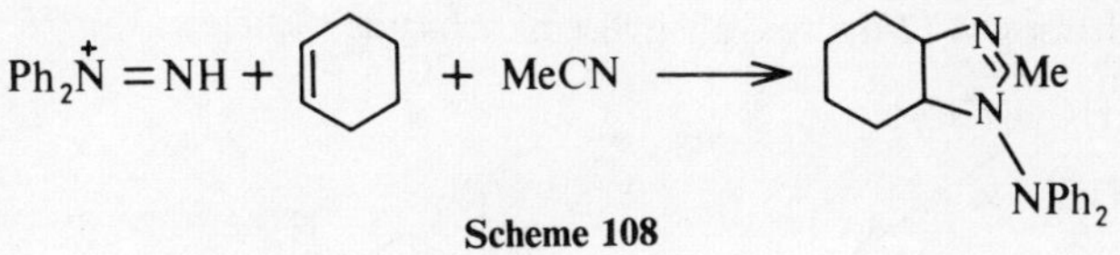

$$Ph_2\overset{+}{N}=NH + \text{[cyclohexene]} + MeCN \longrightarrow \text{[product with NPh}_2\text{]}$$

Scheme 108

which involves the solvent (Scheme 108). The reaction with the but-2-enes is also different:

$$Ph\overset{+}{NR}=NH + MeCH=CHMe \longrightarrow PhNRNHCHMeCH=CH_2 + H^+$$

In the case of the diphenyldiazenium ion it is possible to isolate the intermediate aziridines by neutralizing the acidic medium soon after adding the olefin. The stereochemistry of the olefin is conserved in the aziridine (Scheme 109). Triphenyldiazenium ions condense only with olefins which possess donor groups, and then only

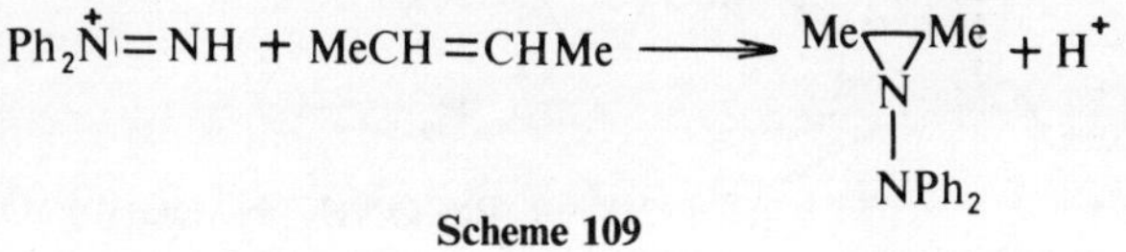

$$Ph_2\overset{+}{N}=NH + MeCH=CHMe \longrightarrow \text{[aziridine with NPh}_2\text{]} + H^+$$

Scheme 109

$$Ph_2\overset{+}{N}=NPh + EtOCH=CH_2 \longrightarrow \text{[product with OEt, NPh, Ph]} + H^+$$

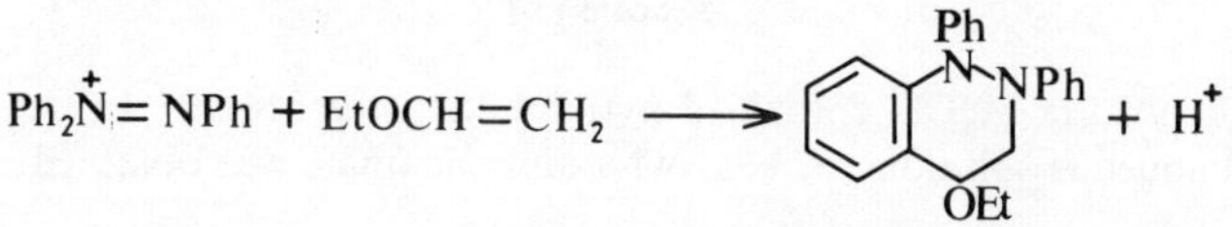

Scheme 110

provided that the base 2,6-lutidine is present (Scheme 110). Other anodic coupling reactions involving organic hydrazines are described in Schemes 111[308] and 112.[309]

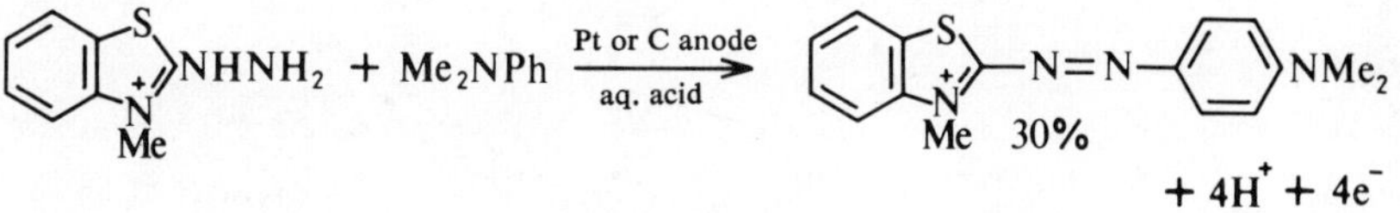

$$\text{[benzothiazolium]-NHNH}_2 + Me_2NPh \xrightarrow[\text{aq. acid}]{\text{Pt or C anode}} \text{[benzothiazolium]}-N=N-\text{[phenyl]}-NMe_2$$
30%
$$+ 4H^+ + 4e^-$$

Scheme 111

[308] G. Henze and E. Keller, *Z. Chem.*, 1974, **14**, 238.
[309] G. Barbey and C. Caullet, *Tetrahedron Letters*, 1974, 1717.

$$\text{ArCH}=\text{N}-\text{NPh}_2 + \text{R} \xrightarrow{\text{LiClO}_4,\ \text{MeCN}} \text{ArC}\overset{+}{\text{R}}=\text{NNPh}_2 + \text{H}^+ + 2\text{e}^-$$

R = pyridine, 2,4-lutidine, triphenylphosphine, or CN$^-$
Ar = Ph, p-MeOC$_6$H$_4$, 2,3,4,6-(MeO)$_4$C$_6$H, 2,4,6-(MeO)$_3$C$_6$H$_2$, 2-furyl, or 2-thienyl

Scheme 112

4-(2-Hydroxyphenyl)semicarbazide undergoes an intramolecular anodic coupling reaction to form 2-benzoxazolinone. A by-product is the dimer 1-(2-hydroxyphenyl)carbamoyl-4-(2-hydroxyphenyl)semicarbazide (Scheme 113).[310]

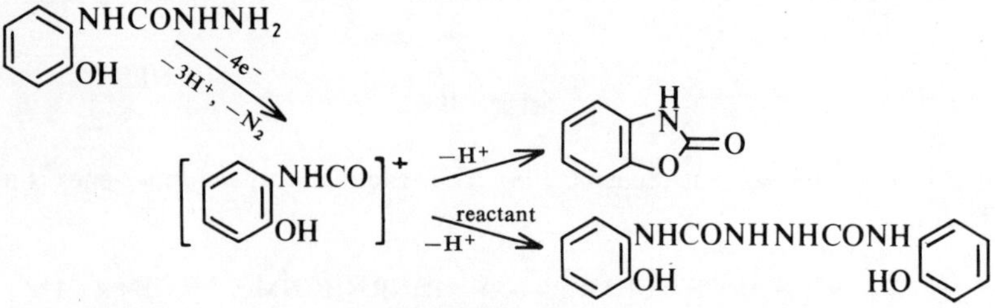

Scheme 113

Millimolar solutions of 4- and 4'-substituted benzanilides in MeCN–0.1M-NaClO$_4$ were oxidized at a vitreous carbon anode. Products depended upon

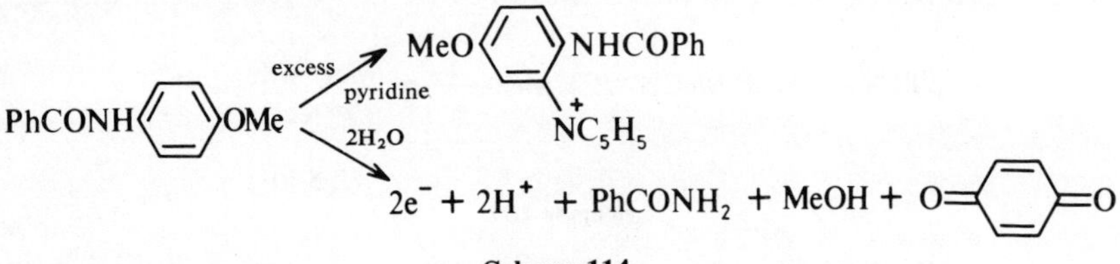

Scheme 114

whether or not the MeCN contained pyridine (Scheme 114).[311] In weakly acidic or neutral aqueous solutions, N-benzoyl-p-quinone imine was obtained.

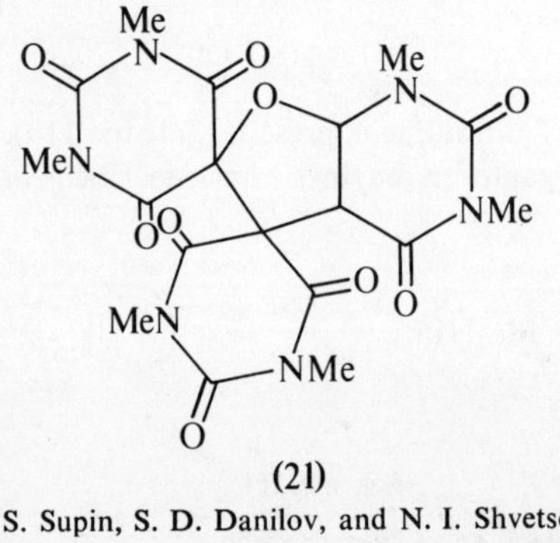

(21)

---

[310] A. A. Shapovalov, C. S. Supin, S. D. Danilov, and N. I. Shvetsov-Shilovskii, *Russ. J. Gen. Chem.*, 1974, **44**, 1535.
[311] S. Ihenoya, M. Nasui, H. Ohmori, and H. Sayo, *J.C.S. Perkin II*, 1974, 571.

1,3-Dimethylbarbituric acid has been oxidized at a pyrolytic graphite anode in 1M-HOAc to give the trimer 5,6-dihydro-1,3-dimethyl-5,6-di-(1′,3′-dimethyl-2′,4′,6′-trioxopyrimid[5,5′]yl)furo[2,3-d]uracil (21).[312]

Intramolecular anodic coupling reactions used to synthesize nitrogen heterocycles include the cyclization of formazans to tetrazolium salts in MeCN in yields exceeding 80% (Scheme 115).[313] Further conversions effected at platinum anodes in MeCN–TEAP in divided cells are shown in Scheme 116.[314]

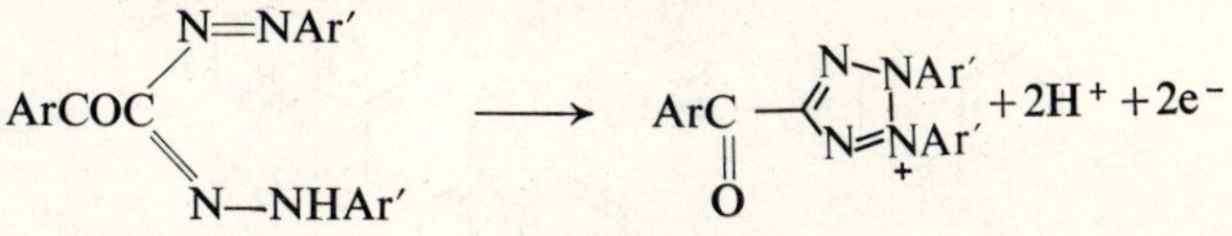

Scheme 115

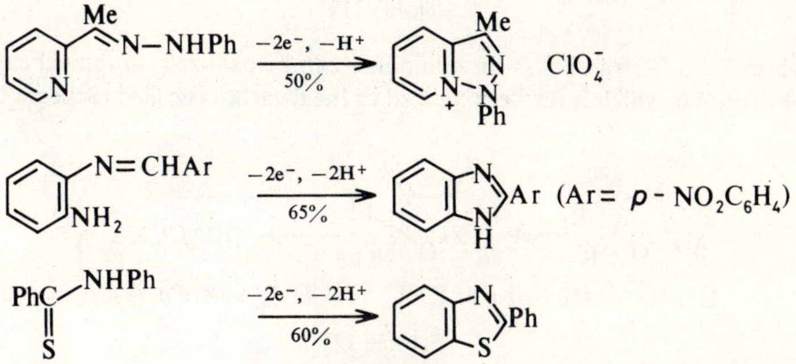

Scheme 116

A similar reaction proceeded only in an undivided cell, where accumulation of acid in the anolyte did not occur (Scheme 117).

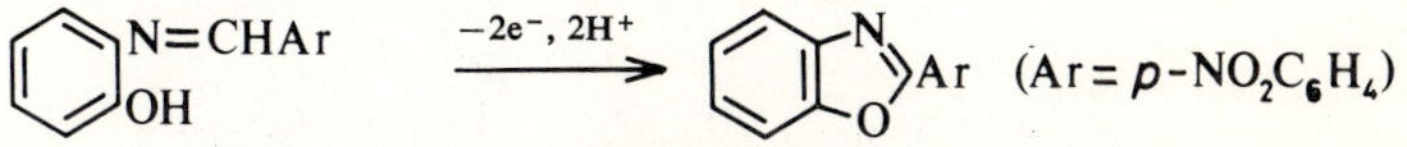

Scheme 117

N-Hydroxy-indoles have been prepared by controlled-potential oxidation of some 2-(o-hydroxylaminophenyl)alkenes in slightly basic media (Scheme 118).[315]

The oxidation of phenazine NN'-dioxide at platinum anodes in benzonitrile has received close study. Radical cations, formed initially, dimerize according to Scheme 119.[316, 317]

---

[312] S. Kato, M. Poling, D. van der Helm, and G. Dryhurst, *J. Amer. Chem. Soc.*, 1974, **96**, 5255.
[313] M. Lacan, I. Tabakovic, and Z. Cekovic, *Tetrahedron*, 1974, **30**, 2911.
[314] M. Lacan, K. Jakopcic, V. Rogic, Sh. Damoni, O. Rogic, and I. Tabakovic, *Synth. Comm.*, 1974, **4**, 219.
[315] R. Hazard and A. Tallec, *Bull. Soc. chim. France*, 1974, 121.
[316] A. Stüwe and H. Baumgärtel, *Ber. Bunsengesellschaft phys. Chem.*, 1974, **78**, 320.
[317] A. Stüwe, M. Weber-Schäfer, and H. Baumgärtel, *Ber. Bunsengesellschaft phys. Chem.*, 1974, **78**, 309.

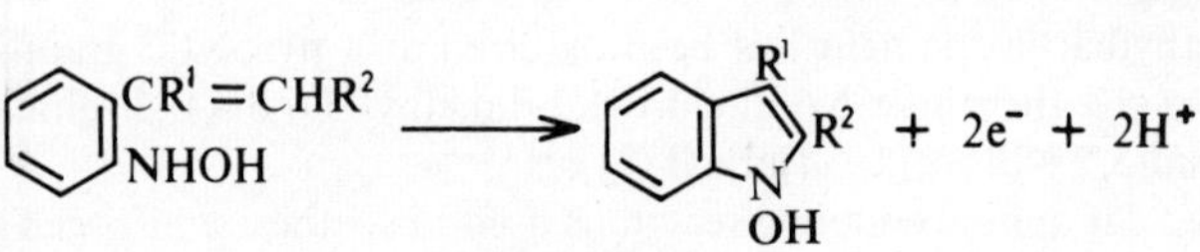

Scheme 118

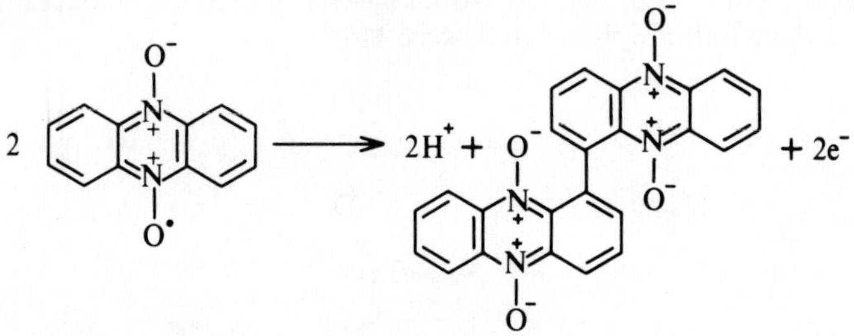

Scheme 119

4-Diazo-2,2,5,5-tetra-alkyl-3-furanidinones can be oxidized *via* radical cations to the 3,4-diketone, which is further oxidized to the dicarboxylic acid (Scheme 120).[318]

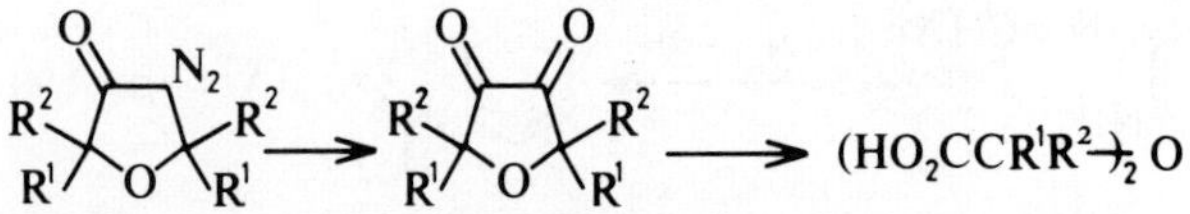

R^1, R^2 = Me or Et or R^1R^2 = $(CH_2)_n$, where n = 4 or 5

Scheme 120

Sulphur Compounds.—Diphenyl sulphoxide has been oxidized at a platinum anode in MeCN. The final product depends upon whether or not benzene is added to the anolyte (Scheme 121).[319] Sulphoxides have been synthesized by oxidation of the corresponding sulphides at C or Pt anodes in aqueous–organic mixed solvents (*e.g.* Me_2S is converted into Me_2SO in 99.5% yield).[320]

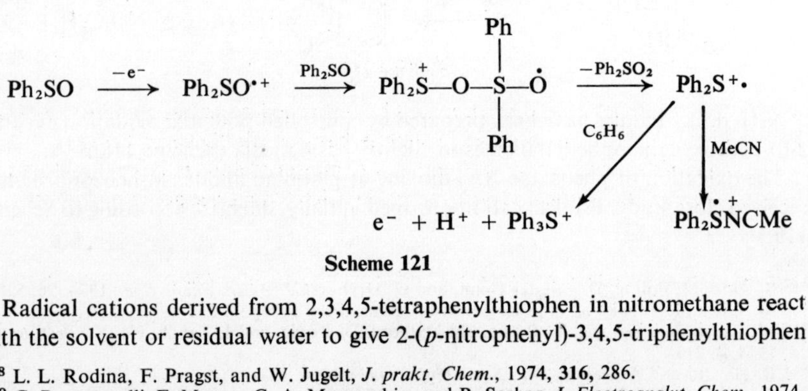

Scheme 121

Radical cations derived from 2,3,4,5-tetraphenylthiophen in nitromethane react with the solvent or residual water to give 2-(*p*-nitrophenyl)-3,4,5-triphenylthiophen

---

[318] L. L. Rodina, F. Pragst, and W. Jugelt, *J. prakt. Chem.*, 1974, **316**, 286.
[319] G. Bontempelli, F. Magno, G. A. Mazzocchin, and R. Seebar, *J. Electroanalyt. Chem.*, 1974, **55**, 109.
[320] C. Thibault and P. Mathieu, U.S.P. 3 808 112 (*Chem. Abs.*, 1974, **81**, 9180).

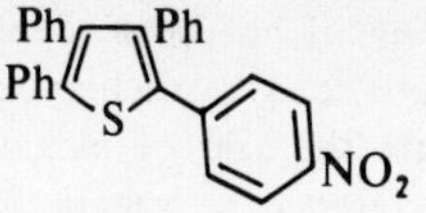

(22) and 1,2,3,4-tetraphenylbut-2-ene-1,4-dione (23), respectively.[321] Yields of (23) and (22) are respectively 40% and 10% in neutral solutions and 20% and 30% in the presence of Na_2CO_3.

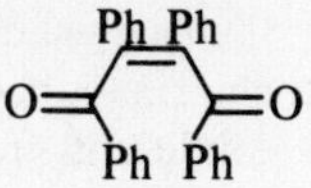

In nominally dry MeCN or $MeNO_2$, some *gem*-polysulphides [*e.g.* (*p*-MeO-$C_6H_4S)_2CH_2$] are oxidized to the corresponding α-disulphides.[322] Even in wet (up to 1 mol l^{-1} of H_2O) MeCN, the result is the same if only 2 F mol^{-1} are passed.[323] Excess charge produces the thiosulphonate in aqueous solutions:[322]

$$(ArS)_2CH_2 \xrightarrow[-CH_2O]{H_2O, -2e^-, -2H^+} ArSSAr \xrightarrow{-4e^-, 2H_2O} ArS-SO_2Ar + 4H^+$$

Anodic oxidation of thiophenols has been used as part of a synthesis of peptides.[324] In the reaction performed in DMF, the disulphide can be constantly regenerated anodically:

$$R^1CO_2H + R^2NH_2 + ArSSAr + R^3_3P \longrightarrow R^1CONHR^2 + 2ArSH + R^3_3PO$$

Typical examples are R^3 = Et, Ar = *p*-$NO_2C_6H_4$ in effecting the reaction:

$$PhCH_2OCO\text{-}Gly\text{-}Leu\text{-}OH + H\text{-}Gly\text{-}OMe \longrightarrow PhCH_2OCO\text{-}Gly\text{-}Leu\text{-}Gly\text{-}OMe$$

Several organic sulphides have been oxidized at mercury anodes with the formation of corresponding mercury compounds:

$$2RSH + Hg \longrightarrow (RS)_2Hg + 2H^+ + 2e^-$$

Examples are 2-mercaptoethylamine[325] and thiosalicylic acid.[326]

Oxidation of the orthothio-oxalate (24) gives rise to sulphonium ions (Scheme 122).[189] 1,2-Dimethylindolizine-3-thial has been oxidized in MeCN to its radical cation, which undergoes dimerization to the dication.[191]

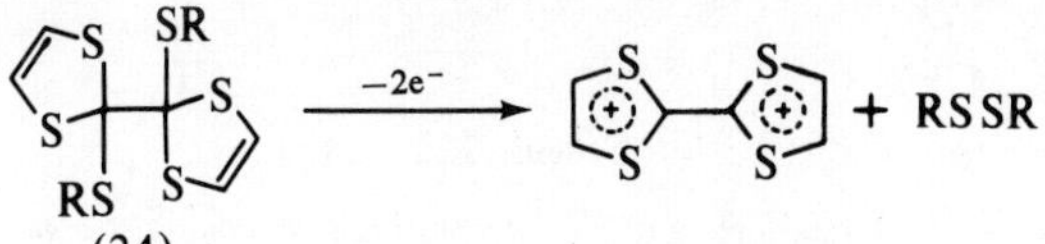

Scheme 122

Halogenation.—The mechanism of electrochemical chlorination of benzene, toluene, chlorobenzene, and *p*-xylene at Pt–Ir alloy anodes in HOAc has been studied by

[321] M. Libert and C. Caullet, *Compt. rend.*, 1974, **278**, C, 439.
[322] J. G. Gourcy, G. Jeminet, and J. Simonet, *J.C.S. Chem. Comm.*, 1974, 634.
[323] N. D. Canfield and J. Q. Chambers, *J. Electroanalyt. Chem.*, 1974, **56**, 459.
[324] Yu. V. Mitui and N. P. Zapevalova, *Russ. J. Gen. Chem.*, 1974, **44**, 2033.
[325] R. S. Saxena and R. Singh, *Z. phys. Chem.* (Leipzig), 1974, **255**, 677.
[326] S. K. Tiwari and A. Kumar, *J. prakt. Chem.*, 1974, **316**, 934.

kinetic measurements; the halogenating agent has been identified as Cl_2^{-}.[327] The advantages of anodic chlorination of ethylene to dichloroethane have been discussed.[328] Optimum conditions include the use of elevated temperatures and graphite or platinum electrodes in preference to magnetite, fused graphite, or RuO–Ti anodes. The ethylene is passed through the concentrated HCl solution being electrolysed.[329] The effect of adsorption of various impurities found in industrial waste gases upon this process has been studied.[330]

A graphite anode has been used in anhydrous methanolic media containing HCl to methoxy-chlorinate and chlorinate olefins admitted to the anolyte.[331] Thus electrogenerated chlorine attacks vinyl chloride to give 1,1,2-trichloroethane and 1,1-dichloro-2-methoxyethane. Similarly, buta-1,3-diene yields 1,2,3,4-tetrachlorobutane and 1,4-dichloro-2,3-dimethoxybutane. Allyl alcohol is chlorinated to 1,2-dichloropropanol in moderate yields.

In aqueous solutions of HCl, hypochlorination of olefins occurs.[332] A PbO_2-on-graphite anode has been used to convert propylene, isobutylene, hept-1-ene, allyl chloride, and other olefins in 3—10% HCl solution with a selectivity >90%. Chlorolignin, used in dyeing and as an additive to plastics, has been obtained by chlorination of lignin suspension in aqueous NaCl (pH 5.0) at a PbO_2-coated Ti anode.[333]

The major effort in anodic halogenation continues to be towards fluorinations. Perfluoro-octanoyl fluoride was formed in 12% yield by oxidation of octanoyl chloride at Ni foam anodes in dry HF.[334] Propane has been perfluorinated at cylindrical carbon anodes in KF,2HF.[335, 336] Perfluoro-alkenes have been converted into perfluoro-alkanes in 75—78% yield in a continuous process based upon a Simon's cell containing HF at 5 °C.[337]

Although aromatic hydrocarbons give anodic waves at Pt or vitreous carbon in liquid HF–0.1M-KF,[338] it has been found that ethanol, hexafluorobenzene, acetic acid, and n-butyric acid do not.[339] It is concluded therefore that industrial fluorina-

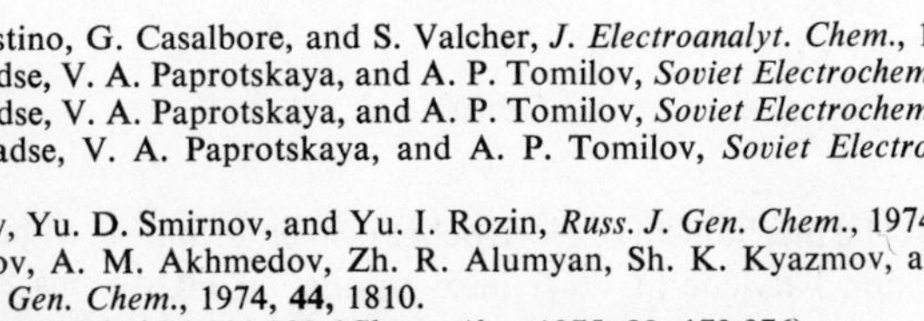

Scheme 123

[327] M. Mastragostino, G. Casalbore, and S. Valcher, *J. Electroanalyt. Chem.*, 1974, **56**, 117.
[328] G. A. Tedoradse, V. A. Paprotskaya, and A. P. Tomilov, *Soviet Electrochem.*, 1974, **10**, 1047.
[329] G. A. Tedoradse, V. A. Paprotskaya, and A. P. Tomilov, *Soviet Electrochem.*, 1974, **10**, 1239.
[330] G. A. Tedoradse, V. A. Paprotskaya, and A. P. Tomilov, *Soviet Electrochem.*, 1974, **10**, 1539.
[331] A. P. Tomilov, Yu. D. Smirnov, and Yu. I. Rozin, *Russ. J. Gen. Chem.*, 1974, **44**, 1990.
[332] D. A. Ashurov, A. M. Akhmedov, Zh. R. Alumyan, Sh. K. Kyazmov, and S. I. Sadykh-zade, *Russ. J. Gen. Chem.*, 1974, **44**, 1810.
[333] M. Ohta, Japan. Kokai 74 131 282 (*Chem. Abs.*, 1975, **82**, 172 876).
[334] F. G. Drakesmith, Ger. Offen. 2 417 860 (*Chem. Abs.*, 1975, **82**, 91 742).
[335] K. L. Mills, U.S.P. 3 806 432 (*Chem. Abs.*, 1974, **81**, 9182).
[336] R. A. Paul and M. B. Howard, U.S.P. 3 840 445 (*Chem. Abs.*, 1975, **82**, 49 313).
[337] S. Benninger, Ger. Offen. 2 302 132 (*Chem. Abs.*, 1974, **81**, 151 510).
[338] J. P. Masson, J. Devynck, and B. Trémillon, *J. Electroanalyt. Chem.*, 1974, **54**, 232.
[339] A. G. Doughty, M. Fleischmann, and D. Pletcher, *J. Electroanalyt. Chem.*, 1974, **51**, 456.

tion does not proceed by way of an initial electron transfer from the organic substrate.

When benzonitriles are perfluorinated at a nickel anode in anhydrous HF, two possible fissions of the nitrile group occur (Scheme 123).[340]

The product mixture obtained by fluorination of 2-methoxyethyl acetate includes CF_3OCF_3, CF_3OCHF_2, CHF_2OCHF_2, $CF_3OC_2F_5$, $C_2F_5OC_2F_5$, $CF_3OC_2F_4OCF_3$, $CF_3OC_2F_4OC_2F_5$, and the perfluorinated dioxolans (25)—(27).[341]

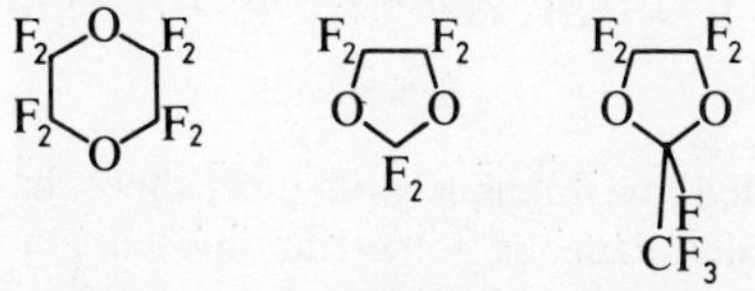

(25)   (26)   (27)

Perfluoroacyl fluorides were obtained by anodic fluorination of alkoxy-carboxylic acids.[342] Yields were superior to those achieved by fluorination of unsubstituted carboxylic acids, and no improvement was found by using the acid anhydrides in place of the acids. The alkoxy-group in the acids was believed to be oxidized to an acyl fluoride. Perfluoro-heterocyclic compounds have been prepared by electrochemical oxidation of corresponding bis(2-hydroxyethyl) compounds in anhydrous HF with nickel electrodes (Scheme 124).[343]

$$(HOCH_2CH_2)_2X \longrightarrow H_2O + \begin{array}{c}F_2\text{—}F_2\\X\quad O\\F_2\text{—}F_2\end{array}$$

(X = S, O, NH, NMe, or NEt)   (X = S, O, NF, NCF$_3$, or NCF$_2$CF$_3$)

Scheme 124

In the perfluorination of THF, the yield of C_4F_8O is increased in the presence of methyl butyrate or ethyl acetate, which prevent polymerization.[344] These esters are themselves fluorinated, to yield the perfluoro-acyl fluorides. Partially fluorinated ethers have been prepared anodically in anhydrous HF from $MeOCCl_3$, $MeOCCl=CCl_2$, and $Cl_2RCOCF_2CFCl_2$.[345] For example, $MeOCCl_3$ gives rise to 13% $MeOCF_2Cl$, in addition to other fluorinated products. Chlorine atoms bound to the α-carbon atoms of the ethers are readily removed, but those attached to the β-carbon atoms are retained.[346] The compounds $Cl(CF_2)_2OCHF_2$, $Cl_2CFCF_2OCF_3$,

---

[340] S. Nagase, H. Baba, K. Kodaira, T. Abe, and M. Yonekura, Japan. Kokai 74 87 647 (*Chem. Abs.*, 1975, **82**, 139 499).
[341] V. V. Berenblit, Yu. P. Doluakov, V. P. Sass, L. N. Senyushov, and S. V. Sokolov, *Zhur. org. Khim.*, 1974, **10**, 2031.
[342] V. V. Berenblit, B. A. Byzov, I. M. Dolgopolskii, and Yu. P. Dolnakov, *Zhur. priklad. Khim.*, 1974, **47**, 2433.
[343] T. Abe, S. Nagase, and H. Baba, Japan. Kokai 74 27 588 (*Chem. Abs.*, 1975, **82**, 170 979).
[344] V. V. Berenblit, V. I. Grachev, I. M. Dolgopolskii, and G. A. Davydov, *Zhur. priklad. Khim.*, 1974, **47**, 590.
[345] S. Nagase, H. Baba, K. Kodaira, and K. Okazaki, Ger. Offen. 2 323 862 (*Chem. Abs.*, 1974, **80**, 82 075).
[346] K. Okazaki, S. Nagase, H. Baba, and K. Kodaira, *J. Fluorine Chem.*, 1974, **4**, 387.

$Cl_2CFCF_2OCHF_2$, $Cl_2CFCF_2OCF_2Cl$, $Cl(CF_2)_2OCF_2Cl$, and $Cl_3CCF_2OCF_3$ have all been prepared in this way.

Five different perfluorinated amino-ethers, used as heat-transfer agents and as dielectrics, have been prepared at nickel plate anodes of a Simon's cell (Scheme 125).[347]

$$R^1_{3-n}N(CH_2CHR^2OCF_2CHFCF_3)_n \longrightarrow (perF-R^1_{3-n})N[CF_2CF(perF-R^2)OC_3F_7]_n$$

$$R^1 = C_4H_9 \text{ or } C_8H_{17}, \quad R^2 = H \text{ or } Me$$

Scheme 125

Various *N*-substituted pyrrolidines and piperidines have been perfluorinated electrochemically.[348] Simultaneous isomerization occurs in most cases (*e.g.* Scheme 126).

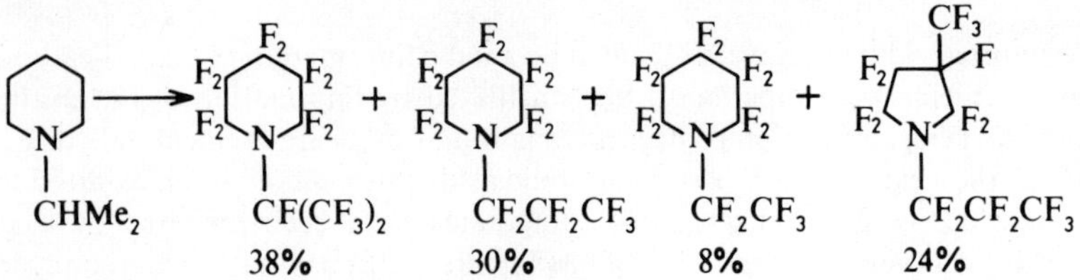

Scheme 126

Trialkylphosphine oxides R_3PO have been perfluorinated successfully,[349] as have octanesulphonyl fluoride and octanoyl fluoride.[350] In these cases,[350] simultaneous adsorption of the organic substance and of fluorine on the NiF_2 surface is recommended in order to optimize yields.

Oxygen Compounds.—In nitromethane, 2,3,4,5-tetraphenylfuran is oxidized to a radical cation, whose decomposition pathway is dependent upon the acidity of the medium.[351] In unbuffered neutral media the dibenzoyl-*cis*-stilbene is obtained, together with the perchlorate of the dication, which, when treated with methanol, yields (28) (Scheme 127). In buffered neutral media, only dibenzoylstilbene is

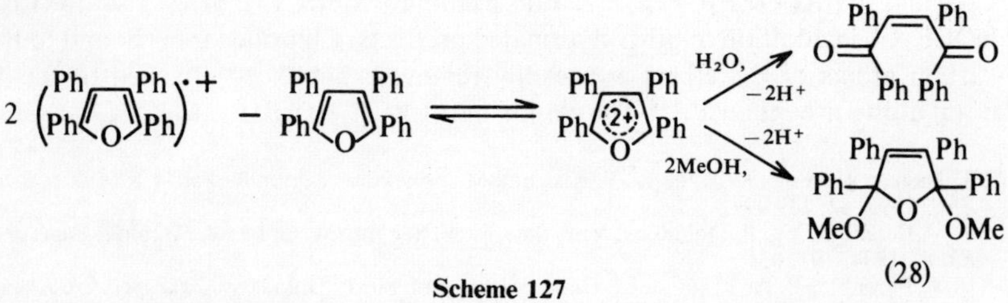

Scheme 127

[347] S. Beninger and T. Martini, Ger. Offen. 2 306 438 (*Chem. Abs.*, 1974, **81**, 169 122).
[348] V. S. Plashkin, L. N. Pushkina, and S. V. Sokolov, *Zhur. org. Khim.*, 1974, **10**, 1215.
[349] L. M. Yagupol'skii and co-workers, Russ. P. 384 349 (*Chem. Abs.*, 1974, **81**, 152 414).
[350] Ch. Comminellis, Ph. Javet, and E. Plattner, *J. Appl. Electrochem.*, 1974, **4**, 289.
[351] M. Libert and C. Caullet, *Bull. Soc. chim. France*, 1974, 805.

obtained, even when water is present. In the presence of perchloric acid, a dimeric product (29) also occurs. When methanol is used as solvent, the yield of (29) is 75%.

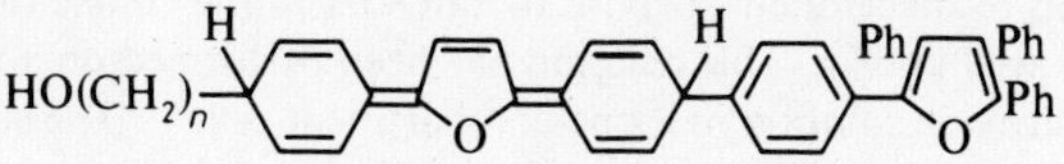

(29) $n = 0$, 1 or 2

The mechanism of oxidation of THF to succinic acid in aqueous media has been studied using platinum, carbon, and PbO_2 anodes.[352] Intermediates include hydroperoxides which dehydrate on PbO_2 anodes to γ-butyrolactone, which is further oxidized to succinic anhydride. The dimer (30) can be formed in high yield at lead in aqueous H_2SO_4 solution by oxidation of butyrolactone or THF.

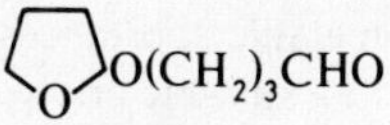

(30)

Oxidation of furfural at PbO_2 anodes in H_2SO_4 solution yields formylacrylic acid.[353]

The neutral radical obtained by the oxidation of the enolate of dibenzoylmethane at a platinum anode in DMSO either abstracts a hydrogen atom from the solvent or, if in high concentration, it dimerizes to 1,1,2,2-tetrabenzoylethane, which on prolonged electrolysis gives tetrabenzoylethylene.[141] The hydrogen-abstraction reaction occurs less readily when benzonitrile is used as the solvent.

Aliphatic ketones have been acetamidated by oxidation at platinum anodes in MeCN–$LiClO_4$. Some rearrangement of the carbon skeleton can occur simultaneously (*e.g.* Scheme 128).[354]

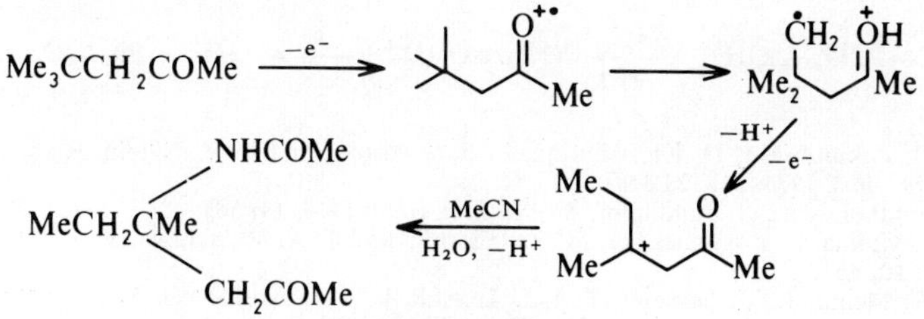

Scheme 128

[352] M. Sugawara, M. Sato, T. Osuda, and H. Yamamoto, *Denki Kagaku Oyobi Kogyo Butsuri Kagaku*, 1974, **42**, 247.
[353] V. A. Smirnov, V. I. Mil'man, and O. B. Krayanskii, Russ. P. 412 176 (*Chem. Abs.*, 1974, **80**, 120 292).
[354] J. T. Becker, L. R. Byrd, and L. L. Miller, *J. Amer. Chem. Soc.*, 1974, **96**, 4718.

In aqueous media, allyl alcohol has been oxidized selectively to acrylic acid by controlling the potential of the platinum electrocatalyst.[355] In acetonitrile, acrolein was formed in 30% current yield.[356]

The oxidation of diacetone-L-sorbose by NiOOH in the course of production of ascorbic acid is well known. This reaction has been performed on a nickel anode in 3% sodium hydroxide solution in a one-compartment cell.[357] The same conversion can be achieved if 1.5% $NiCl_2$ is added to the anolyte.[358] When the unsubstituted sugars L-sorbose or D-glucose are oxidized at platinum or NiOOH anodes in 3% NaOH, they are destroyed by complete oxidation.[359] The effect of electrode material upon the oxidation of diacetone-L-sorbose has shown that in bromide-free electrolytes either nickel anodes must be used or nickel salts must be present in the anolyte.

The preparation of sodium gluconate by the oxidation of glucose in alkaline solution has been patented.[360]

Miscellaneous.—Several organometallic substances have been prepared by anodically dissolving the parent metal in a suitable anolyte. A cell fitted with rotating disc cathodes interleaved with anode sheets has been developed for the production of acetylacetonates of iron, cobalt, or nickel and for the synthesis of metal alkyls.[361, 362]

Si and Si alloys have been dissolved anodically in methyl chloride–lithium chloride–cellusolve electrolytes; Si–Cu and Si–Ca anodes yielded Me_4Si and Me_2SiCl_2.[363]

Trialkylboranes R_3B (R = alkyl, aralkyl, or cyclo-alkyl) have been oxidized at graphite anodes.[364] When the electrolysis was conducted in MeOH–NaOMe–$NaClO_4$ the products were ROMe, R_2, and cyclo-alkanes. In NaOAc–HOAc the product was ROAc.

Triphenylarsine has been oxidized at a platinum anode in wet MeCN.[365] In perchlorate or tetrafluoroborate electrolytes, $Ph_3AsOH^+\ ClO_4^-$ and Ph_3AsOBF_3 are formed, respectively. In chloride electrolytes the halide ion is discharged preferentially, and the product is $Ph_3AsOH^+\ Cl^-$. Subsequent reduction of hydroxytriphenylarsonium perchlorate yields triphenylarsine oxide and perchloric acid:

$$2Ph_3AsOH^+ \xrightarrow[-\frac{1}{2}H_2]{e^-} (Ph_3AsO)_2H^+ \xrightarrow{e^-} \tfrac{1}{2}H_2 + Ph_3AsO$$

[355] G. D. Zakumbaeva, D. Kh. Churina, L. B. Suvorova, and V. A. Naidin, Russ. P. 425 897 (*Chem. Abs.*, 1974, **81**, 121 359).
[356] F. Sundholm and G. Sundholm, *Electrochim. Acta*, 1974, **19**, 565.
[357] T. E. Mulina, I. A. Avrutskaya, M. Ya. Fioshin, and T. A. Malakhova, *Soviet Electrochem.*, 1974, **10**, 467.
[358] M. E. Mellor, L. G. Seleznev, F. I. Luknitskii, L. M. Sukhmaneva, M. A. Veksler, T. V. Sazonova, and T. A. Ivanova, *Khim.-Farm. Zhur.*, 1973, **7**, 33.
[359] M. Ya. Fioshin, I. A. Avrutskaya, T. A. Malakhova, and T. E. Mulina, *Soviet Electrochem.*, 1974, **10**, 760.
[360] R. Ohme and P. Kimmerl, East Ger. P. 104 779 (*Chem. Abs.*, 1974, **81**, 107 759).
[361] W. Eisenbach, *Chem.-Ing.-Tech.*, 1974, **46**, 965.
[362] W. Eisenbach, H. Lehmkuhl, and G. Wilke, Ger. Offen. 2 349 561 (*Chem. Abs.*, 1974, **81**, 20 237).
[363] N. Urabe and F. Hori, *Asahi Garasu Kogyo Gijutau Shoreikai Kenkyu Hokoko*, 1973, **23**, 81.
[364] T. Taguchi, Y. Takahashi, M. Itoh, and A. Suzuki, *Chem. Letters*, 1974, 1021.
[365] S. Zecchin, G. Schiavon, G. Cogoni, and G. Bontempelli, *J. Organometallic Chem.*, 1974, **81**, 49.

2
The Interfacial Tension of Solid Electrodes

BY I. MORCOS

1 Introduction

This chapter outlines the principles and experimental results of presently available methods used in the study of the interfacial tension of solid electrodes. The development in this area is closely related to the significant developments in the study of the double-layer structure of solid electrodes by the differential-capacitance[1—7] and radioactive-tracer methods,[8] the thermodynamic treatment[9] applied by the Frumkin school to study the surface phenomena of hydrogen- and oxygen-adsorbing electrodes, and the results obtained by other methods used to investigate the interface, e.g. ellipsometry,[10] voltammetry,[11,12] and surface conductance.[13] These related areas have been the subjects of previous reviews, and will not be dealt with here.

The literature[14—24] on what is conventionally known as the 'electrocapillary phenomena' is concerned with the interfacial tension of liquid metal electrodes. Quantitative and precise measurements of electrocapillary curves can generally be performed on the latter electrodes. A study of the interfacial tension of the solid electrode/electrolyte interface is complicated by many factors, which can perhaps be best understood if they are viewed in the light of the corresponding conditions prevailing at liquid metal electrodes.

[1] D. I. Leikis, K. V. Rybalka, E. S. Sevastyanov, and A. N. Frumkin, *J. Electroanalyt. Chem. Interfacial Electrochem.*, 1973, **46**, 161.
[2] T. N. Andersen, in 'Modern Aspects of Electrochemistry', ed. J. O'M. Bockris and B. E. Conway, Plenum Press, London and New York, 1969, Vol. 5, Ch. 3.
[3] J. O'M. Bockris, S. D. Argade, and E. Gileadi, *Electrochim. Acta*, 1969, **14**, 1259.
[4] M. A. V. Devanathan and K. Ramakrishnaiah, *Electrochim. Acta*, 1973, **18**, 259.
[5] R. Payne, in 'Advances in Electrochemistry and Electrochemical Engineering', ed. P. Delahay and C. W. Tobias, Interscience, New York, 1969, Vol. 7, pp. 1—76.
[6] G. Valette and A. Hamelin, *J. Electroanalyt. Chem. Interfacial Electrochem.*, 1973, **45**, 301.
[7] G. Nguyen-Van-Huong, J. Clavilier, and M. Bonnemay, *J. Electroanalyt. Chem. Interfacial Electrochem.*, 1975, **65**, 531.
[8] N. A. Balashova and V. E. Kazarinov, in 'Electroanalytical Chemistry', ed. A. J. Bard, Dekker, New York, 1969, Vol. 3, Ch. 3.
[9] B. B. Damaskin, O. A. Petrii, and V. V. Batrakov, in 'Adsorption of Organic Compounds on Electrodes', Plenum Press, London, 1971, pp. 387—435.
[10] J. Kruger, in 'Advances in Electrochemistry and Electrochemical Engineering', ed. P. Delahay and C. W. Tobias, Interscience, New York, 1973, Vol. 9, pp. 227—280.
[11] J. P. Hoare, in 'Advances in Electrochemistry and Electrochemical Engineering', ed. P. Delahay and C. W. Tobias, Interscience, New York, 1966, Vol. 6, p. 201.
[12] S. Gilman, 'Electroanalytical Chemistry', ed. A. J. Bard, Dekker, New York, 1967, Vol. 2, p. 133.
[13] W. J. Anderson and W. N. Hansen, *J. Electroanalyt. Chem. Interfacial Electrochem.*, 1973, **43**, 329.

The theory of capillarity, in the form of the Young[25] and Laplace[26] equations, provides the relationship between the shape of the liquid and its surface tension:

$$\Delta P = \gamma_L (1/R_1 + 1/R_2) \tag{1}$$

where γ_L is the liquid's surface tension, ΔP the pressure difference across the two sides of its curved surface, and R_1 and R_2 are the main radii of curvature. Most experimental techniques[27,28] developed to measure the surface tension of liquids are based on various forms of this same general fundamental equation. The same principles and experimental techniques can be applied to measure the interfacial tension between two immiscible liquids, such as mercury and water.

A solid, because of its rigidity, does not generally exhibit an equilibrium surface, and the methods of capillarity can therefore not be used to define or determine its surface tension. Theoretical estimates of absolute surface tensions of some relatively simple covalently bonded, ionic, rare-gas, and metallic crystals are discussed[29] in the literature. In a few specific situations, the surface tensions of some solid surfaces have been experimentally obtained, using, for example, the zero-creep and cleavage methods.[29,30] These theoretical and experimental methods are designed for the solid/gas interface, and are mostly incompatible for use at room temperature or in the presence of a solvent. Consequently, they cannot be applied to study the solid/liquid or solid electrode/electrolyte interfaces.

When a liquid surface is stretched, its surface area changes, but its surface tension remains constant. The independence of surface tension γ on the surface area A follows from the important thermodynamic relationship:[31]

$$\gamma = G/A \tag{2}$$

where G is the Gibbs free energy. Since the most stable system is the one with the lowest value of G, a liquid drop, in accordance with equation (2), is spherical. The

---

[14] G. Lippmann, *Ann. Chim. Phys.*, 1875, 494; also 1877, **12**, 265.
[15] G. Gouy, *Ann. Chim. Phys.*, 1903, **29**, 145.
[16] D. C. Grahame, *Chem. Rev.*, 1947, **41**, 441.
[17] R. Parsons, in 'Modern Aspects of Electrochemistry', ed. J. O'M. Bockris and B. E. Conway, Academic Press, New York, 1954, Vol. 1, Ch. 3.
[18] D. M. Mohilnes, in 'Electroanalytical Chemistry', ed. A. J. Bard, Dekker, New York, Vol. 1, Ch. 4.
[19] P. Delahay, 'Double Layer and Electrode Kinetics', Interscience, New York, 1965, Ch. 2.
[20] J. A. V. Butler, 'Electrocapillarity', Chemical Publishing Co., Inc., New York, 1940; also 'Electrical Phenomena at Interfaces', The Macmillan Company, New York, 1951.
[21] M. A. V. Devanathan and B. V. K. S. R. A. Tilak, *Chem. Rev.*, 1965, **65**, 635.
[22] M. J. Sparnaay, in 'The International Encyclopedia of Physical Chemistry and Chemical Physics', ed. D. D. Eley and F. C. Tompkins, Pergamon Press, Oxford, 1972, Ch. 3.
[23] H. H. Bauer, in 'Electrodics', ed. K. Niedenzu and H. Zimmer, Wiley, New York, 1972, Ch. 3.
[24] A. N. Frumkin, *Svensk Kem. Tidskr.*, 1965, **6**, 77.
[25] T. Young, in 'Miscellaneous Works', ed. G. Peacock, J. Murray, London, 1855, Vol. 1, p. 418.
[26] P. S. de Laplace, 'Mécanique Céleste', Supplement to Book, Paris, 1806, p. 45.
[27] A. W. Adamson, 'Physical Chemistry of Surfaces', Interscience, New York, 1967, Ch. 1.
[28] A. E. Alexander and B. Hayter, in 'Physical Methods of Chemistry', ed. A. Weissberger and B. W. Rossiter, Wiley, New York, 1971, Vol. 1, Part V, Ch. 9.
[29] R. G. Linford, in 'Solid State Surface Science', ed. M. Green, Dekker, New York, 1973, Vol. 2, Ch. 1; also ref. 27, Ch. 5.
[30] R. A. Oriani and C. A. Johnson, in 'Modern Aspects of Electrochemistry', ed. J. O'M. Bockris and B. E. Conway, Plenum Press, London and New York, 1969, Vol. 5, Ch. 2.
[31] J. W. Gibbs, 'The Collected Works of J. W. Gibbs', Longmans Green, New York, 1931, p. 315.

same relationship is valid for solids, provided that the formation of the new surface is not accompanied by surface deformation, as is the case when the new surface can be formed by cleavage.[29,30] When the solid surface is stretched, there is a time lag between the formation of the new surface and the rearrangement of atoms into new equilibrium positions. Thus new bond angles and bond lengths are formed, and the surface will be under stress. Using the principle of virtual work, Shuttleworth[30,32] showed that the relationship between surface stress and surface tension of a solid is given by:

$$\sigma_s = \gamma + A\, d\gamma/dA \qquad (3)$$

or

$$\sigma_s = \gamma + d\gamma/d\varepsilon_s \qquad (4)$$

where σ_s and ε_s are the surface stress and surface strain, respectively. Different orientations of a crystal have different stresses, and therefore the equilibrium shape[33-35] of a crystal is not necessarily a sphere.

Average surface stresses for the solid surfaces of very small nuclei have been determined[36] from changes in the lattice constant as a function of the particle radius. The experimental procedure involves the use of electron microscopy and electron diffraction for measuring the radii of the nuclei and the lattice constants, respectively. This approach is appropriate for the study of temperature dependence of surface stress in a gaseous atmosphere, but cannot be applied to study the potential dependence of surface stress in an electrochemical cell.

The total or volume stress in a very thin evaporated or electrodeposited film can be obtained from the observed strain by means of the laws of elasticity. The theory and the experimental techniques are discussed in the literature.[37-42] This approach may be applied to study the interfacial tension of solid electrodes because the strain experienced by an electrode of an electrochemical cell is measurable *in situ*.

With the exception of some rare situations, the absolute interfacial tension at the solid/liquid interface is inaccessible by either experiment or theoretical estimation. It is relatively simple, however, to obtain the difference between the interfacial tensions of the solid/gas and solid/liquid interfaces from certain experimentally measurable parameters characterizing the thermodynamic equilibrium between the solid and liquid phases. The most important of these is the contact angle, which is related to the previously mentioned difference in interfacial tensions by the Young[43] and Dupré[44] equation:

$$\gamma_L \cos\theta = \gamma_{SV} - \gamma_{SL} \qquad (5)$$

[32] R. Shuttleworth, *Proc. Phys. Soc. (London)*, 1950, **63A**, 444.
[33] G. Wulff, *Z. Krist.*, 1901, **34**, 449.
[34] C. Herring, *Phys. Rev.*, 1951, **82**, 87.
[35] W. M. Mullins, *Phil. Mag.*, 1961, **6**, 1313.
[36] J. S. Vermaak and D. Kuhlmann-Wilsdorf, *J. Phys. Chem.*, 1968, **72**, 4150.
[37] G. G. Stoney, *Proc. Roy. Soc.*, 1909, **A32**, 172.
[38] J. D. Wilcock and D. S. Campbell, *Thin Solid Films*, 1969, **3**, 3.
[39] R. W. Hoffman, in 'The Use of Thin Films in Physical Investigations', ed. J. C. Anderson, Academic Press, London and New York, 1966, p. 261.
[40] R. W. Hoffman, 'Physics of Thin Films', Academic Press, New York, Vol. III, p. 211.
[41] A. Brenner and S. Senderoff, National Bureau of Standards, Research Paper RP1954, 1949, Vol. 42, p. 105.
[42] H. P. Murbach and H. Wilman, *Proc. Phys. Soc.*, 1953, **B70**, 905.
[43] T. Young, *Phil. Trans. Roy. Soc.*, 1805, **95**, 84; in 'Works', ed. G. Peacock, Murray, London, 1855, Vol. 1, p. 432.
[44] A. Dupré, 'Théorie mécanique de la chaleur', 1869, p. 393.

where γ_L is the liquid's surface tension and γ_{SV} and γ_{SL} are the solid/vapour and solid/liquid interfacial tensions. Equation (5) has been used in the study of the solid/liquid interfacial tension in the absence of an electric field. The extensive literature[45—47] available on this subject provides a useful background for the development of techniques to be used in the presence of an electric field. Inversely, the study of wetting as a function of the electric field should be most useful in understanding the contribution of electrostatic and non-electrostatic forces to the interfacial tension. In equation (5), γ_{SV} is not equivalent to the solid-surface tension because of the presence at the solid surface of an equilibrium adsorbed film[48] of vapour. The difference between the two quantities is given by:

$$\pi = \gamma_S - \gamma_{SV} \tag{6}$$

where π is the film pressure.

In spite of many theoretical and experimental difficulties associated with the phenomenon of the contact angle, the validity of equation (5) is generally accepted.[49,50] The most rigid proof of this equation is that based on thermodynamics, which applies the principle of minimum surface energy.[30,49,50] A proof based on a force balance has been criticized on the ground that it involves an unbalanced force, which is the vertical component, $\gamma_L \sin \theta$, of the liquid's surface tension.[51] This may suggest that the contact angle is determined by a balance of surface stresses or stretching tensions, but as Adamson and Ling[49] convincingly argue, such a hypothesis contradicts the experimental observation that liquid drops on a crystalline surface of low symmetry are of circular cross-section. If the γ_{SV} and γ_{SL} terms in the Young and Dupré equation are stretching tensions, their non-isotropic nature would have prevented the formation of circular liquid drops. On the basis of this and other observations, the authors state that, in the context of a contact-angle situation, the surface free energy is a more rational quantity than surface force.[49] The vertical component of γ_L is a real problem with a soft, deformable solid surface.[52] Such is probably not the case, however, with a rigid solid.

Equation (5) was derived for the ideal rigid, smooth, homogeneous surface. Empirically modified versions of the same derivation were attempted for rough heterogeneous surfaces. For a rough surface, the Wenzel equation[53—55] was suggested:

$$\gamma_L \cos \theta_{app} = r(\gamma_{SV} - \gamma_{SL}) \tag{7}$$

[45] A. W. Adamson, 'Physical Chemistry of Surfaces', Interscience, New York, 1967, Ch. 7, pp. 352—375.
[46] J. J. Bikerman, in 'Physical Surfaces', ed. E. L. Lobel, Academic Press, New York, 1970, pp. 239—299.
[47] T. J. Davies and E. K. Rideal, 'Interfacial Phenomena', Academic Press, New York, 1961.
[48] D. H. Bangham and R. J. Razouk, *Trans. Faraday Soc.*, 1937, **33**, 1459.
[49] A. W. Adamson and I. Ling, in 'Advances in Chemistry, No. 43', ed. R. F. Gould, American Chemical Society, Washington, D.C., 1964, Ch. 3.
[50] R. E. Johnson, jun., *J. Phys. Chem.*, 1959, **63**, 1655.
[51] J. J. Bikerman, Proceedings of the 2nd International Congress on Surface Activity, London, 1957, Vol. III, p. 125.
[52] A. S. Michaels and S. W. Dean, jun., *J. Phys. Chem.*, 1962, **66**, 1790.
[53] R. N. Wenzel, *Ind. and Eng. Chem.*, 1936, **28**, 988; *J. Phys. Colloid. Chem.*, 1949, **53**, 1466.

where θ_{app} is the apparent contact angle and r is the roughness factor. From equation (7) it follows that, if θ is greater than 90°, roughness will increase the contact angle, while if θ is less than 90°, roughness will enhance the wetting.

For a heterogenous surface of two types of patches occupying fractions f_1 and f_2 of the surface, the following modification[45,56,57] of the Young and Dupré form has been used:

$$\gamma_L \cos\theta_{app} = f_1(\gamma_{S_1V} - \gamma_{S_1L}) + f_2(\gamma_{S_2V} - \gamma_{S_2L}) \qquad (8)$$

Hysteresis of wetting constitutes the most serious problem in the measurement of contact angles. In recent years much progress has been made in the study of this problem,[58-65] and the original theory of hysteresis due to roughness and heterogeneity has been further refined by Neumann and Good.[59,60] It is widely accepted[58-63] that surface roughness causes local contortion of the liquid–vapour surface, which results in a large number of metastable configurations. Hysteresis depends on the height of the energy barriers. On an energetically heterogeneous surface the phenomenon is explained[59,64,65] by a very similar mechanism. On rough heterogeneous surfaces, advancing contact angles are considered to be a measure of the wettability of the low-energy sections of the solid surface and receding angles a measure of the wettability of the sections of higher energy.[59] On this basis, it is reasonable to assume that advancing contact angles correspond to equilibrium conditions while receding angles do not. Some justification for this is found in the observation that only advancing angles behave according to equation (7).[45]

In addition to surface roughness and heterogeneity, hysteresis of contact angles may result from the presence of adsorbed organic impurities which are removed when the solid is brought into contact with the liquid.[45,66] High-energy surfaces, such as that of a metal, readily adsorb organic gases from either the atmosphere or the liquid phase, and become more hydrophobic in character. Experimental observations also indicate that contact angles increase with the adsorption of oxygen that is present as an impurity in the gas phase.[46,111]

Mercury has a smooth surface, with a well-defined area, and can be conveniently obtained in a reproducible and pure form. On the other hand, because of the problems of surface roughness, heterogeneity, and adsorbed organic impurities, the surfaces of solids are difficult to reproduce. Chemical, mechanical, and electro-

---

[54] A. B. D. Cassie, *Discuss. Faraday Soc.*, 1948, **3**, 11.
[55] R. Shuttleworth and G. L. J. Bailey, *Discuss. Faraday Soc.*, 1948, **3**, 16.
[56] R. H. Dettre and R. E. Johnson, jun., Symposium on the Contact Angle, Bristol, 1966.
[57] S. Baxter and A. B. D. Cassie, *J. Textile Inst.*, 1945, **36**, T67.
[58] R. H. Dettre and R. E. Johnson, jun., in 'Advances in Chemistry, No. 43', ed. R. G. Gould, American Chemical Society, Washington, D.C., 1964, pp. 112, 136.
[59] A. W. Neumann and R. J. Good, *J. Colloid. Interface Sci.*, 1972, **38**, 341.
[60] J. D. Eick, R. J. Good, and A. W. Neumann, *J. Colloid, Interface Sci.*, 1975, **53**, 235.
[61] R. Shuttleworth and G. L. J. Bailey, *Discuss. Faraday Soc.*, 1948, **3**, 16.
[62] J. J. Bikerman, *J. Phys. Chem.*, 1950, **54**, 653.
[63] R. J. Good, *J. Amer. Chem. Soc.*, 1952, **74**, 5041.
[64] D. C. Pease, *J. Phys. Chem.*, 1945, **49**, 107.
[65] R. E. Johnson, jun. and R. H. Dettre, *J. Phys. Chem.*, 1964, **68**, 1744.
[66] N. K. Adam, 'The Physics and Chemistry of Surfaces', Dover, New York, 1968, Ch. 5.

chemical methods commonly used to polish or smooth the surface may further complicate the problem by contaminating it with foreign particles. Consequently, interfacial tension–potential curves of solid electrodes are far less reproducible than the corresponding curves obtained with mercury.

The interpretation of electrocapillary data obtained with liquid metals (mostly mercury) is greatly facilitated by the characteristics of ideal polarizability. For an ideally polarizable electrode, the relationship between the interfacial tension, the surface charge density, and the electrode potential is described by the Lippmann equation:

$$q = -(\partial \gamma / \partial \phi)_\mu \qquad (9)$$

where γ is the term generally used for the interfacial tension at the liquid–metal electrode/electrolyte interface, q is the surface charge density, ϕ is the electrode potential, and μ is the chemical potential of the electrolyte. Compared to mercury, solid electrodes are mostly unpolarizable surfaces. The deviation from ideal polarizability results from the occurrence of irreversible faradaic and pseudofaradaic electrochemical processes which consume the charge that otherwise would be available for charging the double layer. On the basis of differential capacitance measurements, it was demonstrated that some metals of low melting point, such as lead, cadmium, and bismuth, exhibit ideal polarizable characteristics similar to those of mercury, but within a narrower potential region.[1]

Deviation from polarizability will certainly complicate the interpretation of interfacial tension vs. potential studies of solid electrodes, and in some cases will introduce certain limitations on the use and validity of some experimental techniques. Thus, because there is no available general theoretical relationship between the interfacial tension and other double-layer parameters in the presence of irreversible electrochemical reactions, the interpretation of the data on solid electrodes will be mostly qualitative in nature. Furthermore, even the most qualitative interpretation of the interfacial tension vs. potential plots will require some familiarity with the type of electrochemical activity occurring across the interface. Voltammetric curves are particularly useful in this regard, although data obtained by other electrochemical and non-electrochemical techniques may be required. The most important factor in the interpretation of the results is to know what parameter is being measured.

2 Principles and Experimental Results

General Criteria and Classification of Methods.—Recent attempts to study the interfacial tension of solid electrodes fall into two main categories: measurement of the potential dependence of contact angle[67–86] established by the liquid phase

---

[67] G. Möller, *Ann. Physik.*, 1908, **27**, 665; *Z. phys. Chem.*, 1908, **65**, 226.
[68] A. Frumkin, A. Gorodetskaya, B. Kabanov, and N. Nekrassov, *Zhur. fiz. Khim.*, 1932, **3**, 351.
[69] I. P. Tverdovskii and A. N. Frumkin, *Zhur. fiz. Khim.*, 1947, **21**, 819.
[70] B. N. Kabanoff, 'Electrochemistry of Metals and Adsorption', Nauka, Moscow, 1966.
[71] A. N. Frumkin, *Phys. Chem.*, 1938, **12**, 337.

on the solid surface and the measurement of the variation in surface stress[87—98] experienced by the solid as a function of potential.

Contact angles can either be measured directly[67—74] or obtained indirectly,[75—86] by measuring the rise of a liquid meniscus at a partially immersed plate or the capillary rise inside a metal capillary and then obtaining the contact angle from the relationship correlating either of the latter parameters to the contact angle. Variation in surface stress may either be measured directly,[87—93] with a piezo-electric element, or be obtained indirectly,[94—98] by measuring the potential dependence of the strain (*i.e.* electrode deformation) and then obtaining the variation in stress from the appropriate form of Hooke's law.

Because of the enormous difference in mobility of the liquid and solid phases, the relative change in contact angle for a given change in the potential is much greater than the corresponding change in the relative surface stress of a solid. Consequently, measurement of the latter requires much more sensitive and sophisticated devices than the measurement of the former. This point can be illustrated by the following example. In the case of gold, the variation in meniscus rise between the potential of zero charge and -0.15 V on the rational scale in 0.1M-KI solution is approximately 0.1 cm. The value of the corresponding variation in the length of a gold ribbon obtained by Beck[96] in the same solution is approximately 3.0×10^{-6} cm. At the sensitivity level required to measure such a small variation in the relative stress, the experimental data are seriously influenced by side effects from

---

[72] C. A. Smolders, *Rec. Trav. chim.*, 1961, **80**, 635, 650, 699.
[73] M. J. Sparnaay, *Surface Sci.*, 1964, **1**, 213.
[74] Y. Nakamura, K. Kamada, Y. Katoh, and A. Watanabe, *J. Colloid. Interface Sci.*, 1973, **44**, 517.
[75] I. Morcos and H. Fischer, *J. Electroanalyt. Chem. Interfacial Electrochem.*, 1968, **17**, 7.
[76] I. Morcos, *J. Electroanalyt. Chem. Interfacial Electrochem.*, 1969, **20**, 479.
[77] I. Morcos, *J. Electroanalyt. Chem. Interfacial Electrochem.*, 1975, **62**, 313; also unpublished data by I. Morcos.
[78] I. Morcos, *Coll. Czech. Chem. Comm.*, 1971, **36**, 689.
[79] I. Morcos, *J. Colloid. Interface Sci.*, 1971, **37**, 410.
[80] I. Morcos, *J. Chem. Phys.*, 1972, **56**, 3996.
[81] I. Morcos, *J. Phys. Chem.*, 1972, **76**, 2750.
[82] I. Morcos, in 'Proceedings of the Symposium on Oxide–Electrolyte Interfaces', ed. R. S. Alwitt, The Electrochemical Society, Soft-bound Symposium Series, Princeton, N.J., 1973, pp. 143—154.
[83] I. Morcos, *J. Electrochem. Soc.*, 1974, **121**, 1417.
[84] M. Bonnemay, G. Bronoel, P. J. Jonville, and E. Levart, *Compt. rend.*, 1965, **260**, 5262.
[85] M. Shimokawa and T. Takamura, *J. Electroanalyt. Chem. Interfacial Electrochem.*, 1973, **41**, 359.
[86] H. Dahms, *J. Electrochem. Soc.*, 1969, **116**, 1532.
[87] A. Ya. Gokhshtein, *Soviet Electrochem.*, 1966, **2**, 1204.
[88] A. Ya. Gokhshtein, *Soviet Electrochem.*, 1968, **4**, 590.
[89] A. Ya. Gokhshtein, *Soviet Electrochem.*, 1968, **4**, 551.
[90] A. Ya. Gokhshtein, *Electrochim. Acta*, 1970, **15**, 219.
[91] A. Ya. Gokhshtein, *Soviet Electrochem.*, 1970, **6**, 946.
[92] A. Ya. Gokhshtein, *Soviet Electrochem.*, 1971, **7**, 3.
[93] A. Ya. Gokhshtein, *Russ. Chem. Rev.*, 1975, **44**, 921.
[94] T. R. Beck, *J. Phys. Chem.*, 1969, **73**, 466.
[95] T. R. Beck and K. W. Beach, in 'Proceedings of the Symposium on Electrocatalysis', ed. M. W. Breiter, The Electrochemical Society, Soft-bound Symposium Series, Princeton, N.J., 1974, pp. 357—364.
[96] L. F. Kin and T. R. Beck, *J. Electrochem. Soc.*, 1976, **123**, 1145.
[97] R. A. Fredlein, A. Damjanovic, and J. O'M. Bockris, *Surface Sci.*, 1971, **25**, 261.
[98] R. A. Fredlein and J. O'M. Bockris, *Surface Sci.*, 1974, **46**, 641.

fluctuations in temperature, Joule and overvoltage heating effects, hydrostatic forces due to the liberation of gases, and the variation in bulk stress. For example, the coefficient of thermal expansion of gold is 14.2×10^{-6} K^{-1}, which is significantly larger than the variation in length of the gold ribbon referred to above. Consequently, stringent precautions must be taken to isolate the variation in surface stress that is due to the change in the interfacial tension from any other side-effects. When the variation in surface stress is determined indirectly from electrode deformation, however, the influence of these side-effects does limit the use of this approach to a few idealized systems.

Small fluctuations in room temperature have an insignificant (or no) effect on contact angles.[99-102] A simply constructed air thermostat with control over the temperature to $\pm 1°C$ can effectively isolate the dependence of the contact angle on potential from its temperature dependence.

The variation in the contact angle with potential involves a change in the shape of the liquid side of the interface, but does not involve stretching or applying a force to the solid phase. Bulk stress therefore has no influence on the measurement.

The thermodynamic validity of contact angles does not depend on whether or not the electrode is ideally polarizable, provided that the measured angle corresponds to the electrochemical steady state. Under such conditions, the contact angle is a function of the solid/electrolyte interfacial tension, as given by the Young and Dupré equation. On the other hand, the determination of the variation of surface stress does not necessarily provide knowledge of the variation in interfacial tension unless the electrode is ideally polarizable.

The potential dependence of contact angles can be measured on solids and polarizable liquid metals such as mercury. This provides an opportunity to test the general validity of the method's principles. Methods based on stress measurements cannot be used with a liquid metal.

The potential dependence of contact angles can be measured on metals, semiconductors, and semimetallic electrodes of both single-crystal and polycrystalline type. The potential dependence of stress can be measured only on polycrystalline metals.

Contact-angle methods are not without their disadvanatages. The most important of these are: (1) in a three-phase system there is a greater likelihood of surface contamination and oxidation from organic and oxygen impurities present in the gas phase; (2) hysteresis complicates the interpretation of the data, and reduces the flexibility of the measurement; (3) the wetting of such metals as gold and platinum is still a subject of controversy[103-111] among those who consider these

[99] A. W. Neumann and D. Renzow, *Z. phys. Chem. (Frankfurt)*, 1969, **68**, 11.
[100] A. W. Neumann and W. Tanner, *J. Colloid. Interface Sci.*, 1970, **34**, 1.
[101] R. E. Johnson, jun. and R. H. Dettre, *J. Colloid. Sci.*, 1965, **20**, 173.
[102] C. L. Sutula, R. Houtala, R. A. Dalla Betta, and I. A. Michel, Abstracts, 153rd Meeting, American Chemical Society, April 1967.
[103] M. L. White, *J. Phys. Chem.*, 1964, **68**, 3083.
[104] R. A. Erb, *J. Phys. Chem.*, 1965, **69**, 1306.
[105] F. M. Fowkes, in 'Advances in Chemistry No. 43', ed. R. F. Gould, American Chemical Society, Washington, D.C., 1964, p. 43.
[106] F. M. Fowkes, *Ind. and Eng. Chem.*, 1964, **56**, 40.
[107] A. C. Zettlemoyer, *J. Colloid Interface Sci.*, 1968, **28**, 343.
[108] W. Harkins and A. Feldman, *J. Amer. Chem. Soc.*, 1927, **44**, 2665.

metals to be hydrophobic in nature[103—107] and others who report low or zero contact angle;[108—111] (4) the contact angle–potential curves cannot be easily measured by recording techniques. On the other hand, manually drawn curves are tedious and time-consuming to produce.

Indirect Measurement of Surface Stress.—Beck[94—96] attempted to determine variations in surface stress as a function of potential by using an extensometer which measures the corresponding variation in the length of a very thin metal ribbon. The ribbon is kept under an approximately constant force throughout the experiment by mounting it axially inside a glass tube, with its upper end attached to a spring that has a small spring constant compared with the stiffness of the ribbon. The change in length of the ribbon is measured by comparing the capacitances of a pair of differential capacitors formed between a grounded aluminium bobbin and two aluminium plates fixed in the extensometer head. The aluminium bobbin is mounted on a quartz spindle, which is attached to the upper end of the ribbon. A change in the length of the ribbon as a function of potential causes a change in the position of the bobbin; this creates a difference between the values of the two capcitances. The capacitors are connected to the input of an electronic capacitance sensor which develops a d.c. potential that is directly proportional to the position of the bobbin. The linearity between the output voltage $V_{\Delta L}$ and the change in the length of the ribbon is described in the relationship:[96]

$$V_{\Delta L} = K_1 \left[(1/C_1) - (1/C_2) \right] = K_2 \Delta L \tag{10}$$

where C_1 and C_2 are the two differential capacitances and K_1 and K_2 are proportionality constants.

The calibration of the instrument was carried out by measuring the output voltage as a function of the known weight fixed at the bottom of the glass tube. The output voltage was found experimentally to vary linearly with the weight in accordance with equation (10) and the equation:[96]

$$\Delta L_t = L_t W / A_t E_t \tag{11}$$

where ΔL_t is the change in the length of the tube, L_t its length from the support point to the weight hanger, A_t the cross-sectional area of the tube, and E_t the elastic modulus of the tube.

Thermal expansion constitutes the most serious problem in the extensometer method.[94—96] The error due to thermal expansion can be reduced by using an outer supporting tube with a thermal expansion coefficient of the same order of magnitude as that of the ribbon, and by using materials such as invar and quartz, with low thermal expansion coefficients, for the ΔL-measuring head above the ribbon. The potential must be swept with a frequency sufficiently high to provide adequate resolution from residual thermal drift.[94—96] The extensometer is kept inside a box that is insulated with foam plastic and the measurements are made only after the thermal drift has become negligible.[94—96] None of these measures, however, can effectively eliminate thermal expansion due to the Joule and over-

---

[109] K. W. Bewig and W. A. Zisman, *J. Phys. Chem.*, 1965, **69**, 4238.
[110] M. K. Bernett and W. A. Zisman, *J. Colloid Interface Sci.*, 1968, **28**, 243.
[111] M. E. Schrader, *J. Phys. Chem.*, 1970, **74**, 2313.

voltage heating effects. Beck therefore suggests that the method be limited to systems with low electrochemical activity, but unless the effect of electrochemical activity on thermal expansion can be quantitatively accounted for, the results of the extensometer method cannot be conclusively interpreted.

The variation in surface stress, $\Delta\sigma_s$, can be obtained from the change in the ribbon length ΔL by an equation developed by Beck:[96]

$$\Delta\sigma_s = -(AE/PL)\Delta L \quad (12)$$

where A and P are the cross-sectional area and periphery of the ribbon and E is its elastic modulus. To derive equation (12), Beck[96] made the following assumptions:

(1) the spring constant of the ribbon remains constant after the change in ΔL;
(2) the force F on the ribbon is balanced by two separable surface and bulk stresses which are linearly additive, as shown by:

$$F = \sigma A + \sigma_s P \quad (13)$$

where the surface stress is assumed to be located in the first atomic surface layer;
(3) both the bulk and surface elastic moduli are constants as a function of the applied potential.

If the instrument is designed such that $m_r \gg m_s$, where m_r and m_s are the mass of the ribbon and of the spring respectively, one obtains:

$$\Delta F = -m_s \Delta L \quad (14)$$

By applying the assumptions mentioned above, using equations (13) and (14) as well as Hooke's law in the form:

$$\Delta E = E\,\Delta L/L \quad (15)$$

and dropping negligible terms, one can obtain equation (12).

Another attempt to obtain the potential dependence of surface stress from measurement of the strain or of electrode deformation as a function of potential has been reported by Fredlein et al.[97,98] These authors use the bending-beam technique, which has frequently been discussed in the literature dealing with the elastic properties of thin films.[37–42] The electrode consists of a thin metal film deposited on a strip of infinitely hard substrate such as glass. The electrode is positioned vertically in the solution, with the upper end firmly clamped. The potential dependence of the angular deflection of the free end of the electrode is measured by a laser-optical lever.[97]

The principles of the bending-beam method were first stated by Stoney,[37] who derived an equation relating the stress in the film to the radius of curvature of the beam:

$$\sigma = E\delta^2/6rf \quad (16)$$

where σ is the stress in the film, f the film thickness, δ the substrate thickness, E the modulus of elasticity of the substrate, and r the radius of curvature of the substrate. Stoney[37] used the elastic theory of simple beams to derive equation (16). More accurately, however, the bending of substrates by films should be treated as the bending of a thick plate. To account for this, equation (16) has been modified to:[38—40]

$$\sigma = E\delta^2/6\,rf(1-v) \qquad (17)$$

where v is the Poisson's ratio for the substrate. By combining equation (17) with expression $\theta = l/r$, where l is the length of the electrode and θ the angular deflection, the variation in stress of the substrate can be related to the angular deflection by the equation:

$$\Delta\sigma = 6\,lf(1-v)\,\theta/E\,\delta^2 \qquad (18)$$

The equations used by Fredlein et al.[98] neglect the term f and show $(1-v)$ in the denominator. The validity of equation (18) requires[38] (a) that the elastic moduli of the metal film and the substrate be of the same order of magnitude, (b) that the substrate thickness be much greater than the film thickness, (c) that the substrate curvature be much greater than the substrate thickness, and (d) that the length of the substrate be more than twice its width. Fredlein et al.[97,98] also assume that the variation in stress, $\Delta\sigma$, given by equation (18) corresponds to the variation in surface stress, $\Delta\sigma_s$, as a function of potential. This assumption may be justifiable, provided that the elastic modulus and bulk stress of the thin metal film remain constant during the measurement.

The measuring apparatus consists[98] of a 1 mW He/Ne laser, optical components, and a detector. After reflection from the tip of the electrode, the laser beam is split by a right prism and deflected through diffusion discs to two photoconducting CdS cells. The bending of the electrode displaces the beam and causes one photocell to receive less light than the other. The output voltage from the photocells at a constant potential of the electrode is observed as a function of time on an x–t recorder. It is to be noted that the angular deflection in this case is measured point by point, in contrast to the continuous stress vs. potential plots obtained by Beck.[94—96]

Deposited metal films are known to exhibit considerable intrinsic stresses, and the slow relaxation of the latter during the measurement constitutes a serious source of error.[98] Fredlein et al.[98] found that gold deposits could be stabilized by exposing them for four weeks in 10^{-3}M-KCl at 90 °C. These authors[98] have also drawn attention to the possibility that the Young's modulus of glass may change when the metal is deposited on its surface. The variation in the pH of the solution will change the dissociation rate of the surface hydroxy-groups on the glass reverse surface, leading to a change in both the surface charge density and the surface stress. To avoid such error, the authors[98] recommend the use of constant pH when studying the effect of change of electrolyte concentration.

The authors[97,98] mention no measures taken to control the temperature of the substrate during evaporation and during the measurement of angular deflection. The difference between the two temperatures must be kept constant, so as to obtain a reproducible thermal stress in the electrode, as shown by the equation:[39,40]

$$l_T = (\alpha_f - \alpha_s)\,\Delta T \qquad (19)$$

where α_f and α_s are the average coefficient of expansion for the film and the substrate and l_T is the strain resulting from the thermal stress. A variation in l_T means that l in equation (18) is not constant.

Both Fredlein et al.[98] and Beck[96] have shown independently, by mathematical arguments, that, within the limits of elastic deformation, the term $d\gamma/d\varepsilon_s$ in equation (4) is constant. Accordingly, the variation in surface stress is equal to the variation in interfacial tension:

$$\Delta\sigma_s = \Delta\gamma \qquad (20)$$

In both cases it is assumed that the surface elastic modulus remains constant. Such an assumption is probably valid for systems in which ions are electrostatically adsorbed, but not for systems in which ions are specifically adsorbed.

Plots showing the variation of surface stress as a function of potential for a gold ribbon electrode in 1M-$HClO_4$ and various 0.1M-potassium salts have been reported by Beck and co-workers.[96] Stress–potential plots for platinum and titanium ribbons in 0.1M-H_2SO_4 and 0.1M-K_2SO_4, respectively, have been reported by Beck and Beach.[95] The results[95,96] were accompanied in many cases by simultaneously obtained cyclic voltammograms, showing the variation in the electrochemical activity at the same potential scan rate as was applied in measuring the stress–potential plots. Some of the results obtained by Beck et al.[95,96] are shown in Figures 1—4.

In the case of gold, the maxima of the stress–potential plots shown in Figure 1 are always negatively shifted from the potential of zero charge obtained by the scraping method.[112] The observed shift is larger with iodide and bromide ions (0.28 and 0.2 V respectively) than with chloride, nitrate, and sulphate ions (0.1, 0.12, and 0.12 V, respectively). For a set of 0.1M-potassium salts, the stress–potential

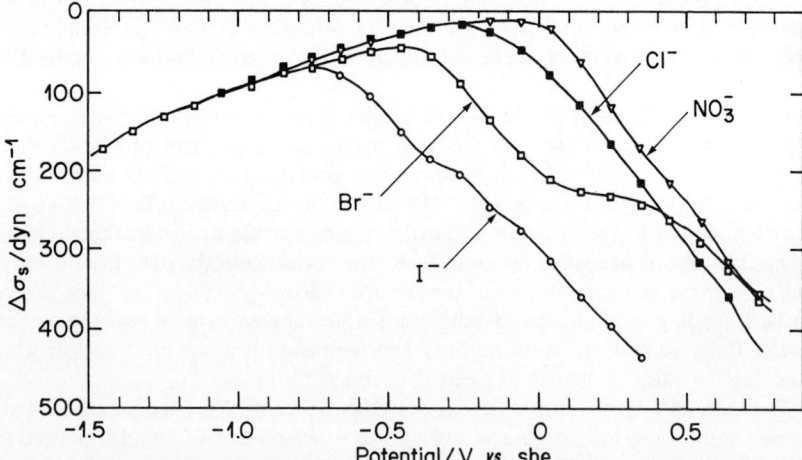

Figure 1 *Anodically swept surface stress–potential curves for gold ribbon with solutions of various potassium salts; sweeping frequency 0.1 Hz; data obtained by the extensometer method.*[96]

[112] D. D. Bodé, jun., T. N. Andersen, and H. Eyring, *J. Phys. Chem.*, 1967, **71**, 79.

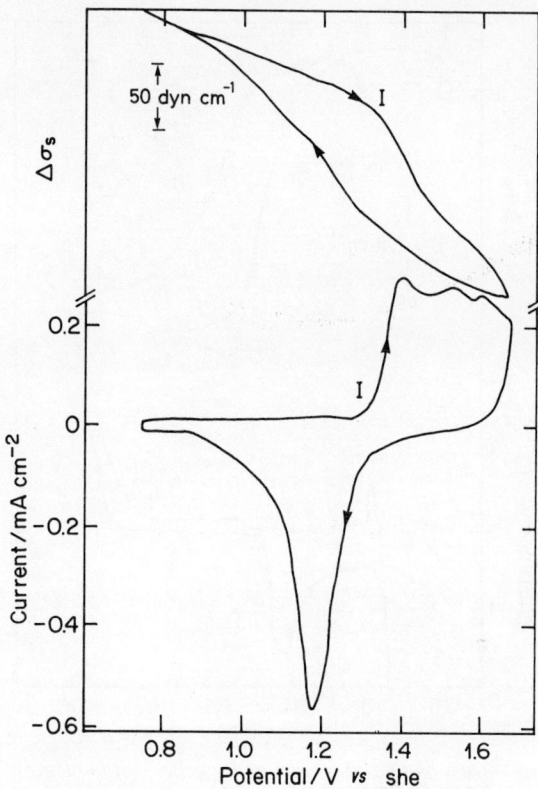

Figure 2 Surface stress–potential curve and cyclic voltammogram for gold ribbon in 1M-$HClO_4$; sweeping frequency 0.1 Hz; data obtained by the extensometer method.[96]

plots have identical slopes at potentials negative to the surface-stress maxima.[96] The change in surface stress at these negative potentials has an order of magnitude similar to the corresponding change in the interfacial tension observed with mercury.[16] At positive potentials, the change in surface stress depends considerably on the nature of the anion and, at a given rational potential, its magnitude may be hundreds of dyn cm^{-1} larger than the corresponding change in interfacial tension with mercury. In 0.1M-perchloric acid, Figure 2 shows that the stress–potential plot exhibits both a marked increase in hysteresis and a steeper slope of the curve at the potential corresponding to the start of anodic oxidation.

Lin and Beck[96] attempted to test the agreement between the Lippmann equation and their experimental results on gold by comparing the presumed charge–potential curve obtained by the numerical differentiation of the stress–potential plot with that obtained by the numerical integration of the anodically swept current–potential plot. Agreement between these two curves in the case of 0.1M-K_2SO_4 and within

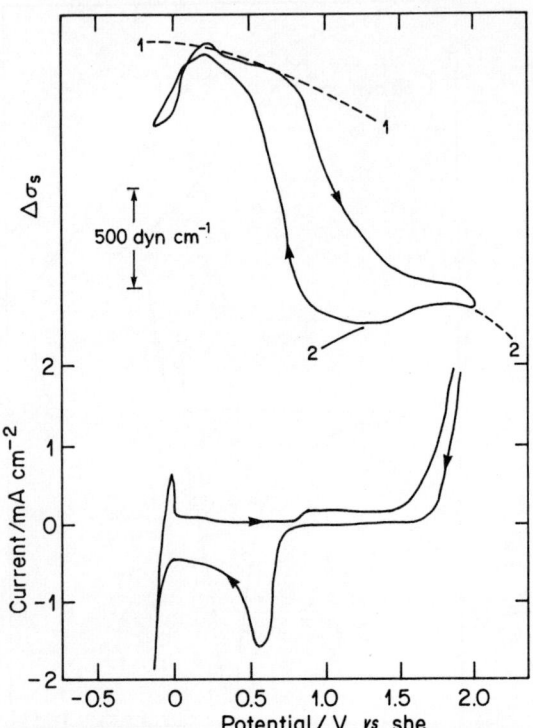

Figure 3 Surface stress–potential curve and cyclic voltammogram for platinum ribbon in 0.1M-H_2SO_4; sweeping frequency 0.01 Hz; data obtained by the extensometer method.[95]

the rational potential region of −0.5 to 0.3 V led the authors to conclude that the assumptions made in deriving equation (12) were justifiable. Such a conclusion may be criticized, however, on the grounds that the agreement between the two curves was demonstrated for what appears to be one isolated case. More conclusive evidence would require similar agreements in the case of other weakly adsorbed anions such as nitrate and perchlorate. Furthermore, there is insufficient justification for the assumption that a current–potential plot obtained at a frequency of 0.1 Hz is purely capacitative in nature.

The disagreement between the Lippmann equation and stress–potential plots observed with gold in other solutions and potential regions has been attributed[96] to the inapplicability of the above-mentioned equation in what the authors consider to be irreversible systems. However, this conclusion is justifiable only if there is sufficient evidence that the measured variation in stress actually corresponds to a variation in interfacial tension, which is not the case here. For example, the formation of an adsorbed oxygen film should result in a decrease in the surface charge density and an increase in the interfacial tension; this is not consistent with

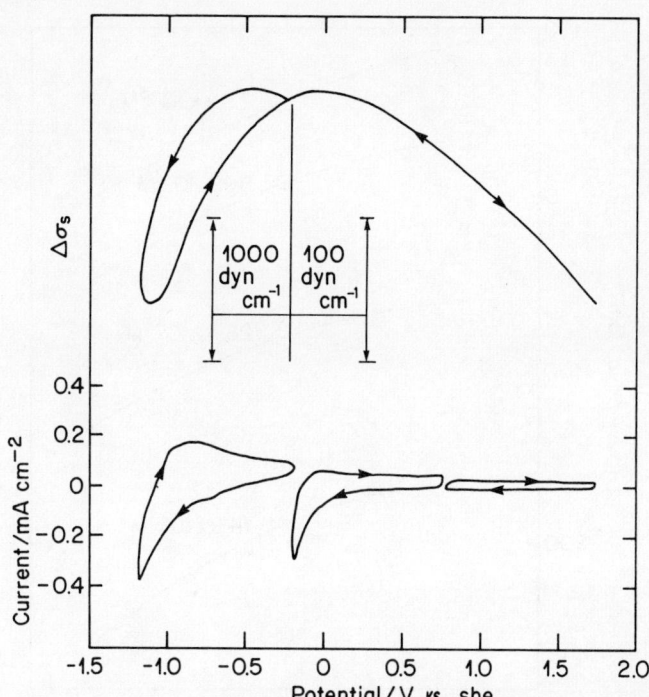

Figure 4 *Surface stress–potential curve and cyclic voltammograms for titanium ribbon in 0.1M-K_2SO_4; pH 4.5 and sweeping frequency 0.1 Hz; data obtained by the extensometer method.*[95]

the sharp decrease observed in stress. Such a large decrease in stress has more probably resulted from a change in thermal expansion due to the large Joule heating effects associated with the electrochemical current and/or from a continuous change in the surface elastic modulus. A change in the elastic modulus as a function of potential violates the condition made in deriving equation (20), with the result that the changes in surface stress and interfacial tension are not equal.

In the case of platinum,[95] Figure 3 shows that the variation in stress is approximately 1500 dyn cm^{-1} for the potential region between 0.3 and 1.5 V. The interaction of oxygen with platinum is stronger than with gold. Consequently, the influence of Joule heating on thermal expansion and of the variation in the Young's modulus on surface stress will be more prominent than with gold.

The surface stress–potential plot obtained with titanium[95] is shown in Figure 4. The maximum of the plot appears to separate an ideally polarizable titanium oxide region at potentials anodic to 0 V (she) from a non-polarizable region in which the titanium oxide undergoes a reduction to lower oxides. The difference in the magnitude of variation in surface stress and in the hysteresis on both sides of the maximum indicates that the maximum most probably defines the potential

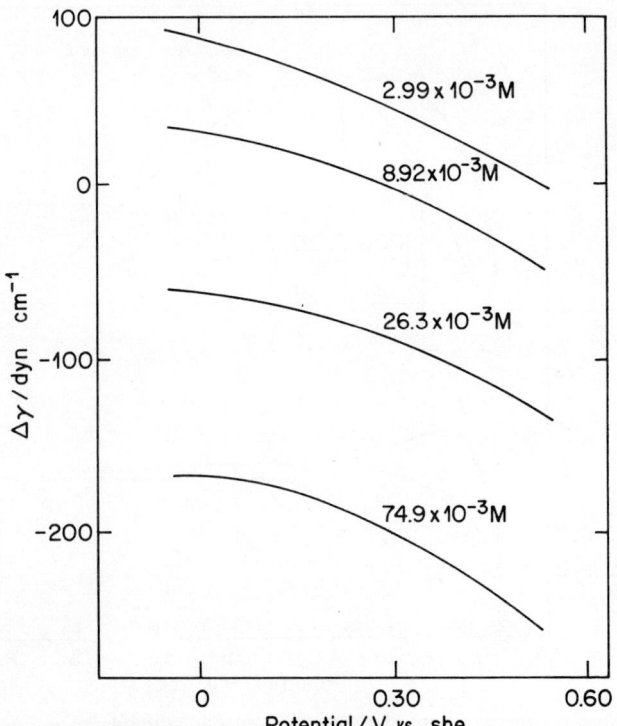

Figure 5 *Interfacial tension–potential curves for gold in different $HClO_4$ concentrations; data obtained by the bending-beam method.*[98]

at which the variation in the surface elastic modulus and Joule heating effects starts to influence the data.

Experimental results obtained by the bending-beam technique on gold and platinum have been reported.[97,98] Plots of interfacial tension–potential for gold[98] in solutions containing different $HClO_4$ concentrations and in the potential region between zero and 0.55 V (she) are shown in Figure 5. Because of the inaccuracy of the value for Young's modulus used in the computation of the data, the uncertainty is high.[98] Assuming that equations (18) and (20) are valid, the analysis of the results indicates that the surface excesses are an order of magnitude higher than those encountered with the mercury–perchlorate solution interface, although the surface charge density is of the same order of magnitude.[98] Caution must be exercised, however, in accepting these conclusions until the validity of the method has been proven with other solutions and with other electrodes.

The results obtained with platinum[97] are shown in Figure 6, and appear similar to those obtained with the extensometer method.[95] This agreement is interesting, and shows that the sweeping of potential in the extensometer method has little effect on the results.

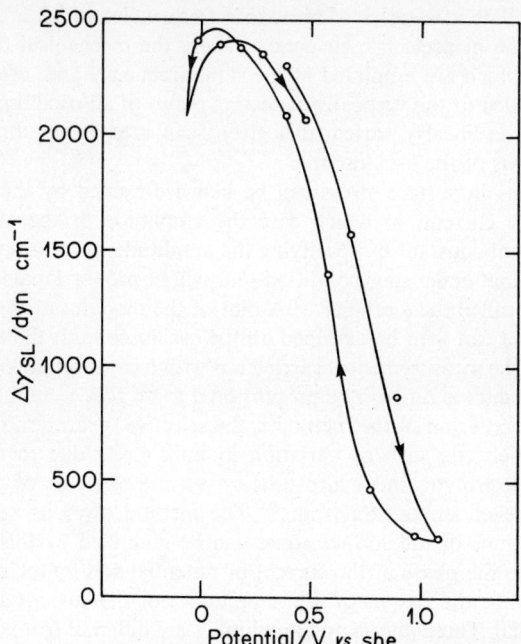

Figure 6 *Interfacial tension–potential curve for platinum in 8.3×10^{-4} M-H_2SO_4; data obtained by the bending-beam method.*[97]

A conclusive test of the assumptions associated with the extensometer and bending-beam methods might have been possible if the methods had been applied to measure the stress–potential plots for semi-polarizable electrodes such as lead, cadmium, and bismuth. Unfortunately, it appears that the large thermal expansion coefficients of these metals have discouraged investigators from performing the experiments.

Direct Measurement of Surface Stress by a Piezoelectric Element.—The only direct method developed to measure surface stress as a function of the potential is due to Gokhshtein.[87—93] The method is direct in the sense that it is the variation in surface stress rather than the variation in the electrode deformation that is registered directly by a piezoelectric element. According to Gokhshtein,[93] no variation in the dimensions of the electrode is allowed to take place during the measurement because the latter is carried out by balancing the measured force against another force, which is the response of the piezoelectric element.

The main features of the method[87] can be outlined as follows. The test electrode is rigidly connected, in a special manner, to a highly sensitive piezoelectric element. The applied potential consists of a mean component upon which is superimposed a high-frequency component. Oscillations with an amplitude $\Delta\phi$ in the potential will result in oscillations with an amplitude $\Delta\sigma_s$ in the surface stress, which in turn set up forces of inertia that excite vibrations in the entire electrode–piezoelement unit. Because the piezoelectric element is mechanically coupled to the test electrode, the two can oscillate at a series of resonance frequencies which are applied in the measurement. The piezoelectric element converts the mechanical oscillations into electrical ones, which are amplified at the same frequency and, after rectification, the signal is applied to the vertically deflecting plates of an oscilloscope. The mean potential, which is linearly varied at a given scan rate, is simultaneously fed to the horizontal axis of the oscilloscope.

The oscillations in surface stress can be likewise excited by the application of a high-frequency current, in which case the amplitude of the electrode charge density, Δq, is kept constant by specifying the amplitude of the alternating current. It can be shown that under such conditions $\Delta \sigma_s$ will be proportional to the derivative $\partial \sigma_s / \partial q$, which is called the q-estance.[93] A plot of the modulus of the estance $\partial \sigma_s / \partial q$ as a function of ϕ can then be obtained on the oscilloscope. If the electric variable is the potential, the measurement is carried out with a constant potential amplitude, under which conditions $\Delta \sigma_s$ will be proportional to $\partial \sigma_s / \partial \phi$.

An important criterion of the method is the selective separation of surface stress from other side-effects, such as variation in bulk stress due to diffusion, Joule heating of the electrolyte, and hydrostatic forces due to traces of gases and other substances deposited on the electrode.[87] The method owes its selectivity to the fact that oscillations of the surface stress can be identified by their characteristic phases relative to the phase of the current or potential and by their specific amplitudes in relation to the frequency and amplitudes of the current and the average electrode potential. These phases and amplitudes are different from those characterizing bulk stresses because of the diffusion of foreign atoms into the electrode lattice and thermal expansion due to heating effects, and they can therefore be identified and recorded separately.[84–93]

Because of the dynamic features of the method, the recorded variation in surface stress does not always correspond to equilibrium conditions.[89] For an ideally polarizable electrode, such as lead, equilibrium is reached during the measurement, because the time for charging or discharging the double layer is shorter than the period of oscillation of surface tension. For an electrode such as platinum, the period of oscillation of the surface stress is shorter than the time necessary for adsorption of hydrogen or oxygen to reach equilibrium.[89] The results, in the latter case, depend on the frequency of oscillation as well as on the scan rate of the mean potential.[89] Surface stress–potential plots corresponding to different frequencies may be simultaneously recorded at a single sweep of the mean potential.[90] For this purpose, the alternating component of the applied current or potential consists of different sinusoids, with different frequencies, rather than just one sinusoid, as in the case of single-frequency measurement. The different frequencies selected are equal to the natural frequencies of the combined mechanical system of the electrode and the piezoelectric element.[90] The complex signal received from the piezoelectric element is resolved into its components by a specially tuned amplifier, operating at the corresponding frequencies.

Typical results[93] by Gokhshtein with lead and platinum are shown in Figures 7 and 8, respectively. In both Figures, the modulus of estance, $|\partial \sigma_s / \partial q|$, is plotted against the applied mean potential ϕ. The plots of Figure 7 with lead were obtained at the same frequency of 5.0 kHz, while those of Figure 8 for platinum were obtained at the two simultaneously applied frequencies of 22.5 and 2.6 kHz. Each plot on

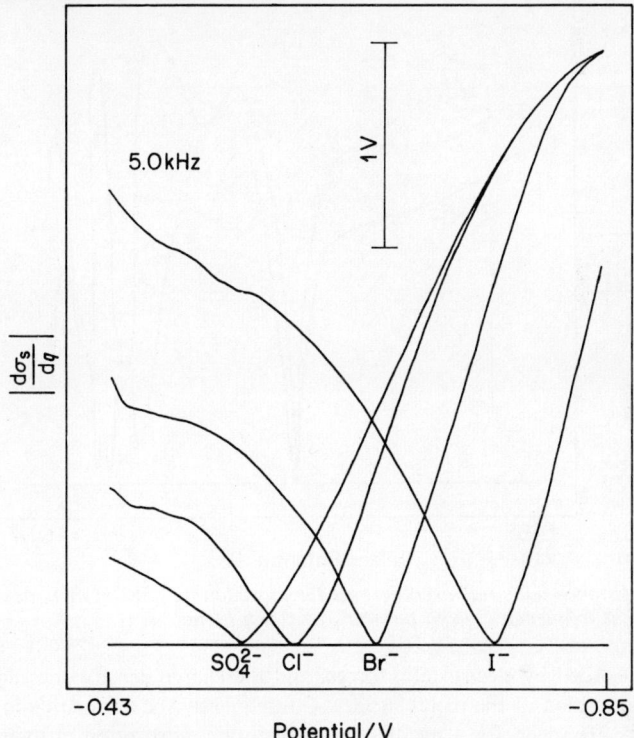

Figure 7 *Estance–potential oscillograms for lead in* 0.5M-Na_2SO_4 + *different* 0.15M *potassium halides; from the work of Gokhshtein.*[93]

Figure 7 has one minimum, and the potentials of the minima are influenced by the surface-active anion, as expected for an ideally polarizable surface. The potentials of the minima also agree with the zero-charge potentials of lead in the given solution. This indicates that the variation in surface stress in the potential region of the minimum corresponds to the variation in interfacial tension or in the surface free energy as defined by the Lippmann equation.

Figure 8 shows a complex behaviour pattern for the variation of surface stress with the potential. This is a case where the variation in surface stress is not equivalent to the variation in interfacial tension as described by the Lippmann equation. To interpret the latter behaviour, Gokhshtein[92] used a modified form of the Lippmann equation which takes into account the effect of surface deformation on the surface charge density:

$$\partial \sigma_s / \partial \phi = -q - \partial q / \partial \theta \qquad (21)$$

where σ_s is surface stress, q is the surface charge density, and θ is the ratio of the solid surface area after deformation to the area before deformation. Equation (21) follows from the relationship $\partial \sigma / \partial q = \partial \phi / \partial \theta$ which describes the results of some experiments on elastic discharging.[92] With an ideally polarizable surface,

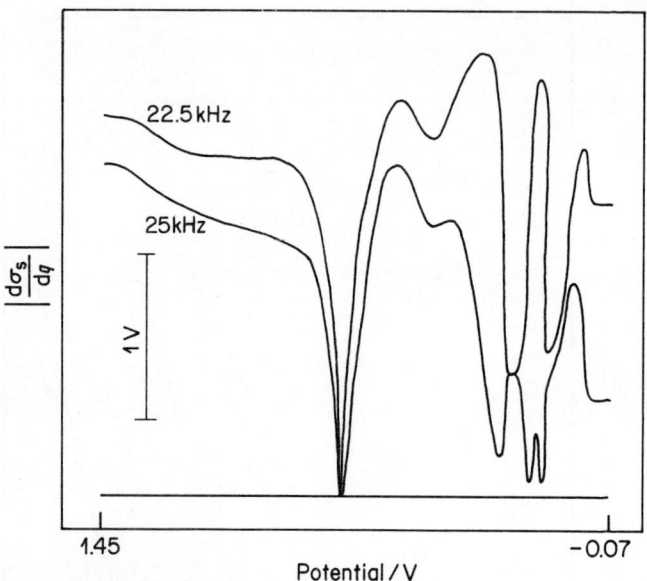

Figure 8 *Estance potential–oscillograms for platinum in 0.5M-H_2SO_4 at two different frequencies; from the work of Gokhshtein.[93]*

ions are adsorbed by electrostatic forces, and the charge density does not depend on the deformation of the metal surface. Consequently the term $\partial q/\partial \theta$ in equation (21) is zero. However, for a metal such as platinum, adsorption of hydrogen and oxygen significantly reduces the role of electrostatic adsorption. Adsorbed particles of hydrogen and oxygen are localized at the surface, and therefore their surface density will be influenced by the deformation of the metal surface. The different maxima and minima of Figure 8 therefore result from the variation in the term $\partial q/\partial \theta$ with potential.[92]

Gokhshtein does not distinguish between surface stress and surface tension, and describes all his data in terms of the second parameter. He does, however, distinguish between surface tension and free surface energy. In the present chapter, however, the term 'surface tension' is used in the same sense as it is applied to liquids, *i.e.* in systems where the surface tension and the surface free energy are identical. Regardless of the term used, the experimental results of Gokhshtein show that, in the absence of side-effects from bulk and thermal stresses, there are two different potential-dependent surface parameters. One parameter is the quantity defined by the Lippmann equation, and the other is the stretching tension of a deformed surface.

Direct Measurement of Contact Angles.—The study of the potential dependence of contact angles started with the work of Möller.[67] The contact angle measured was that of a gas bubble resting on a solid metal completely immersed in the solution. The bubble is formed at the bottom of a tube brought near the metal surface. The maxima of the contact angle–potential curves obtained for different solid metals were found to occur at the same potential. This result most probably reflects the inaccuracy inherent in the method due to hysteresis and the poorly defined nature of the different phases of the system.

Frumkin and co-workers[68—71] used Möller's method to obtain contact angle–potential curves for mercury. The electrolyte was 0.5M-Na_2SO_4 and the gaseous phase was hydrogen. The curves are similar in shape to electrocapillary curves obtained by direct measurements of surface tension, but the variation of the interfacial tension with potential is significantly less than predicted by the equation:

$$d\gamma_{SL}/d\phi = -\gamma_L d\cos\theta/d\phi \qquad (22)$$

Frumkin and co-workers concluded that the disagreement between the experimental results and equation (22) results from the presence of an electrolyte film between the mercury surface and the gas bubble.[69,71] This implies that γ_{SV} (or more precisely π) is not constant, and that equation (22) cannot be used to determine the surface charge density from the curves.

Smolders[72] used the same gas-bubble method to measure contact angles for the mercury/aqueous Na decanesulphonate/hydrogen system as a function of the electrolyte concentration and in the absence of an applied potential. The experimentally obtained values of the contact angle were in good agreement, over a range of concentrations of Na decanesulphonate, with the values calculated from separately measured metal/electrolyte, metal/gas, and electrolyte/gas interfacial tensions by the capillary-electrometer and sessile-drop methods. For calculating contact angles, Smolders[72] used a modified form of equation (5) in which the vertical component of γ_L was compensated for. The form used is given by:

$$\cos\theta = (\gamma_{SV}^2 - \gamma_{SL}^2 - \gamma_{LV}^2)/2\gamma_{SL}\gamma_{LV} \qquad (23)$$

Nakamura and co-workers[74] measured the contact angle between the sessile drop of an electrolyte and a mercury surface as a function of potential. The electrolyte drop is placed on the mercury surface by a micrometer syringe. A vertical capillary touches the top of the drop and connects the drop *via* a salt bridge to a counter-electrode. No reference electrode could be used in such a system, and correction had therefore to be made for the ohmic drop. On the basis of a comparison between the experimental data obtained with the mercury/0.5M-Na_2SO_4/air system and directly measured electrocapillary curves, the authors reached the conclusion already drawn by Frumkin *et al.*[69,71] with regard to the variation of π with the potential.

Sparnaay[73] used the sessile-drop method to study the potential dependence of the contact angle between an electrolyte droplet and an etched germanium surface. Very little potential dependence of the contact angle was reported, which is not surprising in view of the expected strong dipole–dipole interaction between water and the layer of germanium oxide.

Indirect Measurement of Contact Angles by Meniscus-rise Techniques.—Because hysteresis is inevitable in the measurement of contact angles on solid surfaces, the method used to study the potential dependence of the contact angle on such surfaces must permit the separate measurement of both advancing and receding contact angles. Accuracy, establishment of a system of three well-defined phases, and suitability for use in the presence of a reference electrode are other important requirements for obtaining meaningful data. All these requirements can be satisfied in a system in which the test electrode is partially immersed in the cell and the contact angle is indirectly obtained from the observed height of the liquid meniscus at the partially immersed surface. The partially immersed electrode can be in the form of an infinitely wide plate,[75—83] a metal capillary,[84] a rod,[85,125] or a cylinder[86] made of transparent material.

Most of the available data on the potential dependence of contact angles on solid electrodes have been obtained by the method of meniscus rise at partially immersed plates,[75—83] in which the height of the meniscus above or below the level of the electrolyte surface in the cell is related to the contact angle at the three-phase line by the exact equation:[113]

$$\sin\theta = 1 - (\rho g/2\gamma_L) h^2 \tag{24}$$

where γ_L and ρ are the surface tension and density of the liquid and g is the acceleration of gravity. The derivation of equation (24) involves the assumption that the plate is infinitely wide. This is implicit in the fact that equation (24) is obtained by solving equation (1) for the two-dimensional case where h is not a function of the width of the plate. Neumann[113] used equation (24) to obtain contact angles at the organic solid/organic liquid interface, and reported that the theoretical requirement of infinite width can be satisfied by using a plate approximately 2 cm wide.

Since h is the difference between the level of the meniscus edge at the solid surface and the level of the liquid surface in the cell, the heights of both levels must be experimentally measured. This can be accurately and conveniently performed by a cathetometer. The cell used for this purpose should be sufficiently wide and built of a material that cannot be wetted by the solvent. The cell is completely filled until the liquid climbs upward, forming a horizontal surface. The height of the liquid surface level is measured with a glass fibre that is immersed vertically in the liquid at a point sufficiently far away from the working, counter, and reference electrodes. The potential of the working electrode is adjusted by a potentiostat. The potential dependence of both the advancing and the receding meniscus heights can be obtained by measuring the latter in successive steps at different pre-fixed potentials after the electrode at each potential has been lowered (in the case of advancing height) or raised (in the case of receding height) a certain distance with respect to the solution.

For both ideally polarizable and non-polarizable solid electrodes the variation in the solid/electrolyte interfacial tension with potential can be determined from the experimentally measured meniscus height–potential curves by:

$$\gamma_{SL} = \gamma_{SV} - \gamma_L (2Kh^2 - K^2 h^4)^{\frac{1}{2}} \tag{25}$$

where K is a constant for each liquid and replaces $\rho g/2\gamma_L$ in equation (24). Equation (25) is obtained[80] by combining equation (5) with equation (24). When both equations are combined[80] with the Lippmann equation, the result is:

$$q = \gamma_L (2K)^{\frac{1}{2}} [(1 - Kh^2)/(1 - \tfrac{1}{2}Kh^2)^{\frac{1}{2}}] (\partial h/\partial \phi)_\mu \tag{26}$$

which is a useful expression for the graphical determination of the charge density

[113] A. W. Neumann, *Z. phys. Chem. (Frankfurt)*, 1964, **41**, 339.

from the meniscus height–potential curves on ideally polarizable solid electrodes.

In order to test the principles of the meniscus-rise method, some of the early work was carried out on an ideally polarizable surface in the form of a mercury-plated gold plate.[75,78—82] Solutions of inorganic salts in water and mixed water–DMF solvents were used in the study. The meniscus height on this type of electrode varies reversibly with the potential, *i.e.* hysteresis is negligible, and the meniscus height–potential curves can be measured at the same fixed electrode position.

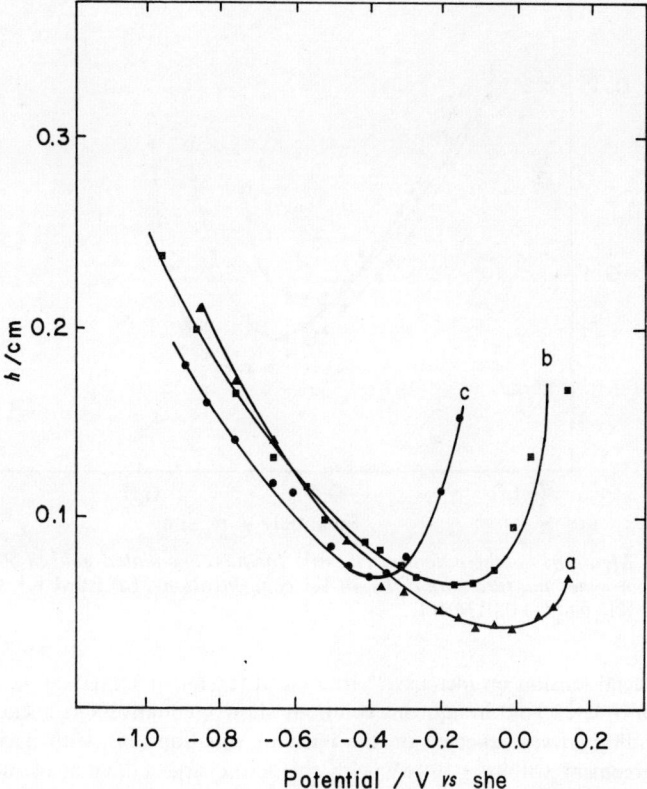

Figure 9 *Meniscus height–potential curves for mercury-plated gold in 50% DMF+ water mixture with (a) 0.1M-LiCl, (b) 0.1M-LiBr, and (c) 0.1M-KI.*

Typical results with the mercury-plated gold electrode in 0.1M solutions of LiCl, LiBr, and KI in 50% DMF–water mixture and in solutions of different concentrations of KI in the same solvent are shown in Figures 9 and 10, respectively. The potentials of the minima of these curves show excellent agreement with the zero-charge potentials of mercury obtained in the same solutions by conventional electrocapillary measurements.[114] The results plotted in Figure 11 show quantitative agreement between the interfacial tension–potential curves obtained in one case from the meniscus-rise method and, in the other, from direct measurement of

[114] V. D. Bezuglyi and L. A. Korshikov, *Soviet Electrochem.*, 1965, **1**, 1422.

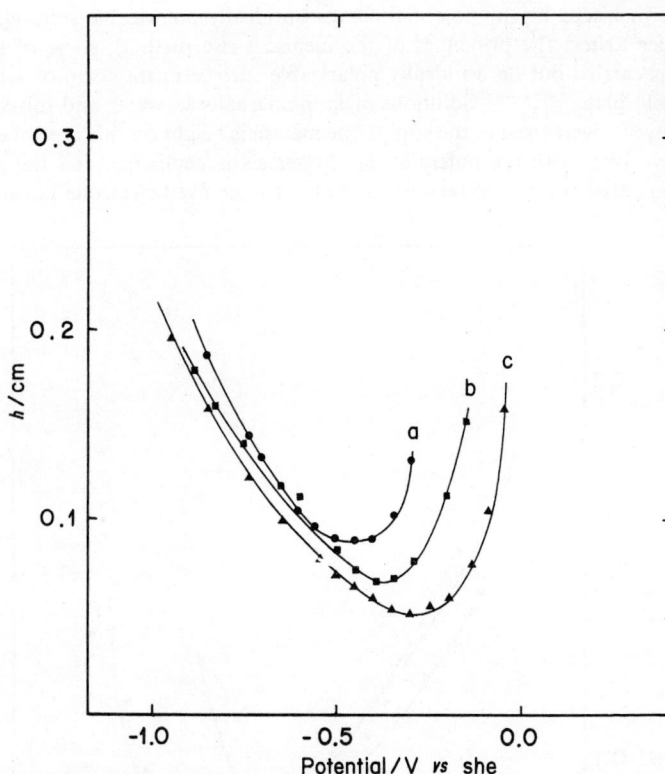

Figure 10 *Meniscus height–potential curves[80] for mercury-plated gold in 50% DMF + water mixture with different KI concentrations; (a) 1.0M-KI, (b) 0.1M-KI, and (c) 0.01M-KI.*

the interfacial tension on mercury.[80] Interfacial tension–potential curves obtained on mercury-plated gold in aqueous solutions show qualitative agreement with the corresponding curves expected on the basis of equation (26), with maxima that are in agreement within ±10 mV with the zero-charge potential of mercury in the same solution. A typical example is shown in Figure 12 for 1.0M-KI solution. In general, the surface charge density obtained from the meniscus height–potential curves at a given potential in aqueous solutions is less than the corresponding charge density calculated from conventional electrocapillary curves in the same solution. The difference between the two quantities diminishes with the increase in the rational potential and the specific adsorption of the anion. These observations strongly suggest that the lower charge densities associated with meniscus height–potential curves in aqueous solutions are at least partly due to surface contamination by organic impurities; this is a more serious problem in a three-phase than in a two-phase system. The possibility that the results are influenced by the presence and stability of a thin film of the electrolyte[69,71] above the proper meniscus cannot be ruled out.

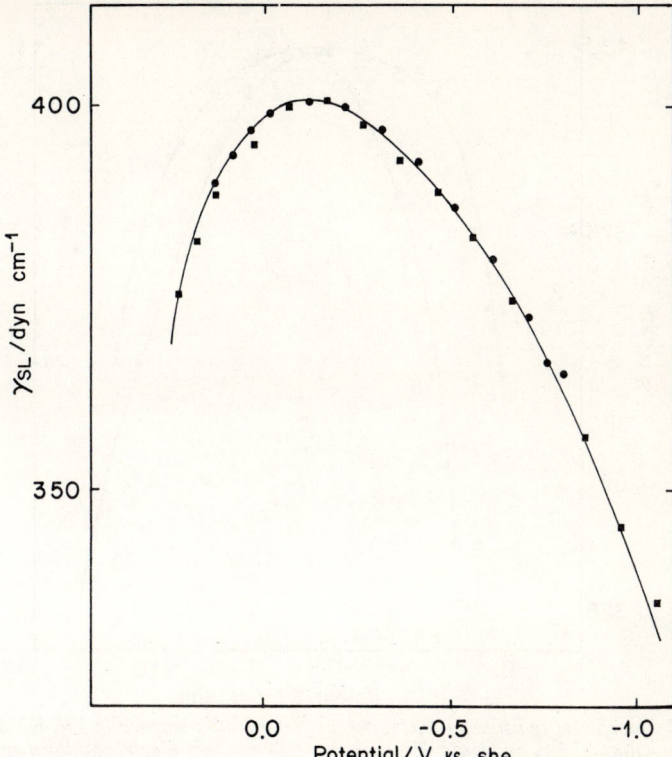

Figure 11 *Interfacial tension–potential curve for mercury-plated gold in 25% DMF + 75% water mixture and 0.1M-LiCl; data calculated from meniscus height–potential curves.*[80]

The meniscus-rise method has been used to study the interfacial phenomena at the silicon electrode.[82] The study was carried out on single-crystal (111) silicon plates, as a function of semiconductor type, resistivity, and electrolyte concentration, using two different illumination intensities of 45 and 260 fc (footcandles). The potential dependence of the *advancing* meniscus height on intrinsic, n-type, and p-type silicon electrodes in 0.5M-H_2SO_4 and with 260 fc illumination intensity is shown in Figure 13. The corresponding potential dependence of the interfacial tension obtained from Figure 13 by means of equation (25) is shown in Figure 14. The curves are strikingly similar in shape to electrocapillary curves of mercury. The variation of interfacial tension with potential is far less with silicon because most of the applied potential is used across the silicon-oxide and the space-charge layers. However, the shape of the curves indicates that the silicon-oxide layer formed is both chemically and electrochemically stable in H_2SO_4 and within the applied potential region. The voltammetry curves shown in Figure 15 exhibit no trace of electrochemical activity between 2.0 and −0.6 V. The electrochemical stability of the oxide has been confirmed by Turner.[115]

[115] D. R. Turner, in 'The Electrochemistry of Semiconductors', ed. D. J. Holmes, Academic Press, New York, 1962, p. 188.

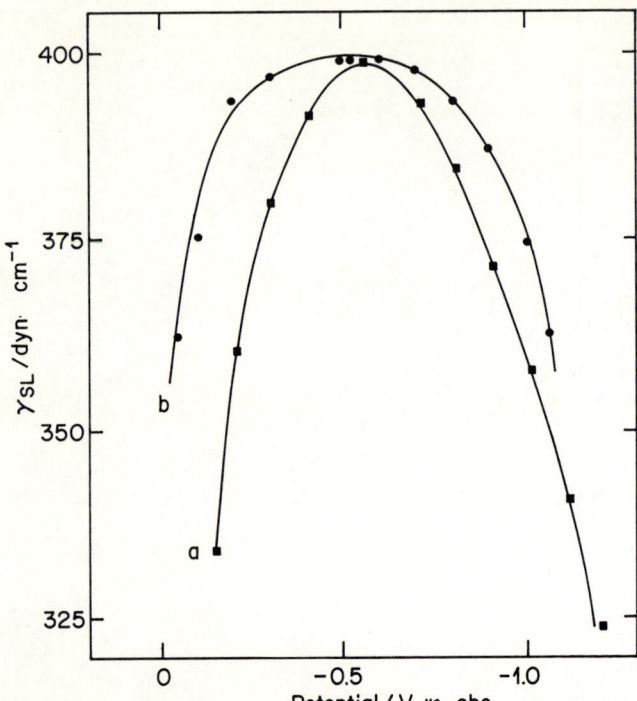

Figure 12 *Interfacial tension–potential curves for mercury surface in* 1M-KI *aqueous solution; (a) on liquid mercury, by conventional electrocapillary measurement,[16] and (b) from meniscus height–potential curve on mercury-plated gold.[80]*

Because of its characteristics of chemical and electrochemical stability,[116] the cleavage orientation of high-pressure stress-annealed pyrolytic graphite provides nearly ideal conditions for the study of interfacial tension by the meniscus-rise method. The potential dependence of the interfacial tension computed by means of equation (25) from the meniscus height–potential curves[81] on this surface is shown in Figure 16. The maxima of the curves and the slight potential dependence of the interfacial tension in aqueous solutions (plots a and b) of 1M-LiCl and -KBr agree with the results of differential capacitance reported on the same surface and in the same solutions by Randin and Yeager.[117] These two authors have attributed the slight potential dependence of the differential capacitance to the semiconducting properties of the graphite used. A different interpretation of the slight potential dependence of both the interfacial tension and the differential capacitance, that takes into account available values for the surface tension of the cleavage graphite surface, has been provided by Morcos.[81,118] Since the cleavage orientation of graphite has a low surface tension (γ_s of *ca.* 35 dyn cm^{-1}), thermodynamics predicts that

[116] I. Morcos and E. Yeager, *Electrochim. Acta*, 1970, **15**, 953.
[117] J. P. Randin and E. Yeager, *J. Electrochem. Soc.*, 1971, **118**, 711.
[118] I. Morcos, *J. Chem. Phys.*, 1972, **57**, 1801.

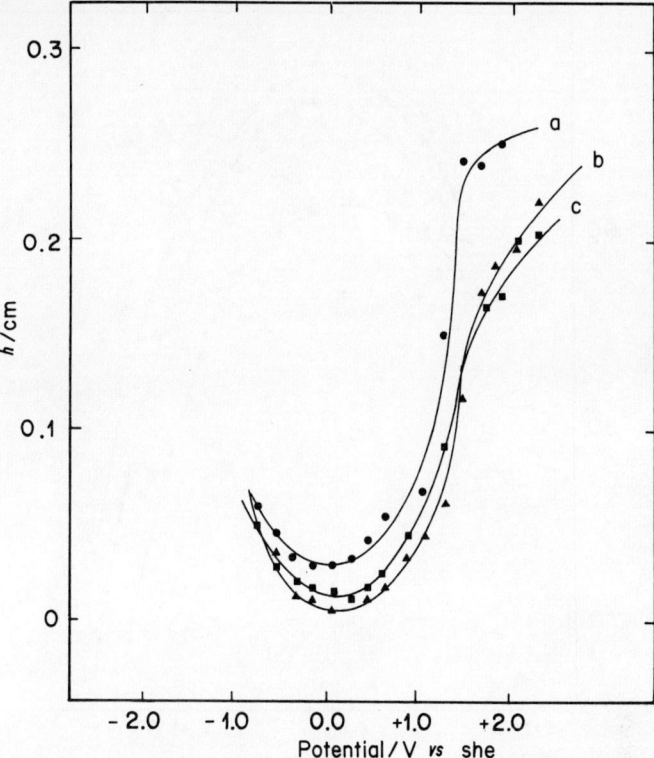

Figure 13 *Meniscus height–potential curves[82] for single-crystal (111) orientation silicon in 0.5M-H_2SO_4 and with an illumination intensity of 260 fc; (a) intrinsic silicon (b) n-type silicon, and (c) p-type silicon.*

water (with a higher surface tension of 72 dyn cm^{-1}) will be very weakly adsorbed on its surface. This in turn will reduce ionic adsorption, particularly if the ions are strongly solvated. However, in the presence of an aprotic solvent, such as DMF, there is less ionic solvation and, consequently, greater surface ionic activity.[81] This should, and [as observed in plot (c) of Figure 16] does, result in a higher potential dependence of the interfacial tension.

Because ideal polarizability is not a requirement for the thermodynamic validity of contact angles, the meniscus height–potential curves provide useful information on the influence of potential-dependent chemisorption on the interfacial tension. An instructive example is found in the case of lead, shown in Figure 17. The interfacial tension–potential curves are calculated[83] from meniscus-rise data obtained on a chemically polished lead electrode in solutions containing different concentrations of Na_2SO_4. The potentials of the maxima agree with the minima of differential capacitance–potential plots reported by Carr *et al.*[119] The concentration dependence of the interfacial tension at potentials negative to the potential of zero

---

[119] J. P. Carr, N. A. Hampson, S. N. Holley, and R. Taylor, *J. Electroanalyt. Chem. Interfacial Electrochem.*, 1971, **32**, 345.

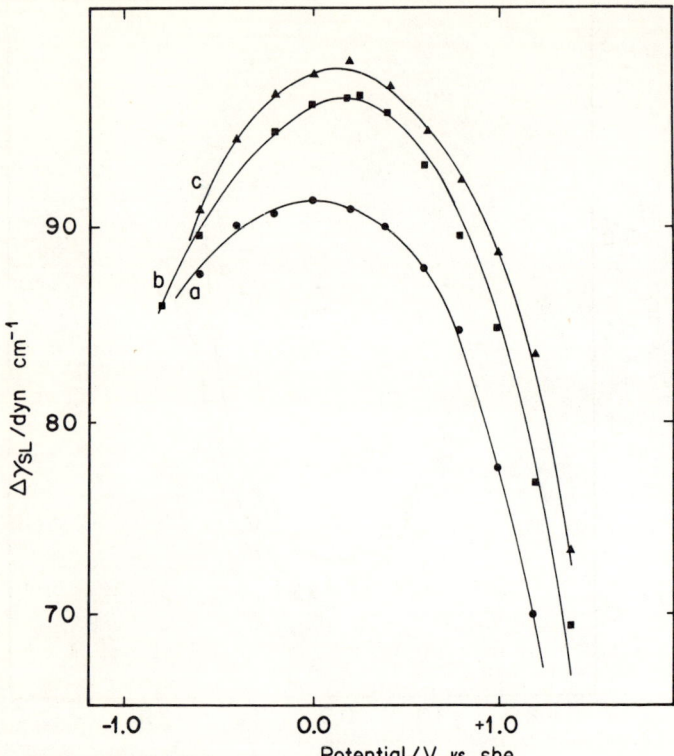

Figure 14 *Interfacial tension–potential curves*[82] *for single-crystal* (111) *orientation silicon in* 0.5M-H_2SO_4 *and with an illumination intensity of* 260 fc; *data calculated from Figure 13.*

charge indicates qualitatively that the lead electrode has some ideally polarizable characteristics. At -1.1 V (Hg/$HgSO_4$), there is a sudden increase in the interfacial tension, signalling a deviation from ideal polarizability which is most probably associated with formation of a new surface phase. The current peak of the anodically swept voltammetric curves[120] for lead in a sulphate medium is observed at 1.0 V (Hg/$HgSO_4$) and the rise in the current associated with same curves starts at -1.1 V.

Since gold is the only metal that does not form a thermodynamically stable bulk oxide at room temperature,[121] it is an important metal for studies of interfacial tension. The potential dependence of the interfacial tension obtained[77] from the meniscus-rise data of a gold electrode in 0.1M-KCl is shown in Figure 18, curve (a). The maximum of the curve agrees with the potential of zero charge from the open-circuit scrape method. The moderate deviation from polarizability observed at the anodic section of the curve probably results from the formation of such covalent

---

[120] J. P. Carr, N. A. Hampson, and R. Taylor, *J. Electroanalyt. Chem. Interfacial Electrochem.*, 1971, **33**, 109.
[121] D. O. Hayward and B. M. W. Trapnell, 'Chemisorption', Butterworths, London, 1964, p. 75.

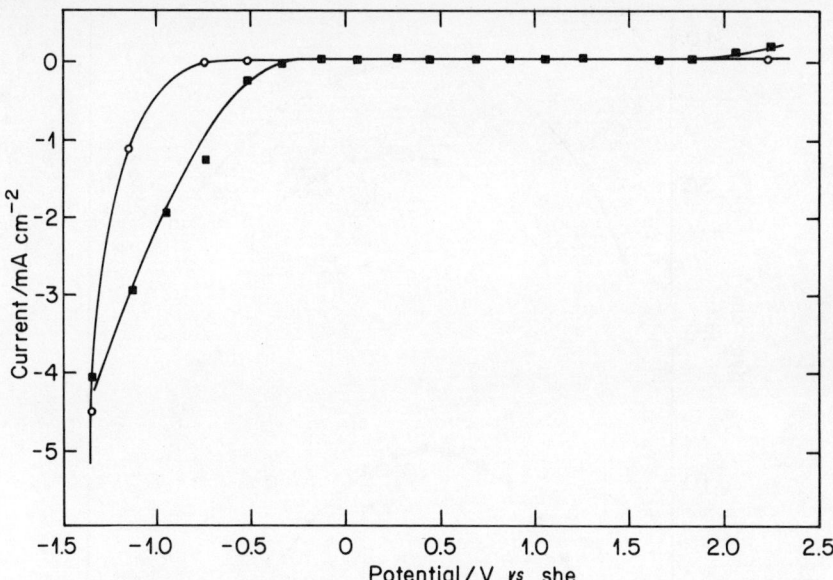

Figure 15 *Steady-state current–potential curves*[82] *for single-crystal* (111) *orientation silicon in* 0.5M-H_2SO_4 *and with an illumination intensity of* 260 fc; ○, *p-type silicon and* ■, *n-type silicon.*

structures as Au—O and/or Au—OH. Similar interpretations have been proposed by Sirohi and Genshaw[122] to explain their ellipsometric data, and by Schmid and O'Brien[123] and Carr and Hampson[124] to explain their capacitance curves. Similar behaviour is observed with a 0.1M-$KClO_4$ solution, as illustrated by curve (b) in Figure 18. A much stronger deviation from ideal polarizability is observed with a 0.1M-KI solution, as shown by curve (c) in Figure 18. In the latter case, the deviation from ideal polarizability most probably results from a strong adsorption of iodide and the subsequent formation of a new surface phase.

Very little work has been carried out at partially immersed electrodes other than those in the form of plates. Bonnemay et al.[84] have reported some data on the potential dependence of the capillary rise inside a glass capillary lined with a metal film deposited from the vapour phase. The relationship between the contact angle θ and the capillary rise h is given by:

$$h = 2\gamma_L \cos\theta / \rho g r \qquad (27)$$

where r is the radius of the capillary. Assuming a constant value of r, the variation in γ_{SL} can be calculated from the variation in h by means of:

$$dh = (2/\rho g r)\, d\gamma_{SL} \qquad (28)$$

which is obtained[84] by combining equation (27) with equation (5). Although the

---

[122] R. S. Sirohi and M. A. Genshaw, *J. Electrochem. Soc.*, 1969, **116**, 910.
[123] C. M. Schmid and R. N. O'Brien, *J. Electrochem. Soc.*, 1964, **111**, 832.
[124] J. P. Carr and N. A. Hampson, *J. Electrochem. Soc.*, 1972, **119**, 325.

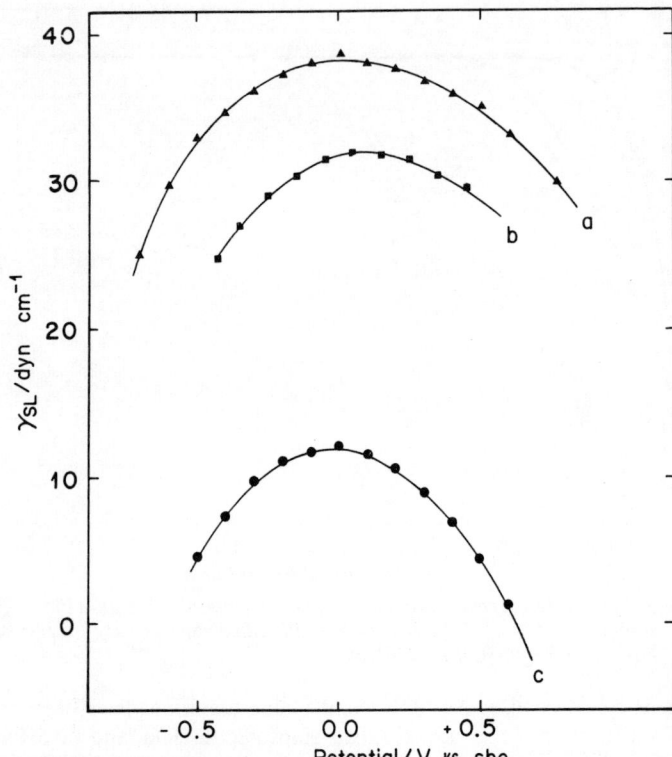

Figure 16 *Interfacial tension–potential curves for cleavage orientation of high-pressure stress-annealed pyrolytic graphite in (a)* 1M-LiCl, *(b)* 1M-KBr, *and (c)* 1M-LiCl *in* 50% *DMF* + 50% *water; data calculated from meniscus height–potential curves.*[81]

principles of this method are straightforward, its use is limited by the requirement of using X-radiography to measure the capillary rise and by the difficulty in obtaining metal capillaries of uniform radius. Figure 19 shows some of the results obtained by Bonnemay *et al.*[84] for the Cu/Na$_2$SO$_4$ system. The differential capacitance calculated from the data of Figure 19 by the equation:

$$C_d = \tfrac{1}{2}\rho g r \, d^2 h/d\phi^2 \qquad (29)$$

gives a value of 5 µF cm^{-2} for the differential-capacity minimum. The value is significantly lower than the experimentally measured differential capacity for copper, namely 40—50 µF cm^{-2}. The low value from the capillary meniscus-rise data was attributed to a lower surface-roughness factor for the evaporated metal surface.[84] However, the authors have not considered the effect of hysteresis on the data, which probably means that the capillary rise was measured under non-equilibrium receding conditions.

Shimokawa and Takamura[85] measured the potential dependence of the meniscus rise at partially immersed wire electrodes. Equation (24) cannot be used to calculate

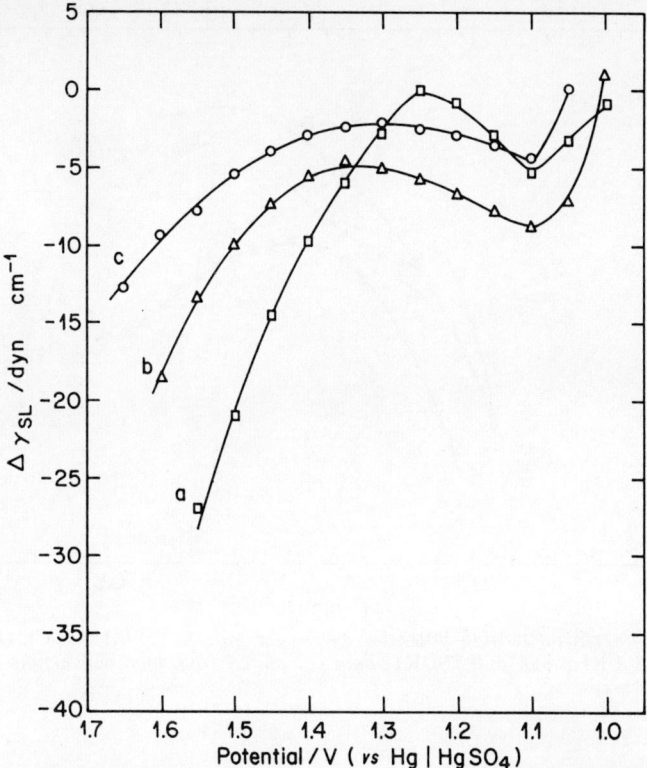

Figure 17 *Interfacial tension–potential curves for chemically polished lead in (a) 0.01M-Na$_2$SO$_4$, (b) 0.1M-Na$_2$SO$_4$, and (c) 1.0M-Na$_2$SO$_4$; data calculated from meniscus height–potential curves.*[83]

the contact angle from the meniscus rise at a wire,[125] and equation (1) cannot be solved analytically for this meniscus shape.[126] In this regard, it was later reported that the meniscus rise depends on the diameter of the wire until the latter reaches at least a few centimetres.[125] The diameter that results in a meniscus rise equivalent to that of an infinitely wide plate increases with the increase in the surface tension of the liquid.[125]

One of the most exciting applications for the method of meniscus rise at partially immersed electrodes was recently investigated by Morcos.[127] The meniscus height of an aromatic hydrocarbon (decylbenzene) at a partially immersed copper rod was measured as a function of the applied voltage between the copper rod and a counter-electrode in the form of a copper cage surrounding the rod. The results are shown in Figure 20. The meniscus height varies reversibly on both sides of a well-defined minimum. The results indicate that there is an electric double layer at the

---

[125] I. Morcos, *J. Electroanalyt. Chem., Interfacial Electrochem.*, 1974, **51**, 211.
[126] C. Hub and L. E. Soriven, *J. Colloid Interface Sci.*, 1969, **30**, 323.
[127] I. Morcos, unpublished work.

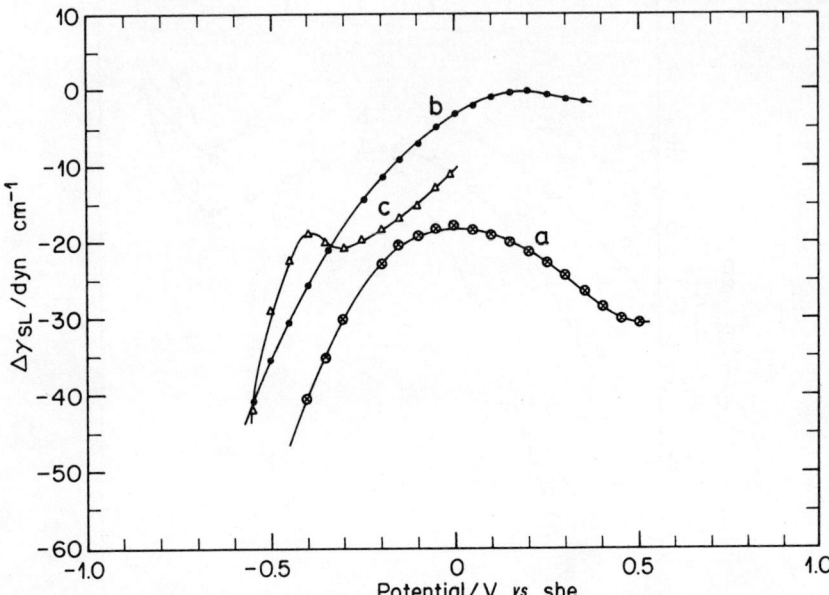

Figure 18 *Interfacial tension–potential curves for gold in* (a) 0.1M-KCl, (b) 0.1M-KClO$_4$, *and* (c) 0.1M-KI; *data calculated from meniscus height–potential curves.*[77]

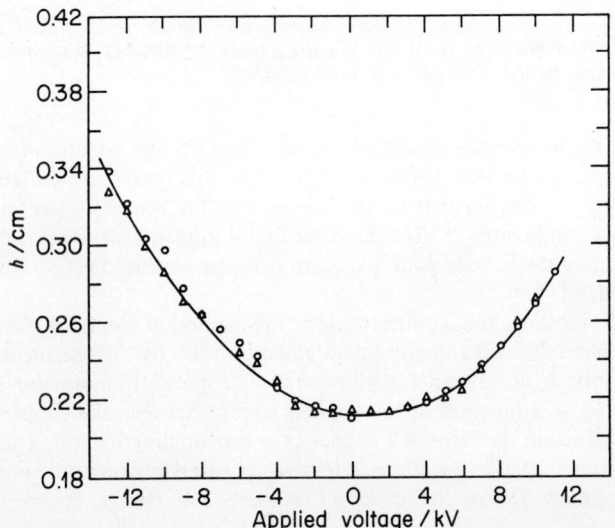

Figure 19 *Potential dependence of meniscus rise inside a partially immersed copper capillary in CuSO$_4$ solution; from the work of Bonnemay et al.*[84]

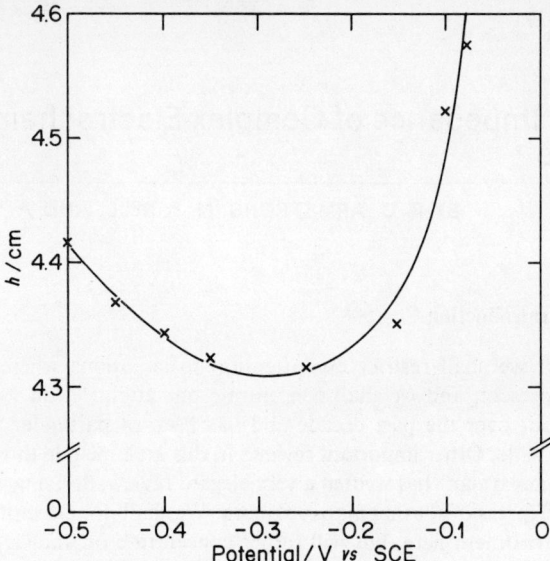

Figure 20 *Meniscus height–voltage curves of two measurements on a copper rod in decylbenzene; unpublished data by Morcos.*

metal/hydrocarbon interface with a surface charge density that depends on the applied voltage. At zero contact angle, equation (24) is reduced to:

$$h_{\theta = 0} = (2\gamma_L/\rho g)^{\frac{1}{2}} \qquad (30)$$

For decylbenzene, with $\gamma_L = 30.7$ dyn cm^{-1} and a density of 0.858 g cm^{-3}, $h_{\theta = 0}$ should be 0.27 cm. Figure 20 shows that at the cathode, values of meniscus height significantly higher than 0.27 cm are reached at applied voltages greater than *ca.* 9 kV. This increase in meniscus height is most probably due to an increase in the surface tension of the decylbenzene under the effect of the electric field. The measurements show that the meniscus-rise method can be used to obtain information that it is impossible to obtain by direct measurement of the contact angle.

3
The A.C. Impedance of Complex Electrochemical Reactions

BY R. D. ARMSTRONG, M. F. BELL, AND A. A. METCALFE

1 Introduction

In this Report we shall restrict our attention to situations where a supporting electrolyte is present, and we shall concentrate our attention on work which has been carried out over the past decade and has been of particular interest to the group at Newcastle. Other important reviews in this area include those by Sluyters[1] and Smith.[2] Rangarajan[3] has written a very elegant review, deriving a large number of theoretical expressions, using matrix algebra. We shall not attempt to give such a sophisticated treatment here, but will rather concentrate on showing the relationships most often met in experimental situations.

2 Historical Summary

In 1940 Dolin and Ershler[4] investigated the adsorption of hydrogen on platinum, and derived equations which describe the frequency dependence of the electrode impedance in terms of a parallel network of a resistance (R_p) and a capacitance (C_p). Subsequently, in 1947, Randles[5] and Ershler[6] considered rapid metal/metal ion and redox reactions in which the time-averaged concentrations at the electrode surface obeyed the Nernst equation. With these assumptions, it was shown that the impedance is given by a series network comprising a resistance (R_s) and a capacitance (C_s), whose values are defined by equations (1) and (2), and it can be represented

$$R_s = \frac{RT}{n^2F^2c}\left[\left(\frac{2}{\omega D}\right)^{\frac{1}{2}} + \frac{1}{k}\right] \quad (1)$$

$$C_s = \frac{n^2F^2c}{RT}\left(\frac{D}{2\omega}\right)^{\frac{1}{2}} \quad (2)$$

by the familiar Randles' equivalent circuit (Figure 1). However this method relies on an 'a priori' separation of the double-layer capacitance (C_{dl}) and the solution resistance (R_{so}), which must be measured separately. Delahay[7] subsequently sug-

---

[1] M. Sluyters-Rehbach and J. H. Sluyters, in 'Electroanalytical Chemistry', ed. A. J. Bard' Dekker, New York, vol. 4, 1970, p. 1.
[2] D. E. Smith, in 'Electroanalytical Chemistry', ed. A. J. Bard, Dekker, New York, vol. 1, 1966, p. 1.
[3] S. K. Rangarajan, *J. Electroanalyt. Chem.*, 1974, **55**, 297, 329, 337, 363.
[4] P. Dolin and D. Ershler, *Acta Physicochim. U.R.S.S.*, 1940, **13**, 747.
[5] J. E. B. Randles, *Discuss. Faraday Soc.*, 1947, **1**, 11.
[6] B. Ershler, *Discuss. Faraday Soc.*, 1947, **1**, 269.
[7] P. Delahay, 'New Instrumental Methods in Electrochemistry', Interscience, New York, 1954.

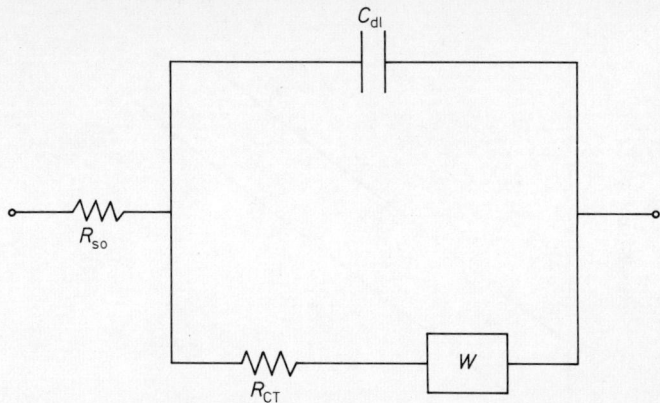

Figure 1 Randles' equivalent circuit

gested that this separation of the faradaic impedance and the double-layer capacitance is not generally valid, particularly in cases where specific adsorption of the electroactive species occurs. In attempts (notably by Gerischer,[8] Grahame,[9] and Delahay[7]) to generalize these equations to potentials other than the equilibrium potential, it was necessary to define the time-averaged concentration at the electrode surface and to solve the necessary equations representing diffusion. One such attempt was the development of the semi-infinite linear diffusion theory by Delahay.[7]

Following this work, expressions for the charge-transfer resistance (R_{CT}) and the Warburg impedance (W) can be derived from the rate equation for a simple one-step reaction and the solution of Fick's laws of diffusion, with the appropriate boundary conditions, to give equations (3a) and (3b), where the Warburg coefficient is defined as shown in equation (4) and $i = \sqrt{-1}$. The Faradaic impedance is given by equations (5) and (6), in which case $R_s = R_{CT} + \sigma\omega^{-\frac{1}{2}}$ and $1/\omega C_s = \sigma\omega^{-\frac{1}{2}}$. The Randles method involves plotting R_s and $1/\omega C_s$ against $\omega^{-\frac{1}{2}}$ (Figure 2) in order to extrapolate out the effects of diffusion.

$$R_{CT} = \frac{RT}{n^2 F^2 k_s} (c_O^s)^\alpha (c_R^s)^{1-\alpha} \tag{3a}$$

$$W = \sigma\omega^{-\frac{1}{2}} - i\sigma\omega^{-\frac{1}{2}} \tag{3b}$$

$$\sigma = \frac{RT}{n^2 F^2 \sqrt{2}} \left(\frac{1}{c_O^s D_O^{\frac{1}{2}}} + \frac{1}{c_R^s D_R^{\frac{1}{2}}} \right) \tag{4}$$

$$Z_f = R_{CT} + W \tag{5}$$

$$= R_s + (1/i\omega C_s) \tag{6}$$

[8] H. Gerischer, *Z. phys. Chem.*, 1951, **198**, 286; *Z. phys. Chem. (Frankfurt)*, 1954, **1**, 278.
[9] D. C. Grahame, *J. Electrochem. Soc.*, 1952, **99**, 370C.

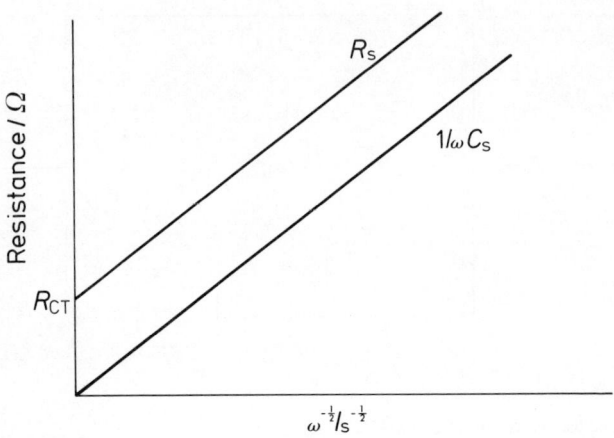

Figure 2 *The Randles method: plots of $1/\omega C_s$ and R_s against $\omega^{-\frac{1}{2}}$*

A second approach, devised by Sluyters[10] and Sluyters-Rehbach,[11] allows the calculation of R_{CT} and σ without prior knowledge of C_{dl}. From Figure 1, it follows that the cell impedance is given by equation (7). From this complicated expression,

$$Z = R_{so} + [1/i\omega C_{dl} + (1/R_{CT} + \sigma\omega^{-\frac{1}{2}} - i\sigma\omega^{-\frac{1}{2}})] \quad (7)$$

two limiting cases can be considered. First, when charge transfer is important (either at high frequencies or when diffusion is unimportant), and when the electroactive species are always at their Nernstian concentration at the electrode surface, equation (7) simplifies to equation (8). If this impedance is plotted in the complex

$$Z = R_{so} + \frac{R_{CT}}{1 + \omega^2 C_{dl}^2 R_{CT}^2} - \frac{i\omega C_{dl} R_{CT}^2}{1 + \omega^2 C_{dl}^2 R_{CT}^2} \quad (8)$$

plane as a function of frequency, a single semicircle (Figure 3) is obtained, having a diameter of numerical value R_{CT} ($= 1/\omega^* C_{dl}$, where ω^* is the frequency/Hz at the maximum of the semicircle).

When charge transfer is small compared with diffusion, the impedance is given by equation (9). In this case, the impedance spectrum in the complex plane is a

$$Z = R_{so} + R_{CT} + \sigma\omega^{-\frac{1}{2}} - i(\sigma\omega^{-\frac{1}{2}} + 2\sigma^2 C_{dl}) \quad (9)$$

straight line of slope 45° (Figure 4). Implicit in this derivation is the assumption that the a.c. diffusion layer is much smaller than the Nernstian diffusion layer. A more exact consideration of the effects of diffusion takes into account the case where these layers are of comparable thickness; the resulting expression for the

---

[10] J. H. Sluyters, *Rec. Trav. chim.*, 1960, **79**, 1092.
[11] M. Sluyters-Rehbach, thesis, Utrecht, 1963.

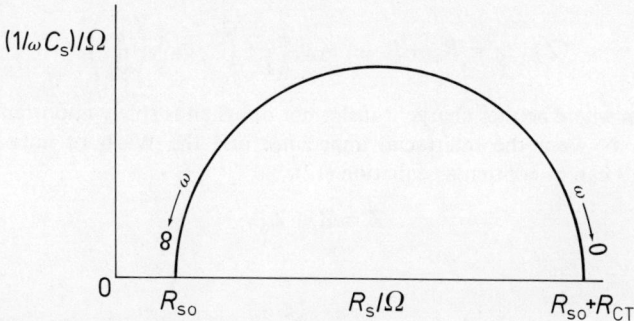

Figure 3 *Complex-plane spectrum for the cell impedance described by Figure 1 when diffusion is unimportant*

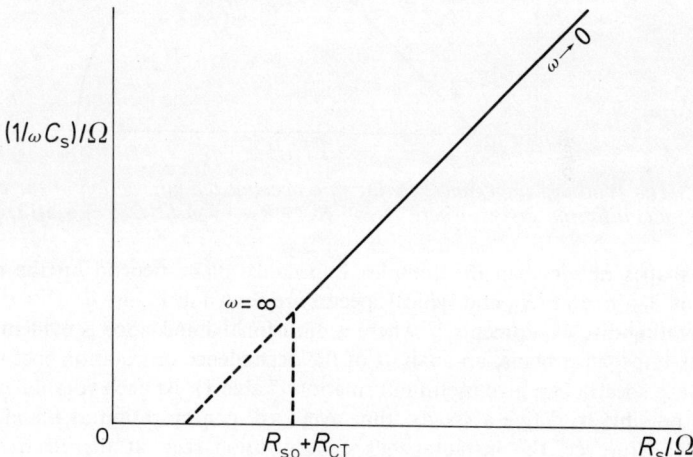

Figure 4 *The Warburg impedance*

diffusional impedance (Z_d) being of the form shown in equation (10). This was originally derived by Llopis,[12] and subsequently re-derived by Sluyters and Yzer-

$$Z_d = \frac{\sigma_O}{\omega^{\frac{1}{2}}}(1-i)\tanh\left[\delta_O\left(\frac{i\omega}{D_O}\right)^{\frac{1}{2}}\right] + \frac{\sigma_R}{\omega^{\frac{1}{2}}}(1-i)\tanh\left[\delta_R\left(\frac{i\omega}{D_R}\right)^{\frac{1}{2}}\right] \quad (10)$$

[12] J. Llopis and F. Colon, in 'Proceedings of the Eighth Meeting of the C.I.T.C.E., 1956', Butterworths, London, 1958, p. 414.

mans,[13,14] Drossbach and Schultz,[15] and Schuhmann.[16] At high frequencies this reduces to equation (9), and it relaxes at low frequencies to the real axis (Figure 5), where the value of Z as $\omega \to 0$ is defined as shown in equation (11).

$$(Z)_{\omega \to 0} = R_{so} + R_{CT} + \sigma_O \delta_O \left(\frac{2}{D_O}\right)^{\frac{1}{2}} + \sigma_R \delta_R \left(\frac{2}{D_R}\right)^{\frac{1}{2}} \quad (11)$$

For cases where neither charge transfer nor diffusion is solely important, there is interaction between the interfacial impedance and the Warburg impedance. In general, this can be written as equation (12).

$$Z = Z_f + Z_d \quad (12)$$

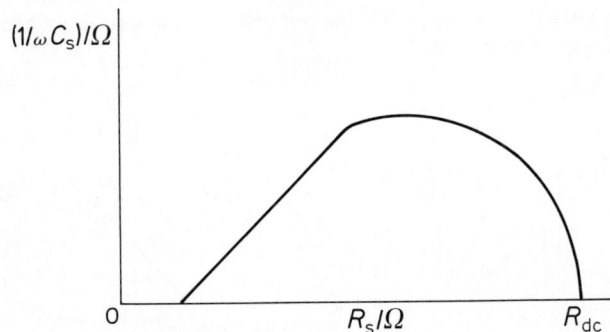

Figure 5 *The Warburg impedance, taking into account the interaction of the a.c. and d.c. diffusion layers, where* $R_{dc} = R_{so} + R_{CT} + \sigma_O \delta_O (2/D_O)^{\frac{1}{2}} + \sigma_R \delta_R (2/D_R)^{\frac{1}{2}}$

The shapes obtained in the complex impedance plane depend on the relative values of R_{CT}, σ, and C_{dl}, and typical spectra are shown in Figure 6.

In rotating-disc experiments,[17] where a diffusional impedance is evident in the complex impedance plane, an analysis of the dependence on rotation speed of the impedance spectra can give useful information (Table 1). At each rotation speed it will be possible to define a steady 'time-averaged' concentration at the electrode interface. However, the instantaneous concentration seen at the electrode will depend upon the frequency of the sinusoidal perturbation. If only the Warburg impedance is dependent on rotation speed, the time-averaged concentration (and not the instantaneous) is always equal to the Nernstian. If, however, neither the time-averaged nor the instantaneous concentrations are equal to the equilibrium concentration, both the interfacial and diffusional impedances will be dependent on the speed of rotation. This implies some irreversibility in the charge-transfer

[13] J. H. Sluyters, thesis, Utrecht, 1956.
[14] A. B. Yzermans, thesis, Utrecht, 1965.
[15] P. Drossbach and J. Schultz, *Electrochim. Acta*, 1964, **11**, 1391.
[16] D. Schuhmann, *Compt. rend.*, 1966, **262**, 1125.
[17] A. A. Pilla, thesis, Paris, 1966.

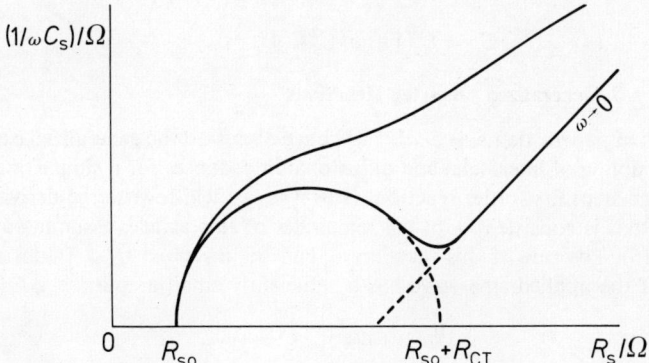

Figure 6 *Diagrams in the complex impedance plane for Figure 1 as diffusion becomes more important*

reaction. In both cases, the Warburg impedance will decrease with increased rotation speed (diffusion processes become less important).

Table 1 *Dependence of impedance spectra on rate of rotation*

Relationships of the surface concentration (c) to the Nernstian concentration (c_N)		Variation of the impedance with increased rate of rotation	
Instantaneous surface concentration	Time-averaged surface concentration	Charge-transfer	Warburg
$c = c_N$	$\bar{c} = c_N$	—	Decreases
$c = c_N$	$\bar{c} \neq c_N$	None	Decreases
$c \neq c_N$	$\bar{c} \neq c_N$	Increases	Decreases

Having now briefly discussed the historical development of the theory of the a.c. impedance technique, the remainder of this Report will be devoted to discussions of various reaction schemes for which generalized expressions have appeared in the literature. Throughout the following discussion, the presence of a geometric capacitance between the working and subsidiary electrodes has been ignored.[18,19] In the simulations given, to avoid any interaction of the time constants of the high- and low-frequency shapes in the complex plane, the double-layer capacitance (C_{dl}) is set equal to zero. These also ignore the solution resistance (R_{so}). In the analysis of some reaction schemes it is more convenient to use a parallel network of resistance (R_p) and capacitance (C_p) rather than the more usually measured

---

[18] R. D. Armstrong and W. P. Race, *J. Electroanalyt. Chem.*, 1971, **33**, 2084; 1972, **34**, 244.
[19] R. D. Armstrong and R. Mason, *J. Electroanalyt. Chem.*, 1973, **41**, 231.

quantities from a series network, R_s and C_s. These can be interconverted by use of equations (13) and (14).

$$R_p = R_s[(1+\omega^2 C_s^2 R_s^2)/\omega^2 C_s^2 R_s^2] \quad (13)$$

$$C_p = C_s(1+\omega^2 C_s^2 R_s^2)^{-1} \quad (14)$$

3 Generalized One-step Reactions

In a series of papers, de Levie *et al.*[20—22] have discussed the generalized expressions for the coupling of interfacial and diffusional impedances for a simple one-step (or pseudo-one-step) first-order reaction. However, we will rewrite the derivation in a notation that is consistent with the remainder of this article, assuming a reaction scheme (15). The rate of this reaction, v, can be linearized by a Taylor series expansion if the applied sine wave has a sufficiently small amplitude, ΔE [equation

$$v_A A_{\text{solution}} \rightleftharpoons v_C C_{\text{solution}} \quad (15)$$

(16)]. The reaction rate can also be related to the mass fluxes of A and C to the electrode surface. Thus equation (17) may be written. Elimination of $\Delta c_A^s \exp(i\omega t)$ and $\Delta c_C^s \exp(i\omega t)$ from equation (16) and subsequent rearranging gives the expres-

$$v = \left(\frac{\partial v}{\partial E}\right)_{c_A, c_C} \Delta E \exp(i\omega t) + \left(\frac{\partial v}{\partial c_A}\right)_{E, c_C} \Delta c_A \exp(i\omega t) + \left(\frac{\partial v}{\partial c_C}\right)_{E, c_A} \Delta c_C \exp(i\omega t) \quad (16)$$

$$\frac{(i\omega D_A)^{\frac{1}{2}}}{v_A} \Delta c_A^s \exp(i\omega t) = -\frac{(i\omega D_C)^{\frac{1}{2}}}{v_C} \Delta c_C \exp(i\omega t) \quad (17)$$

$$Y_f = nF\left(\frac{\partial v}{\partial E}\right)_{c_A, c_C}\left[1 - \frac{v_A}{(i\omega D_A)^{\frac{1}{2}}}\left(\frac{\partial v}{\partial c_A}\right)_{E, c_C} + \frac{v_C}{(i\omega D_C)^{\frac{1}{2}}}\left(\frac{\partial v}{\partial c_C}\right)_{E, c_A}\right]^{-1} \quad (18)$$

$$Z_f = \left[1/nF\left(\frac{\partial v}{\partial E}\right)_{c_A, c_C}\right]\left[1 - \frac{v_A(1-i)}{(2\omega D_A)^{\frac{1}{2}}}\left(\frac{\partial v}{\partial c_A}\right)_{E, c_C} + \frac{v_C(1-i)}{(2\omega D_C)^{\frac{1}{2}}}\left(\frac{\partial v}{\partial c_C}\right)_{E, c_A}\right] \quad (19)$$

$$R_{\text{CT}} = 1/nF\left(\frac{\partial v}{\partial E}\right)_{c_A, c_C} \quad (20)$$

$$\sigma' = \frac{v_C}{(2D_C)^{\frac{1}{2}}}\left(\frac{\partial v}{\partial c_C}\right)_{E, c_A} - \frac{v_A}{(2D_A)^{\frac{1}{2}}}\left(\frac{\partial v}{\partial c_A}\right)_{E, c_C} \quad (21)$$

$$Z_f = R_{\text{CT}} + Z_d \quad (22)$$

$$Z_d = (\sigma'/\omega^{\frac{1}{2}})R_{\text{CT}}(1-i) \quad (23)$$

sion (18) for the admittance, and thus the impedance is given by equation (19). Defining R_{CT} and σ' by equations (20) and (21), equation (19) can be rewritten as equation (22), where Z_d is defined by equation (23). This is described by the same

[20] R. de Levie and A. A. Husousky, *J. Electroanalyt. Chem.*, 1969, **22**, 29.
[21] R. de Levie and L. Pospisil, *J. Electroanalyt. Chem.*, 1969, **22**, 277.
[22] H. Moreira and R. de Levie, *J. Electroanalyt. Chem.*, 1971, **29**, 353.

equivalent circuit as shown in Figure 1 after vectorial subtraction of the double-layer capacitance. Thus defined, the Warburg impedance is different from that at the equilibrium potential in that the charge-transfer resistance is incorporated in the expression. As pointed out by de Levie, the Warburg impedance has the same sign as R_{CT}, and therefore can be negative (Figure 7). This has been experimentally verified for indium in aqueous NaSCN solutions.[20]

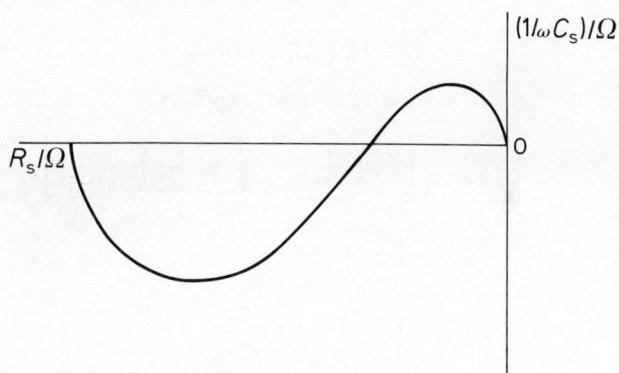

Figure 7 *A schematic representation of a complex-plane impedance plot when the charge-transfer resistance is negative*

4 Adsorption

Adsorption of Neutral Molecules.—The phenomena associated with the adsorption of neutral molecules at an electrode interface have traditionally been studied by analysing electrocapillary curves and differential-capacitance/potential curves. It has also been shown, by various authors, that the adsorption kinetics can be analysed by studying the frequency dependence of the impedance of the electrode.

Frumkin and Melik-Gaikazyan[23] considered only the two limiting cases of these so-called desad processes, in which either diffusion or adsorption is rate-determining. The situation with mixed diffusion and heterogeneous control can be derived, using the same general approach as that outlined in the previous section. If we consider the electrochemical reaction to be (24), then the response to a sinusoidal perturbation of potential is given by equation (25). The reaction rate can also be equated to the flux at the electrode surface by equation (26). The surface concentration on the electrode surface is given by equation (27), and the rate by equation (28).

Substitution of Δc_A^s and $\Delta \Gamma$ from equations (26) and (28) into equation (25) yields, for the faradaic admittance Y_f, equation (29). Rearrangement of this equation by defining the relaxation times for heterogeneously controlled (τ_H) and diffusion-controlled (τ_D) adsorption processes [equations (30) and (31)] and ΔC [equation (32)] gives equation (33) or equation (34). This is equivalent to the expression derived by Lorenz and Möckel,[24] and may be written as equations (35) and (36)

---

[23] A. N. Frumkin and V. I. Melik-Gaikazyan, *Doklady Akad. Nauk. S.S.S.R.*, 1951, **77**, 855.
[24] W. Lorenz, *Z. Electrochem.*, 1958, **62**, 192.

$$v_A A_{\text{solution}} \rightleftharpoons B_{\text{adsorbed}} \tag{24}$$

$$v = \left(\frac{\partial v}{\partial E}\right)_{\Gamma, c_A} \Delta E \exp(i\omega t) + \left(\frac{\partial v}{\partial \Gamma}\right)_{E, c_A} \Delta\Gamma \exp(i\omega t) + \left(\frac{\partial v}{\partial c_A}\right)_{\Gamma, E} \Delta c_A^s \exp(i\omega t) \tag{25}$$

$$v = -\frac{(i\omega D_A)^{\frac{1}{2}}}{v_A}\Delta c_A^s \exp(i\omega t) \tag{26}$$

$$\Gamma = \Gamma_0 + \Delta\Gamma \exp(i\omega t) \tag{27}$$

$$v = \partial\Gamma/\partial t = i\omega\Delta\Gamma \exp(i\omega t) \tag{28}$$

$$Y_f = nF\left(\frac{\partial v}{\partial E}\right)_{\Gamma, c_A^s}\left[1 - \left(\frac{\partial v}{\partial \Gamma}\right)_{E, c_A^s}\frac{1}{i\omega} + \left(\frac{\partial v}{\partial c_A^s}\right)\frac{v_A}{(i\omega D_A)^{\frac{1}{2}}}\right]^{-1} \tag{29}$$

$$\tau_H = -\left(\frac{\partial \Gamma}{\partial v}\right)_{E, c_A} \tag{30}$$

$$\tau_D^{\frac{1}{2}} = -\left(\frac{\partial \Gamma}{\partial c_A}\right)_{v, E}\frac{v_A}{D_A^{\frac{1}{2}}} \tag{31}$$

$$\Delta C = -nF\left(\frac{\partial \Gamma}{\partial E}\right)_{c_A, \omega \to 0} \tag{32}$$

$$Y_f = i\omega\Delta C[1 + i\omega\tau_H + (i\omega\tau_D)^{\frac{1}{2}}]^{-1} \tag{33}$$

$$Z_f = [1 + i\omega\tau_H + (i\omega\tau_D)^{\frac{1}{2}}]/i\omega\Delta C \tag{34}$$

after the addition of the double-layer capacitance, where now ΔC is the difference between the interfacial capacities as $\omega \to 0$ and as $\omega \to \infty$, i.e. $\Delta C = C_0 - C_{dl}$.

$$C_p = \frac{\Delta C[(\omega\tau_D/2)^{\frac{1}{2}} + 1]}{[(\omega\tau_D/2)^{\frac{1}{2}} + \omega\tau_H]^2 + [(\omega\tau_D/2)^{\frac{1}{2}} + 1]^2} + C_{dl} \tag{35}$$

$$\frac{1}{\omega R_p} = \frac{\Delta C[(\omega\tau_D/2)^{\frac{1}{2}} + \omega\tau_H]}{[(\omega\tau_D/2)^{\frac{1}{2}} + \omega\tau_H]^2 + [(\omega\tau_D/2)^{\frac{1}{2}} + 1]^2} \tag{36}$$

A plot of $1/\omega R_p$ vs. C_p is generally called the Cole–Cole[25] plot, as the equations are of the same form as those derived for a dielectric relaxation process with a particular spread of relaxation times. If the reaction is entirely controlled by the rate of the heterogeneous process ($\tau_D = 0$), such a plot takes the form of a semicircle,[26] with centre $C_p = C_{dl} + \Delta C/2, 1/\omega R_p = 0$, whereas, if the kinetics are diffusion controlled ($\tau_H = 0$), the plot is a quarter-circle,[26] intersecting the C_p axis at values of C_{dl} and $(C_{dl} + \Delta C)$. If τ_H and τ_D are of the same order, this plot gives a

---

[25] K. S. Cole and R. H. Cole, *J. Chem. Phys.*, 1941, **9**, 341.
[26] R. D. Armstrong, W. P. Race, and H. R. Thirsk, *J. Electroanalyt. Chem.*, 1968, **16**, 517.

gradual transition from a semicircle (at high frequencies, when there is effectively heterogeneous control) to a quarter-circle (at low frequencies, when there is effectively complete diffusion control).[26] A simulation of such behaviour is shown in Figure 8, together with the equivalent circuits of the three cases. The validity of this approach has been verified experimentally by Armstrong and Race.[26] For comparison, the behaviour for this case in the complex impedance plane is shown in Figure 16.

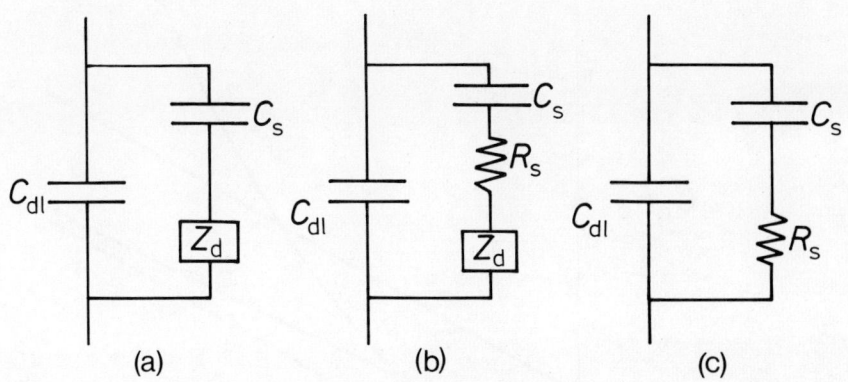

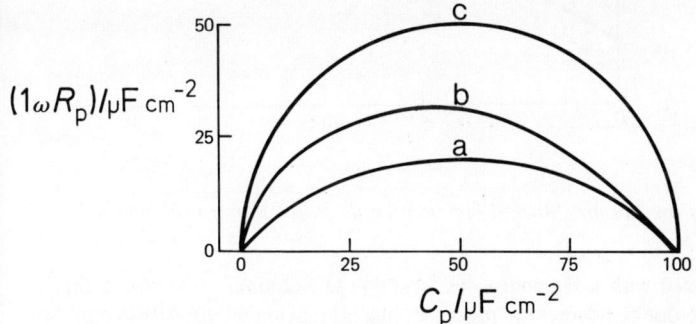

Figure 8 *A simulation of equations (35) and (36) for $\Delta C = 100 \,\mu F \, cm^{-2}$ and (a) $\tau_D = 10^{-7}$ s, $\tau_H = 0$; (b) $\tau_D = 10^{-7}$ s, $\tau_H = 10^{-7}$ s; (c) $\tau_D = 0$, $\tau_H = 10^{-7}$ s, together with their equivalent circuits*

Lorenz and Möckel[27] have analysed the behaviour of the quantity $\omega C_p^* R_p$ [i.e. $\omega (C_p - C_{dl}) R_p$] as a function of $\omega^{-1/2}$. From equations (35) and (36) it follows that one may derive equation (37). Figure 9 shows a comparison of those plots with the

$$\omega C_p^* R_p = \{1 + (2/\omega\tau_D)^{\frac{1}{2}}\}/\{1 + (2/\omega\tau_D)^{\frac{1}{2}}\tau_H\} \tag{37}$$

[27] W. Lorenz and F. Möckel, *Z. Elektrochem.*, 1956, **60**, 507.

cases that were discussed above. As can be seen from this Figure, this function can be used to differentiate between pure reaction control, pure diffusion control, or intermediate cases.

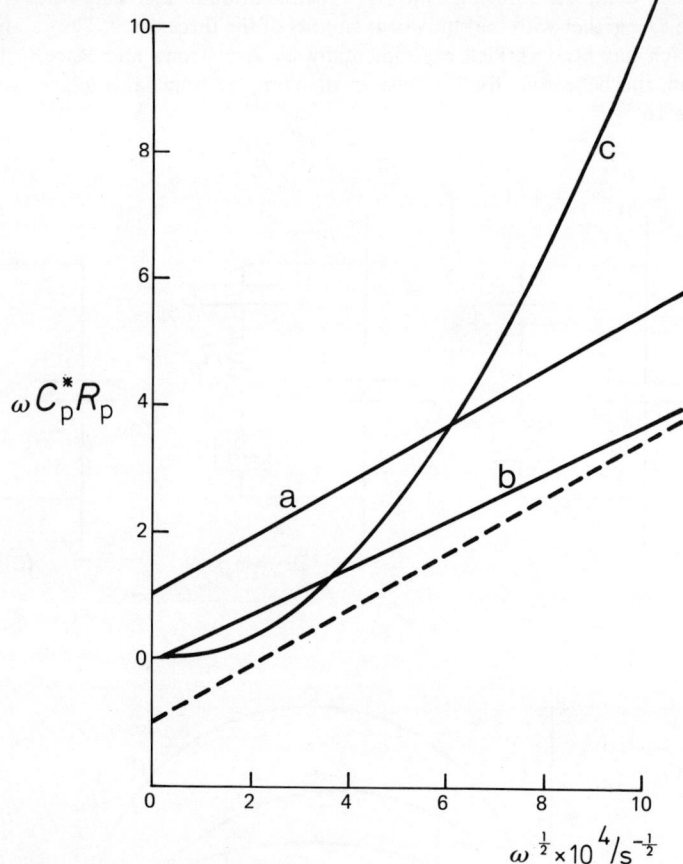

Figure 9 *Plots of the Lorenz–Möckel function for the cases shown in Figure 8*

Adsorption Coupled with a Homogeneous Reaction in Solution.—The above theory of the impedance due to adsorption processes has been extended by Armstrong[28] to the case where adsorption is coupled with a homogeneous reaction in solution. A species A is considered to be always in equilibrium with B adsorbed molecules at the electrode surface, and also reacts in solution to form S [equation (38)]. An example of such a process would be the adsorption of benzoic acid from a solution containing a large excess of benzoate ion. The reaction in solution would then be the addition of a proton to the benzoate ion to form benzoic acid prior to adsorption.

$$S_{solution} \rightleftharpoons A_{solution} \rightleftharpoons B_{adsorbed} \tag{38}$$

[28] R. D. Armstrong, *J. Electroanalyt. Chem.*, 1969, **22**, 49.

The expression (39) has been derived for the electrode admittance, with the assumption that the diffusion coefficients of A and S are equal. This represents a

$$Y_f = \frac{\Delta C i\omega D^{\frac{1}{2}}(k_o+i\omega)^{\frac{1}{2}}\left(1+\dfrac{c_A}{c_S}\right)}{D^{\frac{1}{2}}(k_o+i\omega)^{\frac{1}{2}}\left(1+\dfrac{c_A}{c_S}\right)+\left(\dfrac{\partial \Gamma}{\partial c_A}\right)_E i\omega\left[1+\dfrac{c_A}{c_S}\left(\dfrac{k_o+i\omega}{i\omega}\right)^{\frac{1}{2}}\right]} \quad (39)$$

relaxation process where there are three relaxation times, these being defined by equations (40)—(42). It should be noted that τ_{D1} and τ_{D2} are mean times with a Cole–Cole distribution of values, whilst τ_R, the reaction relaxation time, has a single value.

$$\tau_{D1} = \left(\frac{\partial \Gamma}{\partial c_A}\right)^2 / D \quad (40)$$

$$\tau_{D2} = \left(\frac{\partial \Gamma}{\partial c_A}\right)^2 \left(\frac{c_A}{c_A+c_S}\right) / D \quad (41)$$

$$\tau_R = \left(\frac{\partial \Gamma}{\partial c_A}\right) / k_o^{\frac{1}{2}} D^{\frac{1}{2}} \quad (42)$$

Computer simulations of equation (39) were carried out by Armstrong and Bell[29] with the data shown in Table 2, which give the corresponding admittance spectra shown in Figure 10. It is interesting to note that when τ_R is important it takes the role of τ_H, the relaxation time for a heterogeneously controlled process, and a semicircle is generated in the complex plane. When diffusion is important (*i.e.* τ_{D1} or τ_{D2}), a quarter circle is produced.

Table 2 *Summary of parameters used and the conditions fulfilled for the simulations shown in Figure 10*

τ_{D1}/s	c_A/c_s	τ_{D2}/s	k_o/s^{-1}	τ_R/s	Conditions fulfilled	
10^{-3}	10^{-3}	10^{-9}	10^{-1}	10^{-1}	$\omega \ll k_o$	(τ_{D1})
10^{-3}	10^{-3}	10^{-9}	10^{3}	10^{-3}	$\omega \approx k_o$ and $c_A \ll c_S$	($\tau_{D1}+\tau_R$)
10^{-3}	10^{-3}	10^{-9}	10^{7}	10^{-5}	$\omega \ll k_o$ and $\omega^{\frac{1}{2}}c_S \gg k_o^{\frac{1}{2}}c_A$	(τ_R)
10^{-3}	10^{-3}	10^{-9}	10^{15}	10^{-9}	$\omega \ll k_o$ and $\omega^{\frac{1}{2}}c_S \approx k_o^{\frac{1}{2}}c_A$	($\tau_R+\tau_{D1}$)
10^{-3}	10^{-3}	10^{-9}	10^{19}	10^{-11}	$\omega \ll k_o$ and $\omega^{\frac{1}{2}}c_S \ll k_o^{\frac{1}{2}}c_A$	(τ_{D2})

5 Two-step Reactions

Reactions in which there is an Adsorbed Intermediate.—This situation can be represented by the reaction (43). The equivalent circuit for the reaction can be obtained

$$v_A A_{\text{solution}} \rightleftharpoons B_{\text{adsorbed}} \rightleftharpoons v_C C_{\text{solution}} \quad (43)$$

---

[29] R. D. Armstrong and M. F. Bell, *J. Electroanalyt. Chem.*, 1975, **58**, 419.

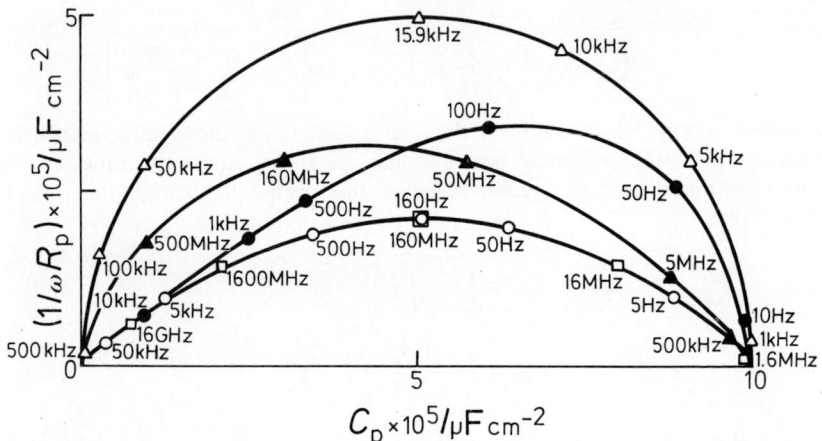

Figure 10 Computer simulations of equation (39) with the data shown in Table 2: (○) $k_0 = 10^{-1}$ s^{-1}; (●) $k_0 = 10^3$ s^{-1}; (△) $k_0 = 10^7$ s^{-1}; (▲) $k_0 = 10^{15}$ s^{-1}; (□) $k_0 = 10^{19}$ s^{-1}

from the work of Gerischer,[30] and is shown in Figure 11. Equations for the impedance which are appropriate to the specific case of the hydrogen-evolution

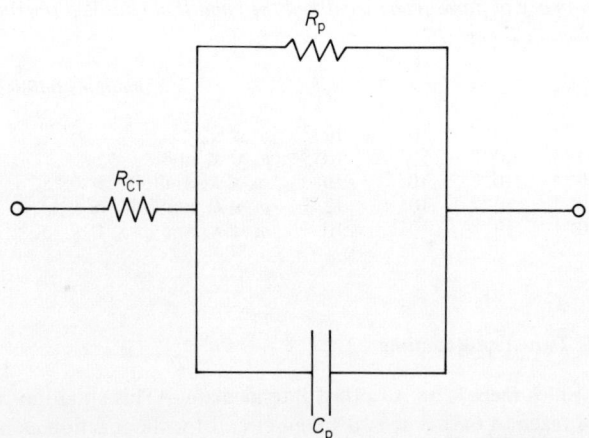

Figure 11 Equivalent circuit proposed by Gerischer[30] for a reaction involving an adsorbed intermediate

[30] H. Gerischer, Z. phys. Chem. (Leipzig), 1952, **201**, 55.

reaction were given by Grahame[31] and by Gerischer and Mehl.[32] Subsequently, an analysis of the situation was given by Epelboin and Keddam,[33] assuming that both steps of the reaction are first-order and that the adsorption of B obeys the Langmuir isotherm. The problem was tackled in a more general fashion, without the above assumption, by Armstrong and Henderson.[34] They derived equation (44) for the electrode admittance, where α_1 and α_2 are defined as shown in equations (45) and (46).

$$Y_f = \frac{n_1 F}{\alpha_1}\left(\frac{\partial v_1}{\partial E}\right)_{\Gamma, c_A} + \frac{n_2 F}{\alpha_2}\left(\frac{\partial v_2}{\partial E}\right)_{\Gamma, c_C} + \frac{\left[\frac{n_1 F}{\alpha_1}\left(\frac{\partial v_1}{\partial \Gamma}\right)_{E, c_A} + \frac{n_2 F}{\alpha_2}\left(\frac{\partial v_2}{\partial \Gamma}\right)_{E, c_C}\right]\left[\frac{1}{\alpha_1}\left(\frac{\partial v_1}{\partial E}\right)_{\Gamma, c_A} - \frac{1}{\alpha_2}\left(\frac{\partial v_2}{\partial E}\right)_{\Gamma, c_C}\right]}{i\omega - \left(\frac{\partial v_1}{\partial \Gamma}\right)_{E, c_A}\frac{1}{\alpha_1} + \left(\frac{\partial v_2}{\partial \Gamma}\right)_{E, c_C}\frac{1}{\alpha_2}} \quad (44)$$

$$\alpha_1 = 1 + \frac{v_A}{(i\omega D_A)^{\frac{1}{2}}}\left(\frac{\partial v_1}{\partial c_A}\right)_{E, \Gamma} \quad (45)$$

$$\alpha_2 = 1 - \frac{v_C}{(i\omega D_C)^{\frac{1}{2}}}\left(\frac{\partial v_2}{\partial c_C}\right)_{E, \Gamma} \quad (46)$$

When there is no influence of an adsorbed intermediate, this expression reduces to equation (18), which was derived for a one-step reaction.[35] Similarly, it can be shown that, when the rate of the second reaction is very fast, the equation becomes identical[35] with equation (29) for the adsorption of neutral molecules.

In the case when diffusion is unimportant, this complicated expression simplifies to equation (47), where R_{CT}, R_0, and τ are defined as an infinite-frequency charge-transfer resistance [equation (48)], an additional resistance at zero frequency [equation (49)], and a time constant [equation (50)].

$$Y_f = \frac{1}{R_{CT}} + \frac{1}{R_0(1+\omega^2\tau^2)} - \frac{i\omega\tau}{R_0(1+\omega^2\tau^2)} \quad (47)$$

$$\frac{1}{R_{CT}} = \left[n_1 F\left(\frac{\partial v_1}{\partial E}\right)_\Gamma + n_2 F\left(\frac{\partial v_2}{\partial E}\right)_\Gamma\right] \quad (48)$$

$$\frac{1}{R_0} = \tau\left[n_1 F\left(\frac{\partial v_1}{\partial \Gamma}\right)_E + n_2 F\left(\frac{\partial v_2}{\partial \Gamma}\right)_E\right]\left[\left(\frac{\partial v_1}{\partial E}\right)_\Gamma - \left(\frac{\partial v_2}{\partial E}\right)_\Gamma\right] \quad (49)$$

$$\tau = \left[\left(\frac{\partial v_2}{\partial \Gamma}\right)_E - \left(\frac{\partial v_1}{\partial \Gamma}\right)_E\right]^{-1} \quad (50)$$

[31] D. C. Grahame, *J. Electrochem. Soc.*, 1952, **99**, 370C.
[32] H. Gerischer and W. Mehl, *Z. Elektrochem.* 1955, **59**, 1049.
[33] I. Epelboin and M. Keddam, *J. Electrochem. Soc.*, 1970, **117**, 1052.
[34] R. D. Armstrong and M. Henderson, *J. Electroanalyt. Chem.*, 1972, **39**, 81.
[35] R. D. Armstrong, R. E. Firman, and H. R. Thirsk, *Faraday Discuss. Chem. Soc.*, 1973, **56**, 244.

112   *Electrochemistry*

A simulation of Y_f leads to three possibilities in the complex plane (Figure 12 a, b, c), depending on the relative values of R_{CT} and R_0 when the time constants for

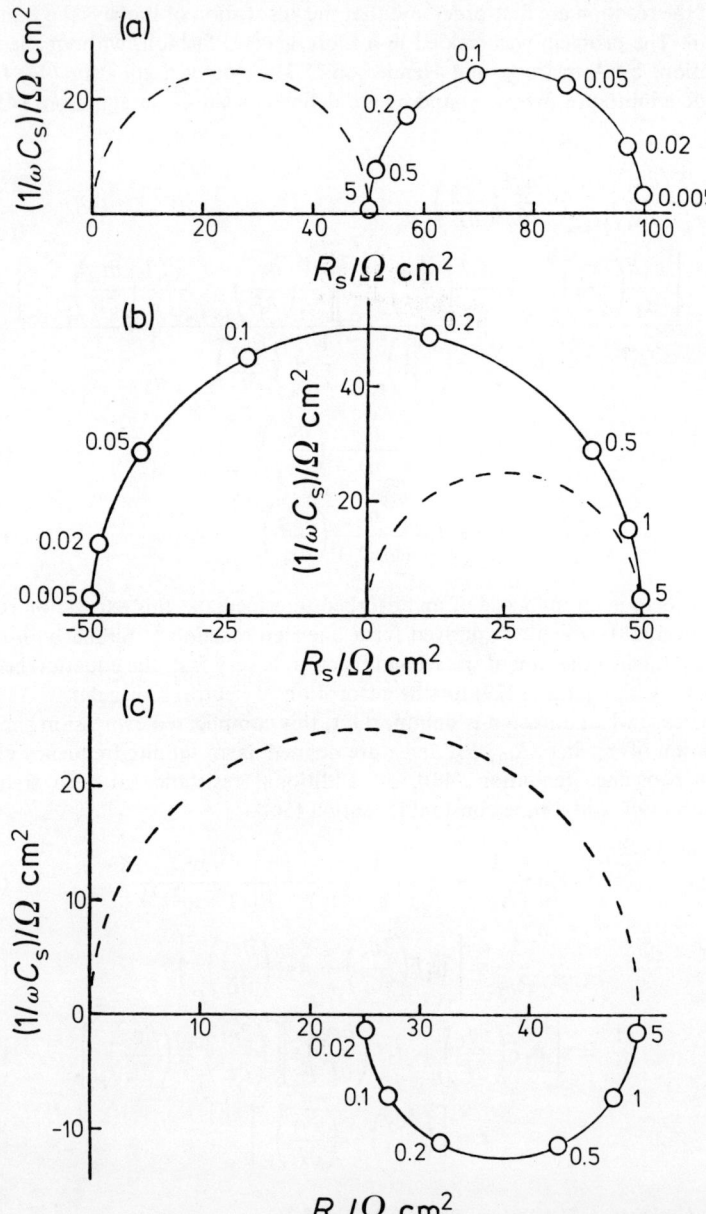

Figure 12 *Simulations in the complex impedance plane for an adsorbed intermediate when diffusion is unimportant, where $R_{CT} = 50\ \Omega\ cm^2$, $\tau = 1$ s, and* (a) $R_0 = -100\ \Omega\ cm^2$; (b) $R_0 = -25\ \Omega\ cm^2$; (c) $R_0 = 50\ \Omega\ cm^2$

the two reactions are sufficiently well separated. From these diagrams it is possible to obtain the resistances R_{CT} and R_{dc}, from which R_0 and τ can be calculated, using equations (51) and (52), where ω^* is the angular frequency at the maximum of the low-frequency semicircle. These quantities can be related to the frequency-independent elements of the equivalent circuit shown in Figure 11 through the equations (53) and (54).

$$\frac{1}{R_0} = \frac{1}{R_{dc}} - \frac{1}{R_{CT}} \tag{51}$$

$$\tau = \frac{1}{\omega^*}\left(\frac{R_0 + R_{CT}}{R_0}\right) \tag{52}$$

$$R_p = -R_{CT}^2/(R_0 + R_{CT}) \tag{53}$$

$$C_p = -R_0\tau/R_{CT}^2 \tag{54}$$

When the effects of diffusion are taken into account, three general situations may arise.[35] If we consider the case where the diffusion of both A and C is important, the typical impedance-plane diagrams simulated from equation (44) and shown in Figure 13 are generated. This shows two extremes; cases where diffusion is fairly unimportant (Figure 13a) and where it is very important (Figure 13c). Similar diagrams are obtained when the diffusion of only one of the species is important.[35] However, when the possibility of negative reaction orders is considered, much more complicated shapes are generated.[35]

In a series of papers, Delahay[36—39] has discussed the separation of the double-layer capacitance and the faradaic impedance when there is specific adsorption of electroactive species, and he has derived expressions for the electrode admittance for fast metal ion/metal amalgam reactions. His expression can be written in its simplest form as equation (55). As pointed out by Delahay, this is a more specific

$$Y = \frac{1}{2\sigma\omega^{-\frac{1}{2}}} + \frac{\omega(C_o - C_{dl})u}{2 + 2u + u^2} + i\left[\frac{1}{2\sigma\omega^{-\frac{1}{2}}} + \frac{\omega(C_o - C_{dl})(2+u)}{2 + 2u + u^2} + \omega C_{dl}\right] \tag{55}$$

case of the derivation advanced by Lorenz. Numerical evaluation of this [equation (55)], and comparison with equation (44) in the case of very rapid reactions, shows that it would be difficult to distinguish experimentally between the two, though it should be noted that, analytically, the equations are different.[40]

Reactions in which there is a Solution-soluble Intermediate.—This is a similar situation to that of the adsorbed intermediate, except that the species B is soluble and is not present on the electrode surface.

This situation was considered to some extent by Smith,[2] for an a.c. polarographic wave, and by Despic[41] for Cd^+ as an intermediate in the $Cd^{2+}/Cd(Hg)$ reaction.

[36] P. Delahay, *J. Phys. Chem.*, 1966, **70**, 2067, 2373.
[37] P. Delahay and G. G. Susbielles, *J. Phys. Chem.*, 1966, **70**, 3150.
[38] P. Delahay, K. Holub, G. G. Susbielles, and G. Tessari, *J. Phys. Chem.*, 1967, **71**, 779.
[39] K. Holub, G. Tessari, and P. Delahay, *J. Phys. Chem.*, 1967, **71**, 2612.
[40] S. K. Rangarajan, personal communication.
[41] A. R. Despic, D. R. Jovanovic, and S. P. Bingulae, *Electrochim. Acta*, 1970, **15**, 459.

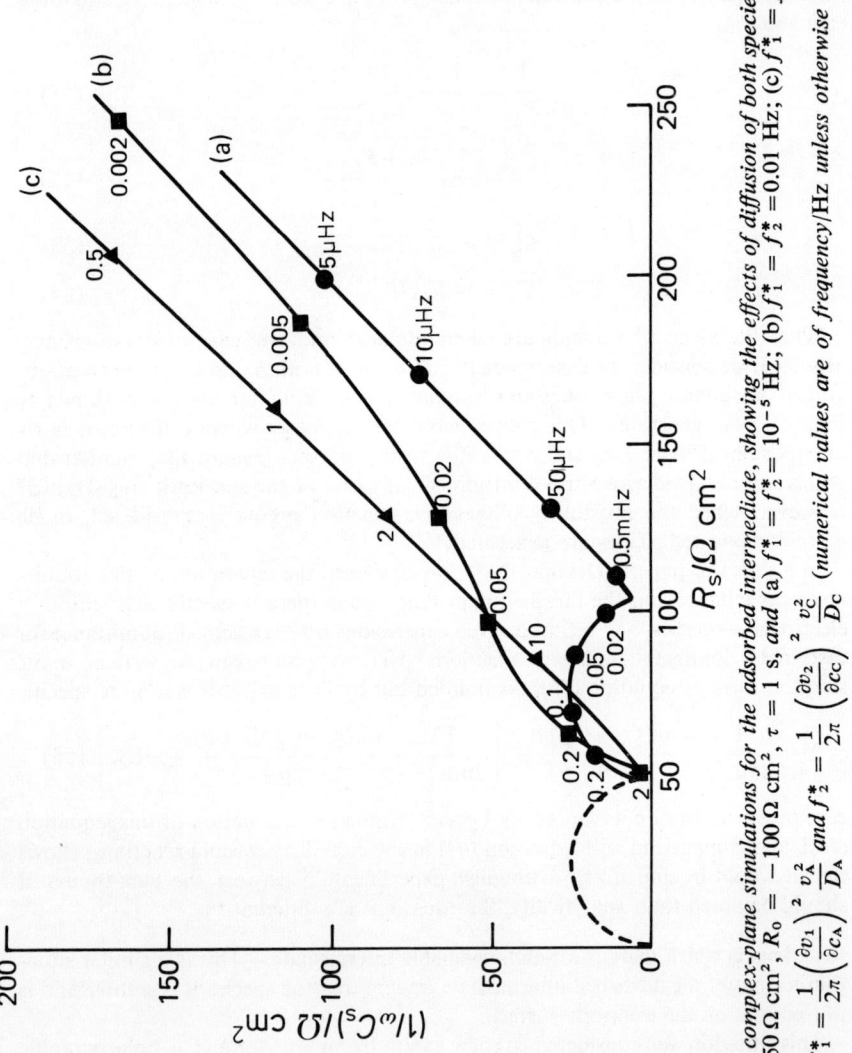

Figure 13 *Typical complex-plane simulations for the adsorbed intermediate, showing the effects of diffusion of both species in solution; $R_{CT} = 50\,\Omega\,cm^2$, $R_0 = -100\,\Omega\,cm^2$, $\tau = 1\,s$, and (a) $f_1^* = f_2^* = 10^{-5}\,Hz$; (b) $f_1^* = f_2^* = 0.01\,Hz$; (c) $f_1^* = f_2^* = 10\,Hz$, where $f_1^* = \frac{1}{2\pi}\left(\frac{\partial v_1}{\partial c_A}\right)^2 \frac{v_A^2}{D_A}$ and $f_2^* = \frac{1}{2\pi}\left(\frac{\partial v_2}{\partial c_C}\right)^2 \frac{v_C^2}{D_C}$ (numerical values are of frequency/Hz unless otherwise marked)*

Subsequently, a derivation similar to that for the adsorbed intermediate was carried out by Armstrong and Firman[42] in which they obtained the expression (56)

$$Y_f = \frac{n_1 F}{\alpha_1}\left(\frac{\partial v_1}{\partial E}\right)_{c_A, c_B} + \frac{n_2 F}{\alpha_2}\left(\frac{\partial v_2}{\partial E}\right)_{c_B, c_C} +$$

$$\frac{\left[\frac{n_1 F}{\alpha_1}\left(\frac{\partial v_1}{\partial c_B}\right)_{E, c_A} + \frac{n_2 F}{\alpha_2}\left(\frac{\partial v_2}{\partial c_B}\right)_{E, c_C}\right]\left[\frac{1}{\alpha_1}\left(\frac{\partial v_1}{\partial E}\right)_{c_A, c_B} - \frac{1}{\alpha_2}\left(\frac{\partial v_2}{\partial E}\right)_{c_B, c_C}\right]}{(i\omega D_B)^{\frac{1}{2}} - \frac{1}{\alpha_1}\left(\frac{\partial v_1}{\partial c_B}\right)_{E, c_A} + \frac{1}{\alpha_2}\left(\frac{\partial v_2}{\partial c_B}\right)_{E, c_C}} \quad (56)$$

for the faradaic admittance, where α_1 and α_2 are as defined earlier [equations (45) and (46)]. It was shown that the impedance becomes a pure Warburg impedance at sufficiently low frequencies, and that at high frequencies two general situations arise. Firstly, when the diffusion of A and C is unimportant, equation (56) reduces to equation (57), where R_{CT}, R_∞, and $\tau[= (\tau_1/D_B)^{1/2}]$ have the same form as defined

$$Y_f = \frac{1}{R_{CT}} + \frac{1}{R_0[1+(i\omega\tau_1)^{\frac{1}{2}}]} \quad (57)$$

earlier [equations (48), (49), and (50)], replacing the surface concentration Γ by c_B, the concentration of the intermediate in solution. The spectra predicted by this equation are shown in Figure 14a, b, c. An unusual situation occurs when R_{CT}/R_0 equals -1, in which case a straight line of unit slope is observed (Figure 14d).[42]

It should be noted that the relaxation time τ_1 is now defined in terms of R_{CT} and R_0 by equation (58).

$$\tau_1 = \frac{1}{\omega^*}\left(\frac{R_0 + R_{CT}}{R_0}\right)^2 \quad (58)$$

When the diffusion of A and C become significant, the low-frequency shape is masked by the appearance of a Warburg impedance (Figure 15).[35]

6 Passive Behaviour of Metals

The Influence of a Catalyst/Inhibitor on an Electrode Reaction.—A number of authors (including Schuhmann,[43] Epelboin et al.,[44, 45] and Armstrong[46, 47]) have considered the effect of adsorption of a catalyst/inhibitor on the electrode reaction, particularly for the case where the active–passive transition on certain metals can be attributed to this model. A typical example is in the dissolution of Ti (A) to Ti^{III} (B), with the inhibition of the reaction by the adsorption of oxygen species.

---

[42] R. D. Armstrong and R. E. Firman, *J. Electroanalyt. Chem.*, 1973, **45**, 3.
[43] D. Schuhmann, *J. Electroanalyt. Chem.*, 1968, **17**, 45.
[44] I. Epelboin, M. Keddam, and Ph. Morel, 'Proceedings of the 3rd International Congress on Metallic Corrosion, Moscow, 1966', vol. 1, Moscow, 1969, p. 110.
[45] I. Epelboin, C. Gabrielli, M. Keddam, and H. Takenouti, *Compt. rend.*, 1973, **276**, C, 145.
[46] R. D. Armstrong, *J. Electroanalyt. Chem.*, 1968, **17**, 45.
[47] R. D. Armstrong, M. F. Bell, and R. E. Firman, *J. Electroanalyt. Chem.*, 1973, **48**, 150.

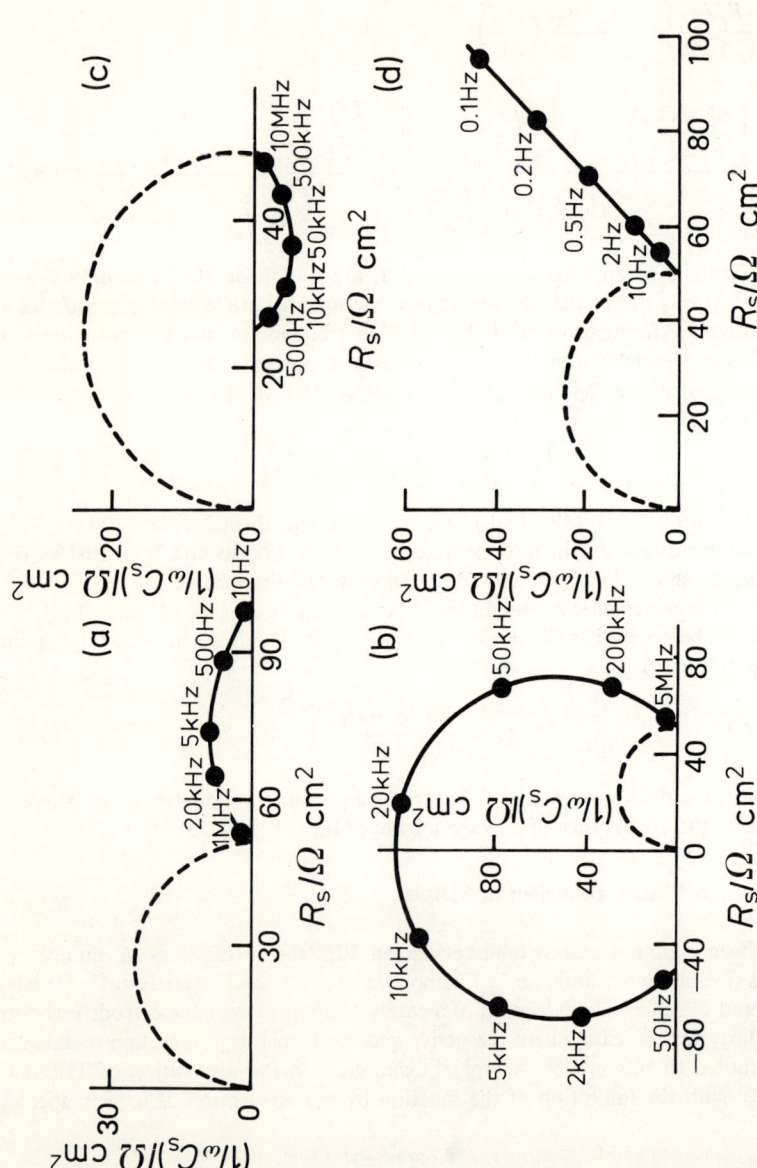

Figure 14 Simulations in the complex impedance plane of equation (56) for the solution-soluble intermediate with the data given in Figure 12 for (a), (b), and (c) and $R_0 = -50\,\Omega\,\text{cm}^2$ for (d).

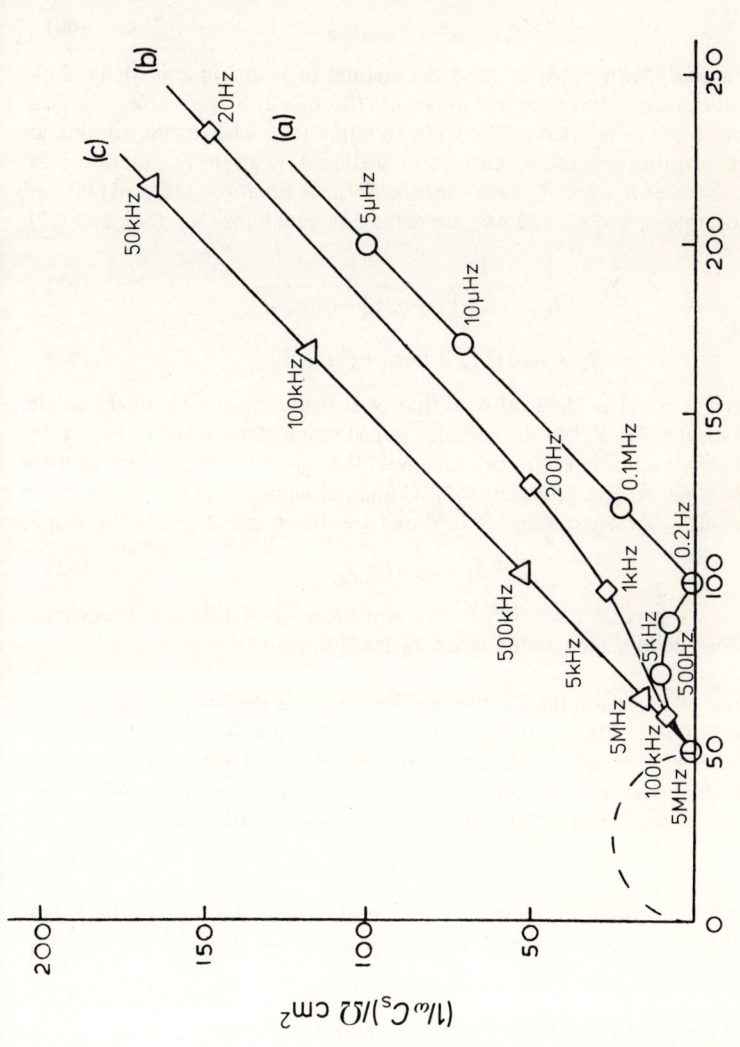

Figure 15 *Typical complex-plane simulations for the solution-soluble intermediate, showing the effects of diffusion in solution of both A and C, with the data of Figure 13 and* (a) $f_1^* = f_2^* = 10^{-5}$ Hz; (b) $f_1^* = f_2^* = 10^2$ Hz; (c) $f_1^* = f_2^* = 10^6$ Hz

The reaction can be written in the general form indicated by equations (59) and (60), and it is assumed that the surface concentration of the adsorbed species

$$v_A A \underset{}{\overset{ne}{\rightleftharpoons}} v_B B \qquad (59)$$

$$N_{\text{solution}} \rightleftharpoons N_{\text{adsorbed}} \qquad (60)$$

varies continuously with potential, and the method of Frumkin and Melik-Gaikazyan[23] for the adsorption of organic molecules (Section 2) is applicable. The faradaic admittance can be written as shown in equation (61), whereas the admittance due to the adsorption process, as derived in Section 4, is given by equation (62), where the resistances R_{CT} and R_0 have the same form as equations (48) and (49) and the time constants τ_H and τ_D, and ΔC, are defined in equations (30), (31), and (32).

$$Y_f = \frac{1}{R_{\text{CT}}} + \frac{1}{R_0}\frac{1}{[1+i\omega\tau_H+(i\omega\tau_D)^{\frac{1}{2}}]} \qquad (61)$$

$$Y_a = i\omega\Delta C/[1+i\omega\tau_H+(i\omega\tau_D)^{\frac{1}{2}}] \qquad (62)$$

The behaviour of Y_f is identical with that of the solution-soluble intermediate, whereas the display for Y_a^{-1} in the complex impedance plane is given in Figure 16. The total electrode admittance for this scheme is then given by the addition of these two equations, *i.e.* to give equation (63). Computer simulations of this equation were carried out by Armstrong and Bell,[47] and are shown in Figure 17. The shapes

$$Y = Y_f + Y_a + i\omega C_{\text{dl}} \qquad (63)$$

generated show a gradual transition from a semicircle when diffusion is unimportant to a more complicated shape when τ_H tends to zero.

The Behaviour of Metals in the Passive and Transpassive Regions.—When a metal that is immersed in solution is covered by a continuous film, the total electrode impedance can be considered as the sum of the impedance of the metal/film ($Z_{\text{M/F}}$) interface, the film (Z_F), and the film/solution interface ($Z_{\text{F/S}}$) [equation (64)]. When the impedance of the film/solution interface is important, it has been shown[48] that

$$Z_T = Z_{\text{M/F}} + Z_F + Z_{\text{F/S}} \qquad (64)$$

the situation is mathematically equivalent to that of the adsorbed intermediate. This is as a result of the exchange of both anions and cations at the film/solution interface.

7 Special Cases Due to Electrocrystallization Effects

In general, it is possible that the impedance for metal deposition/dissolution reactions will be influenced by electrocrystallization effects. In this section we shall consider two possible models for which simulations have appeared in the literature. These are (i) the adatom model and (ii) two-dimensional nucleation and growth.

[48] R. D. Armstrong and K. Edmondson, *Electrochim. Acta*, 1973, **18**, 937.

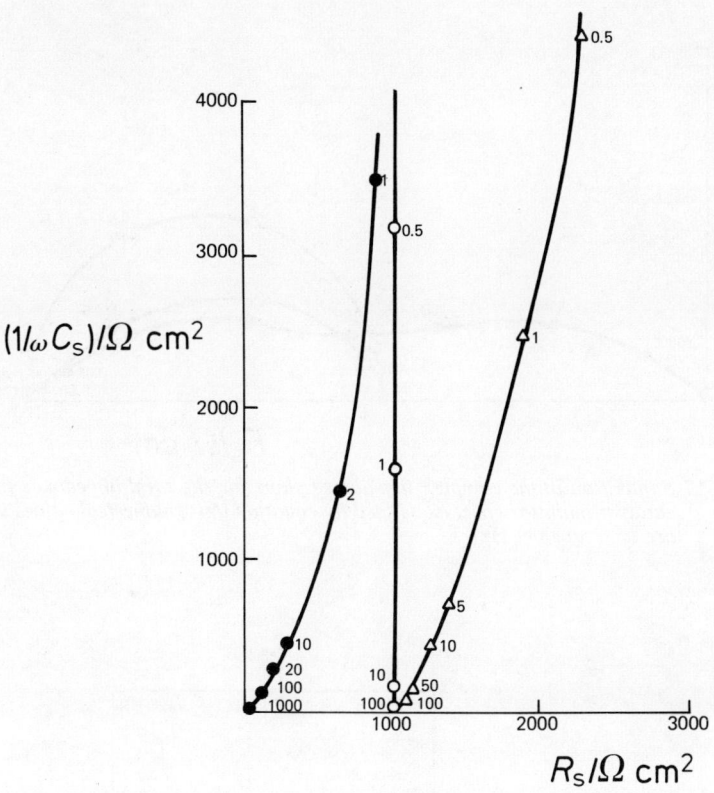

Figure 16 *The influence of τ_H and τ_D on the complex-plane simulations of the adsorption impedance (Y_a^{-1}) for the catalyst/inhibitor case, as defined by equation (62) (numerical values shown are of frequency/Hz)*

Adatom Model. The reaction scheme for this model can be represented by equation (65), and in general the impedance will be a function of charge transfer, surface

$$M^{n+} \underset{\text{transfer}}{\overset{\text{charge}}{\rightleftharpoons}} M_{ad} \underset{\text{diffusion}}{\overset{\text{surface}}{\rightleftharpoons}} (M_{ad})_{\text{stepline}} \underset{\text{incorporation}}{\overset{\text{lattice}}{\rightleftharpoons}} M_{lat} \qquad (65)$$

diffusion, incorporation into the lattice, and diffusion in solution, and it can be represented by the equivalent circuit shown in Figure 18.

This situation was first treated by Lorenz[49] in 1954 for the case when incorporation into the lattice and diffusion in solution are fast. Subsequently, Rangarajan[50] derived expressions which take these quantities into account.

---

[49] W. Lorenz, *Z. Naturforsch*, 1964, **9a**, 716.
[50] S. K. Rangarajan, *J. Electroanalyt. Chem.*, 1968, **17**, 61.

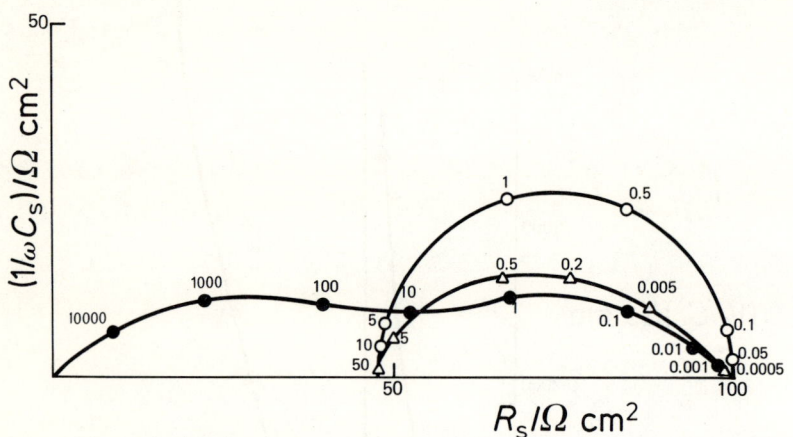

Figure 17 *Simulation in the complex impedance plane for the total impedance for the catalyst/inhibitor case, as defined by equation (64) (numerical values shown are of frequency/Hz)*

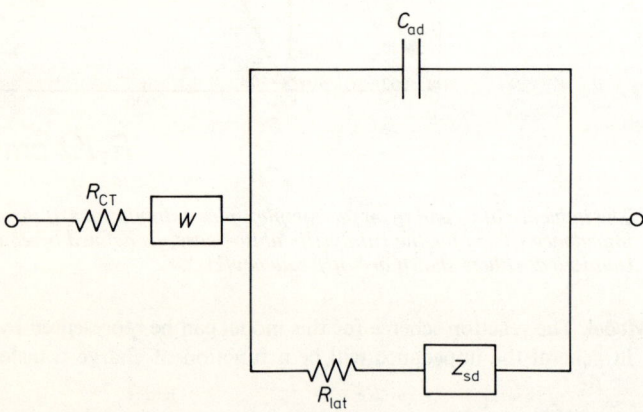

Figure 18 *Equivalent circuit for the adatom model*

The impedance contributions from charge transfer and diffusion in solution may, for this purpose, be defined as shown in equations (66) and (67). The total impedance is the sum of these two quantities and that due to surface effects (Z_s).

$$Z_{\omega \to \infty} = R_{CT} = RT/nFi_0 \qquad (C_{dl} = 0) \tag{66}$$

$$Z_D = k_1 R_{CT}/(i\omega)^{\frac{1}{2}} \tag{67}$$

including surface diffusion (Z_{sd}) and lattice incorporation (R_{lat}) [equation (68)], the definition of Z_s being as shown in the equation (69), where $C_{ad} = 1/k^2 R_{CT}$. The impedance for surface diffusion (Z_{sd}) will contain terms for the distance between two step lines ($2x_0$) and the average distance an adatom may travel before discharge $\lambda_0 = (D_a/k_1)^{1/2}$], and it may be written as shown in equation (70).

$$Z_T = Z_d + R_{CT} + Z_s \tag{68}$$

$$1/Z_s = 1/C_{ad} + 1/(R_{lat} + Z_{sd}) \tag{69}$$

$$Z_{sd} = 2R_{CT} \sum_{m=1}^{\infty} \frac{[k_1 + (i\omega D + D/\tau_m)^{\frac{1}{2}}]k_2}{[k_1 + (i\omega D + D/\tau_m)^{\frac{1}{2}}(i\omega + 1/\tau_m^a)] + k_2(i\omega D + D/\tau_m)^{\frac{1}{2}}} \tag{70}$$

Transferring to the admittance plane, the total expression may, in general, be written as equation (71), where ρ_0 is defined in a similar way to λ_0, as the distance a metal ion may travel before discharge.

$$Y_f = R_{CT}^{-1} f(\omega/k_2, x_0/\lambda_0, \rho_0/x_0, \rho_0 k_2/k_1) \tag{71}$$

Simulations of this situation have been carried out by Armstrong and Firman.[51] They have shown that, when solution diffusion is unimportant, a semicircular shape is generated in the complex impedance plane, the size of which is dependent on the ratio x_0/λ_0 (Figure 19). When x_0/λ_0 is very large (>100), the shape is a true semicircle, whereas when this quantity becomes smaller a deviation is observed at high frequencies, It has also been pointed out that this model is experimentally indistinguishable from that of the adsorbed intermediate.

By varying c_{Mn^+} and D_{Mn^+}, the effect of solution diffusion was examined (Figure 20). It was found that the solution diffusion term Z_w masked any small variations in the other terms.

Two-dimensional Nucleation and Growth: Potentiostatic Case.—This model proposes the growth of discrete centres on electrode surfaces where the centres are nucleated at a rate $A(E,t)$. The nuclei grow outwards in two dimensions with a velocity $v(E,t)$, which under potentiostatic conditions can be assumed to be independent of time. A consideration of this model, taking into account the overlap of the growing centres and the formation of successive layers, leads to the expression (72) for the current due to the $(n+1)^{th}$ layer,[52] where q_m is the charge involved in the formation of a monolayer and $\beta = (\pi v^2 A)/3$. As no analytical solution has yet appeared for equation (72), it is not possible to evaluate the impedance directly. However, Armstrong and Metcalfe[53] have calculated the initial transient behaviour by numerical

$$i_{n+1}(t) = q_m \int_0^t 3\beta(t-t')^2 \exp[-\beta(t-t')^3] i_n(u) du \tag{72}$$

integration of equation (72). They further calculated[54] the frequency response of the impedance by a Fourier transform of the input perturbation and response to a small potentiostatic pulse applied after the steady-state current had been reached.

---

[51] R. D. Armstrong and R. E. Firman, *J. Electroanalyt. Chem.*, 1973, **45**, 257.
[52] R. D. Armstrong and J. A. Harrison, *J. Electrochem. Soc.*, 1969, **116**, 328.
[53] R. D. Armstrong and A. A. Metcalfe, *J. Electroanalyt. Chem.*, 1975, **63**, 19.
[54] R. D. Armstrong and A. A. Metcalfe, *J. Electroanalyt. Chem.*, 1976, **71**, 5.

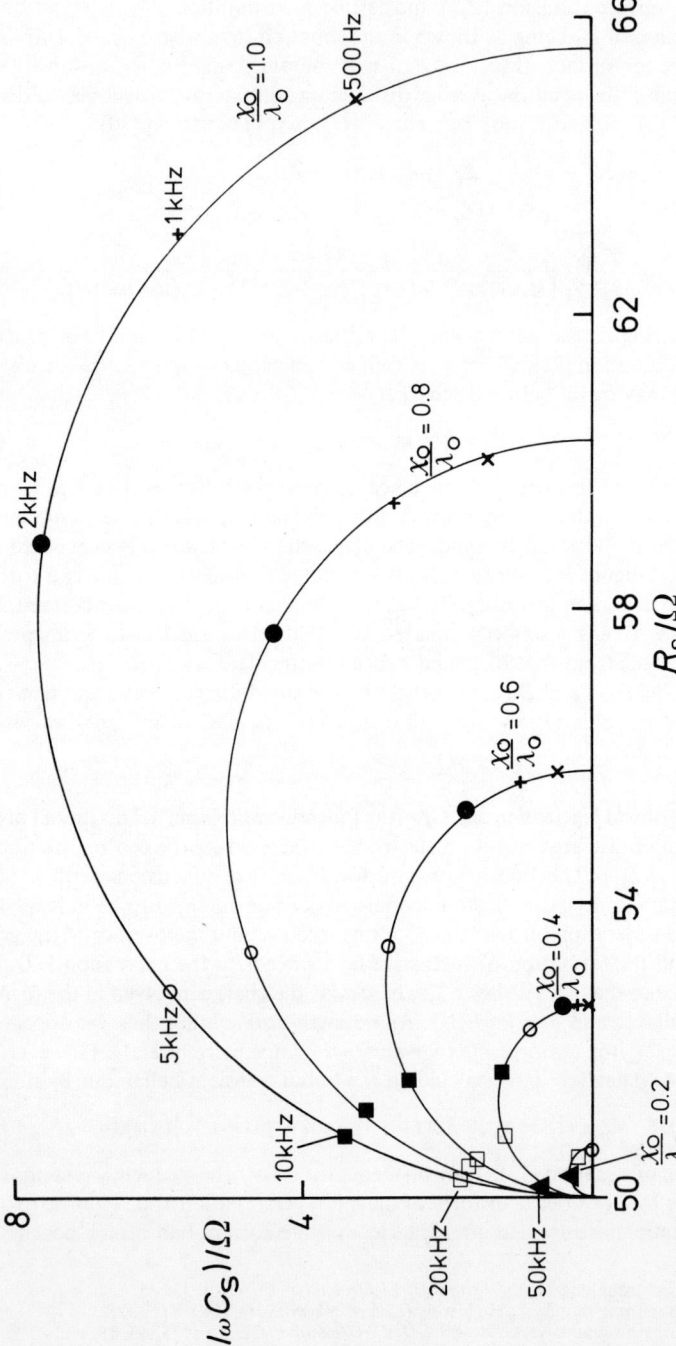

Figure 19 *The influence of the rates x_0/λ_0 on the electrode impedance for the adatom model when solution diffusion is unimportant;* $R_{CT} = 50\,\Omega$, $k_2 = 5327\,s^{-1}$

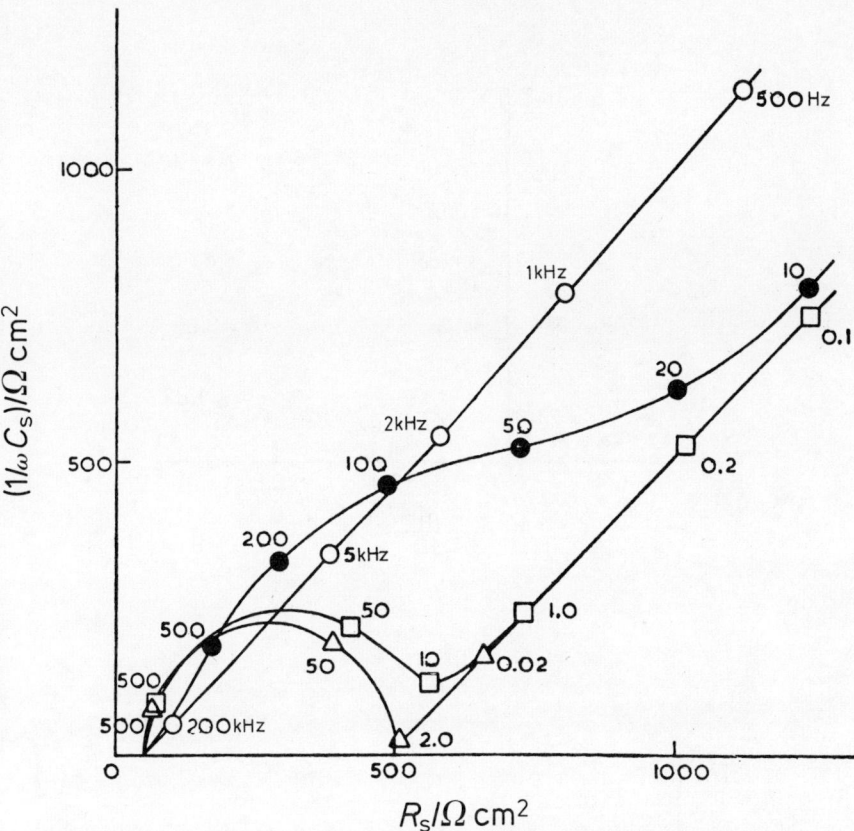

Figure 20 The effect of solution diffusion complex-impedance simulations when $R_{CT} = 5\,\Omega$, $k_2 = 5327\,\text{s}^{-1}$, $x_0/\lambda_0 = 10$, $R_{lat} = 0$, $D_0 = 10^{-5}$; ($\triangle$) $c_{M^{n+}} = 10^{-6}\,\text{mol cm}^{-3}$; ($\square$) $c_{M^{n+}} = 10^{-7}\,\text{mol cm}^{-3}$; ($\bullet$) $c_{M^{n+}} = 10^{-8}\,\text{mol cm}^{-3}$; ($\circ$) $c_{M^{n+}} = 10^{-9}\,\text{mol cm}^{-3}$ (numerical values are of frequency/Hz except where otherwise shown)

Since both A and v are dependent on potential, the perturbation was applied by changing the nucleation rate and/or the rate of advance of an edge. The impedance spectrum for a change in v is shown in Figure 21a. However, for a change in A, the high-frequency limit of the impedance is infinite, and for this reason the admittance spectrum is given (Figure 21b). It is interesting to see that the admittance changes quadrant as the frequency is increased. This is due to the quadratic form of the current rise from the steady-state value. Comparison of Figures 21a and 21b shows that the admittance spectrum for a change in A is very similar to the impedance spectrum for a change in v. One is essentially the complex reciprocal of the other.

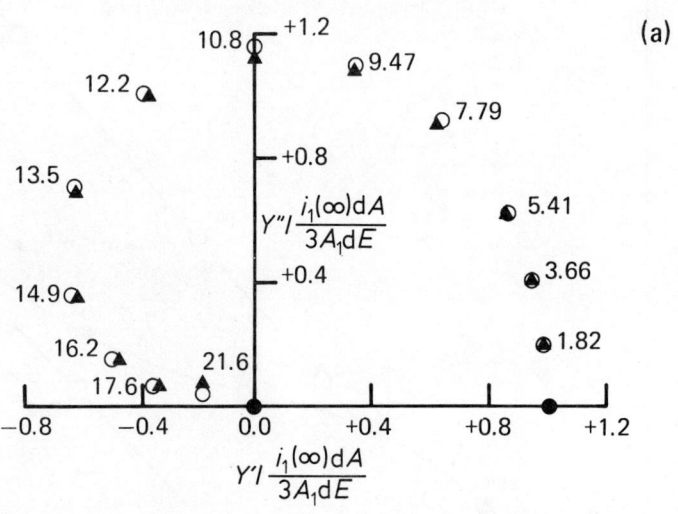
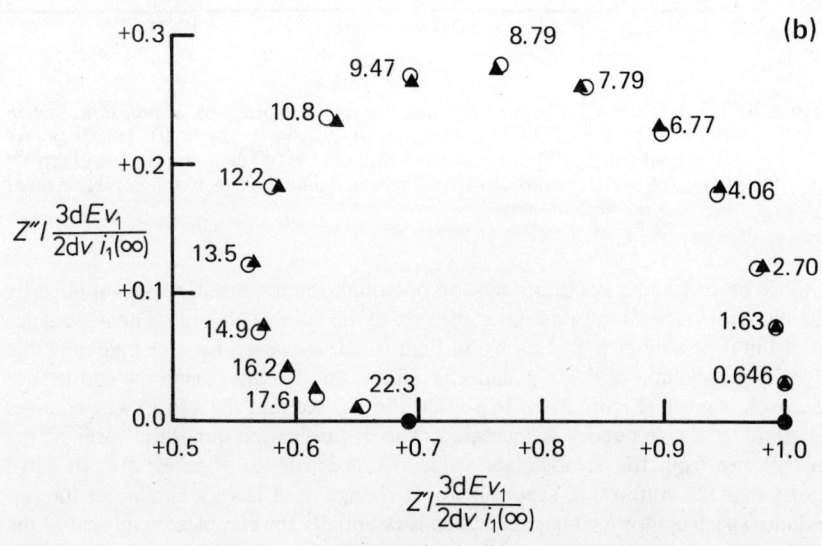

Figure 21 (a), (b)

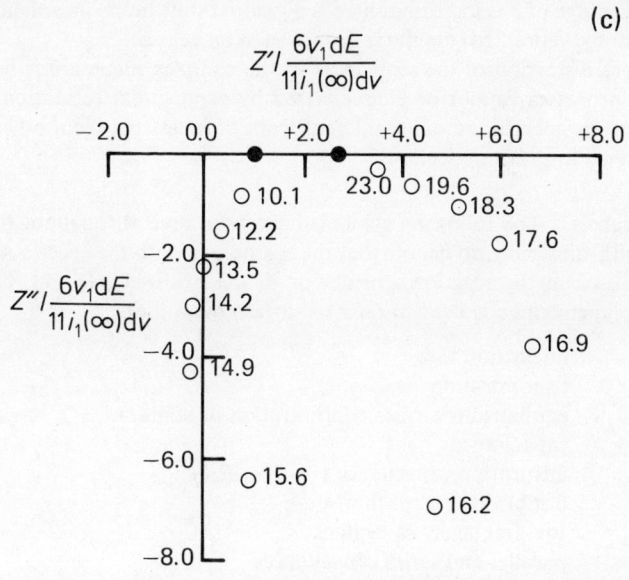
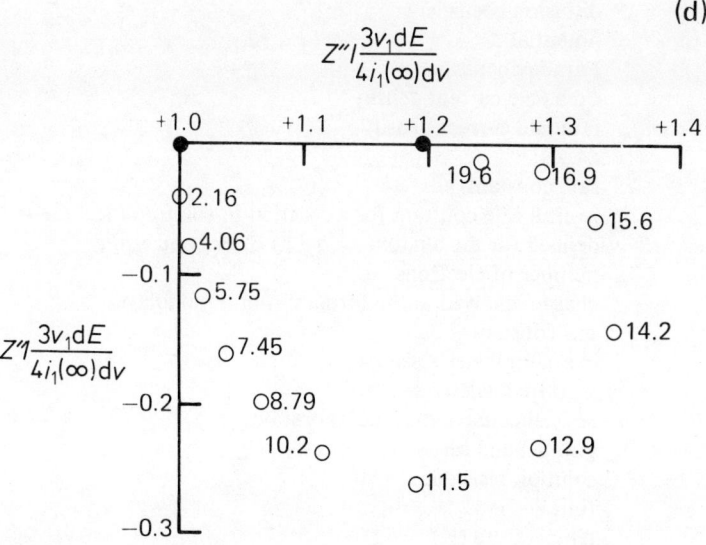

Figure 21 *Plots in the complex plane for two-dimensional nucleation and growth simulated by* (a) *a 3% change in rate of advance of an edge* (v), (b) *a 3% change in the nucleation rate* (A), (c) *a 3% change in A with a 0.5% change in v*, (d) *a 3% change in A with a 2% change in v. Numerical values are of frequency/Hz, and the graphs are plotted dimensionlessly*

Where there is a small contribution from v (Figure 21c), the impedance spectrum occurs in two quadrants, again due to the quadratic nature of the initial current rise. As the percentage of v is increased there is a gradual shift in the impedance towards one quadrant (Figure 21d) (as the influence of A decreases).

The severe distortion of the semicircles in the complex plane arises because the relaxation processes cannot be characterized by exponential relaxation times. In this respect, the spectra are quite different from those arising from other mechanisms described in previous sections.

List of Symbols.—The following symbols have been used throughout this article, together with subscripts, to denote that the symbols refer to the species A, B, C, N, S, and M used in the reaction schemes or to the oxidized (O) and reduced (R) species. A superscript s is used to refer to surface quantities.

A	nucleation rate
c	concentration
c_a	equilibrium surface concentration of adatoms
C	capacitance
C_{ad}	adsorption capacitance ($= 1/k_2 R_{CT}$)
C_{dl}	double-layer capacitance
C_0	low-frequency capacitance
C_p, C_s	parallel and series capacitances
C_p^*	($= C_p - C_{dl}$)
ΔC	($= C_0 - C_{dl}$)
D	diffusion coefficient
E	potential
F	Faraday constant
i_0	exchange current density
i_f	faradaic current density
i	$\sqrt{-1}$
k	rate constant
k_o	overall rate constant for a reaction in solution (Section 4)
k_1, k_2	defined for the adatom model as $i_0 n F c_0$ and $i_0 n F c_a$
n	number of electrons
q_m	charge involved in the formation of a monolayer
R	gas constant
R_{CT}	charge-transfer resistance
R_{dc}	zero-frequency resistance
R_{lat}	resistance for lattice incorporation
R_p, R_s	parallel and series resistances
R_{so}	solution resistance
t, t'	time
T	temperature (K)
u	$\left[\dfrac{(\omega D)^{\frac{1}{2}}}{v} \left(\dfrac{\partial c}{\partial v} \right)_{E, \Gamma} \right]$
v, v_1, v_2	rate of reactions
v	rate of advance of an edge (Section 7)

The A.C. Impedance of Complex Electrochemical Reactions

W	Warburg impedance for a Randles equivalent circuit
$2x_0$	distance between two step lines
Y	admittance
Y_a	admittance for adsorption
Y_f	faradaic admittance
Z	impedance
Z_d	diffusional impedance
Z_f	faradaic impedance
$Z_{M/F}, Z_F, Z_{F/S}$	impedances for the metal undergoing passive dissolution
Z_s	$[1/C_{ad} + 1/(R_{lat} + Z_{sd})]^{-1}$
Z_T	total impedance
α_1, α_2	$\left[1 + \dfrac{v_A}{(i\omega D_A)^{\frac{1}{2}}}\left(\dfrac{\partial v_1}{\partial c_A}\right)_{E,\Gamma}\right], \left[1 - \dfrac{v_C}{(i\omega D_C)^{\frac{1}{2}}}\left(\dfrac{\partial v_2}{\partial c_C}\right)_{E,\Gamma}\right]$
β	$(\pi v^2 A)/3$
Γ	surface excess (time-dependent)
Γ_0	equilibrium surface excess
δ	thickness of the diffusion layer
λ_0	$(D_a/k_s)^{\frac{1}{2}}$, where D_a is the diffusion coefficient of adatoms
v	stoicheiometric number
ρ_0	$(D_0/k_2)^{\frac{1}{2}}$
σ	diffusion coefficient
σ'	diffusion coefficient defined by equation (21)
τ	time constant
$\tau_D, \tau_{D1}, \tau_{D2}$	time constants for diffusional processes
τ_m, τ_m^a	$x_0^2/m^2\pi^2 D_M^{2n+}$, $x_0^2/m^2\pi^2 D_a^2$
τ_H	time constant for a heterogeneous process
τ_R	reaction time constant
ω	frequency/rad s^{-1}
ω^*	frequency/s^{-1} at the maximum of the semicircle

4
The Theory of Electron-transfer Reactions in Polar Media: Part II

BY P. P. SCHMIDT

1 Introduction

This chapter is the second part of a survey of the electron-transfer reaction in polar solution; Part I appeared in Volume 5 of this series of Specialist Periodical Reports on Electrochemistry, Chapter 2. The discussion of principles introductory to this part of the survey is contained in Part I. Hence, occasional reference will be made to passages and formulae already given.

It is the purpose of this part of the survey to continue the discussion initiated in the first part. There, general matters pertaining to the descriptions of various electron-transfer systems were considered. Definitions of types of systems were presented, and the specific case of the simple outer-sphere electron-transfer reaction was examined in detail. In the succeeding sections of this part of the survey, therefore, timbers of the edifice of the theory are erected on the foundations already laid. The structure of the theory cannot be completed at this time. Nevertheless, recent developments certainly constitute a considerable progress toward the aim of establishing a reasonably general theoretical account of most of the commonly encountered electron-transfer reaction systems. It seems, as of this writing, that avenues for further exploration are much better defined. There is justification for the expectation that these systems should admit useful theoretical descriptions.

Apart from the general account of the electron-transfer reaction provided in Part I, there, as mentioned, the simple electron-transfer reaction was considered. The substance of the contents of this part of the survey builds upon that simple treatment. The objective of any general formulation of chemical reactivity, obviously, is to provide some means, at least in principle, if not in fact, to treat any given situation. It is not possible, certainly at this time, actually to be able to carry out this plan in detail. However, based on the use of the most sophisticated current theoretical methods, it seems that considerable progress can be made.

Specifically, in the next section the principles of chemical reactivity are formulated in terms of the linear-response theory,[1] an element of the statistical mechanics of non-equilibrium systems.[2] With the use of the linear-response theory, it is shown that the general semi-classical and quantum-mechanical derivations of the rate constants for electron-transfer reactions,[3,4] due originally to Marcus[5] and to

[1] T. Yamamoto, *J. Chem. Phys.*, 1960, **33**, 281.
[2] R. Kubo, *J. Phys. Soc. Japan*, 1957, **12**, 507.
[3] P. P. Schmidt, *J. Chem. Phys.*, 1973, **48**, 4384.
[4] P. P. Schmidt, *J. Phys. Chem.*, 1974, **78**, 1684.
[5] R. A. Marcus, *J. Chem. Phys.*, 1965, **43**, 679.

Levich and Dogonadze,[6] are contained as limits within one general formulation. This demonstration serves further to substantiate the essential similarity between these two treatments, a subject which was pursued in Part I. Also, in the next section, the question of the degree of adiabaticity of a given reaction is considered.

Following the next section, an extensive discussion of the recent treatments of realistic systems is given. These treatments seek to account for the contributions to the activation energy arising from intramolecular degrees of freedom of the reactant. In large measure, many of these intramolecular contributions were considered in 1965 by Marcus[5] in his general statistical-mechanical analysis. However, at the time his methods were not implemented for the consideration of specific cases; this was due partly to the very general character of Marcus' treatment, and due partly to the fact that, until recently, an adequate method of analysis of the intramolecular vibrational degrees of freedom, especially as they participate in a radiationless relaxation process, was not available. The situation now is considerably different.[7] A machinery of analysis of the vibrational degrees of freedom exists in the general theory of molecular radiationless processes.[8] This machinery is beginning to be applied to the electron-transfer system, with encouraging results.[7,9]

Still not adequately accounted for are the electron-transfer processes which involve gross change in the molecular structure during the course of the transfer of the electron from the donor to the acceptor species. Several simple rectilinear treatments have been provided,[10,11] and with generous appraisal they can be said to be reasonably satisfactory. Nevertheless, these simple treatments suffer from a lack of rigour as well as from a lack of flexibility. More substantial analysis is required. Consequently, in a later section of this survey, an account is given of some recent attempts[12—14] to establish a theory of highly excited molecular vibrational states and the application of this theory to the electron-transfer system. The basis of the treatment is a quantum hydrodynamic approach similar in form, but different in some details, to the theory of the atomic nucleus.[15] This theoretical approach, still in its infancy, seems to offer grounds for optimism. This optimism derives entirely from a knowledge of the success of the approach in nuclear physics.[15] The analogies between the electron-transfer system and the nuclear system are many. It is certainly hoped that the optimism based on analogy will be rewarded.

2 The General Treatment of Electron-transfer Reactions

Introduction.—The electron-transfer reaction is an example of a radiationless relaxation process; this aspect of the reaction system was described in Part I in some detail. In addition, in Part I the simple outer-sphere electron-transfer reaction system was examined. This reaction limit is characterized by changes only in

---

[6] V. G. Levich and R. R. Dogonadze, *Coll. Czech. Chem. Comm.*, 1961, **26**, 193.
[7] J. Ulstrup and J. Jortner, *J. Chem. Phys.*, 1975, **63**, 4358.
[8] F. K. Fong, 'Theory of Molecular Relaxation: Applications in Chemistry & Biology', J. Wiley and Sons, New York, 1975.
[9] W. Schmickler and W. Vielstich, *Electrochim. Acta*, 1973, **18**, 883.
[10] P. P. Schmidt, *Austral. J. Chem.*, 1970, **23**, 1287.
[11] R. R. Dogonadze, J. Ulstrup, and Yu. I. Kharkats, *J.C.S. Faraday II*, 1972, **58**, 744.
[12] P. P. Schmidt, *J.C.S. Faraday II*, 1976, **72**, 1099.
[13] P. P. Schmidt, *J.C.S. Faraday II*, 1976, **72**, 1125.
[14] P. P. Schmidt, *J.C.S. Faraday II*, 1976, **72**, 1144.
[15] J. M. Eisenberg and W. Greiner, 'Nuclear Models', North Holland, Amsterdam, 1970.

the degrees of freedom of the polar solvent during the transfer of an electron from a donor to an acceptor species. The complications which accompany the alteration, formation, and destruction of chemical bonds within the reactants are not considered in the outer-sphere limit.

In order to consider the more complicated and realistic electron-transfer reaction systems, however, it is necessary to determine whether the straightforward postulates of reactivity discussed in Part I are adequate to apply to these general cases. Hence, one purpose of this section is to formulate general approaches to the theoretical examination of an electron-transfer reaction. In principle, we seek representations which can be used to treat the combined difficult problems of the adiabatic and non-adiabatic reaction limits as well as the gross molecular rearrangements which are characteristic of many typical reactions.

In particular, in this section a general formulation of solution-phase reactivity is examined. It is based on the application of the non-equilibrium statistical-mechanical linear response theory[2] to the solution-phase electron-transfer reaction.[3,4] We show, with the use of the rate-constant expression obtained with this theory,[1] that both the semi-classical Marcus[5] and the quantum-mechanical Levich–Dogonadze theories[6] are contained within this single formulation.

Summary of the Outer-sphere Theory.—The requirements of any general theoretical treatment of chemical reactivity, and in particular of the electron-transfer reaction, can best be seen in comparison to the simple approaches taken to account for the outer-sphere reaction. Consequently, a brief summary of the outer-sphere theories is presented here.

As discussed in Part I, the first treatments of the outer-sphere electron-transfer reaction proceeded along two separate routes. The Marcus[16] and Hush[17] theories, chronologically first, were developed, based upon the use of the assumption that the transition state for the reaction is an activated, stationary, quasi-equilibrium state. Therefore, it is possible to define thermodynamic quantities applicable to the system in the activated state, and the rate-constant expression may be written as[16,17] in equation (1), where Z is the collision number in the solution phase, ΔG^* is the

$$k = Z\exp(-\Delta G^*/k_B T) \qquad (1)$$

activation free energy, and k_B is the Boltzmann constant. The determination of ΔG^* depends on the assumption that the activated state may be described in terms of equilibrium thermodynamic quantities. The details of the analyses which reveal the dependence of the activation energy on contributions from the solvent, the transfer electron, and on other factors are given in Part I.

The second approach, introduced by Levich and Dogonadze in 1959,[6,18] depended on the use of first-order time-dependent perturbation theory in order to obtain an expression for the transition probability for the transfer of an electron from a donor to an acceptor species. The expression for the transition probability used by them is written as equation (2). The evaluation of this expression requires

---

[16] R. A. Marcus, *J. Chem. Phys.*, 1956, **24**, 966.
[17] N. S. Hush, *J. Chem. Phys.*, 1958, **28**, 962.
[18] V. G. Levich and R. R. Dogonadze, *Doklady Akad. Nauk S.S.S.R.*, 1959, **134**, 123.

$$w_{\mathrm{fi}} = \left[\sum_i \exp(-E_i/k_B T)\right]^{-1} \sum_{i,f} \exp(-E_i/k_B t) |\langle f| H' | i\rangle|^2 \delta(E_i - E_f) \quad (2)$$

that appropriate initial and final system states be found, based, usually, on the use of the Born–Oppenheimer separation method. That is, it is necessary to investigate the states of the donor and acceptor species dissolved in a polar solvent medium. The Born–Oppenheimer separation, applied to the electron-transfer system, isolates the degrees of freedom of the solvent medium from the internal, solvated degrees of freedom of the electron that is transferred. In a non-adiabatic limit, the separation also provides a number of operator quantities, some of which govern the transfer of the electron from the donor to the acceptor species.

In equation (2) it is necessary to take into account the possibility that the electron transfer can take place from any weighted initial state to any accessible final state. Hence, the expression is summed over all final states, as indicated. Furthermore, it is summed over all the possible statistically weighted initial states. The quantity H' is an effective perturbation which can be assumed to be responsible for the coupling of the initial and final states. The matrix elements of this perturbation operator are a measure of the degree of interaction between the initial and final states, and in

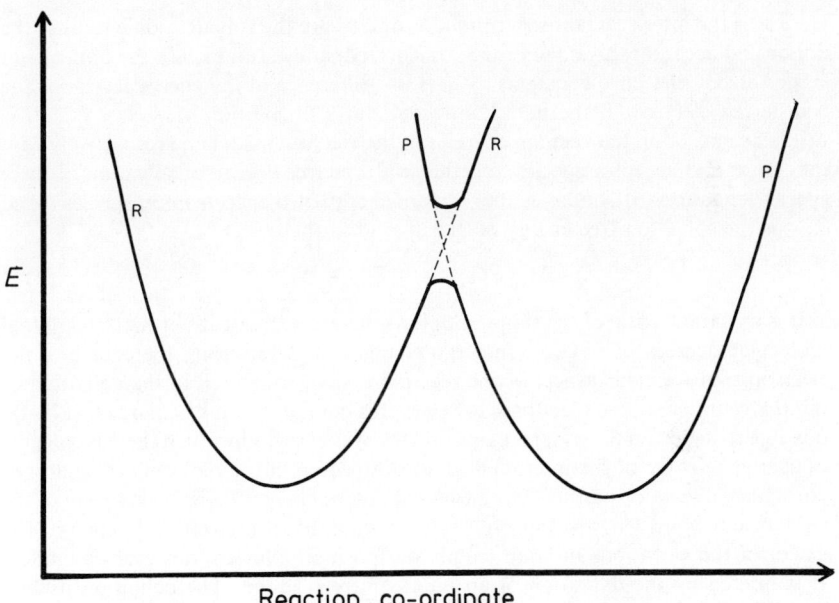

Figure 1 *A representation of the intersection of the potential-energy curves of the initial and final reaction state. R labels the reactant state and P the product state.*

terms of the familiar intersection of potential-energy surfaces (see Figure 1), are measures of the degrees of splitting of the surfaces in the intersection region.

In both the classical and statistical-mechanical treatments of Marcus[16] and Hush[17] and in the quantum statistical-mechanical treatment due to Levich and Dogonadze,[6,18] there is a stipulation that energy must be conserved in the electron-transfer step. Therefore, in terms of the representations of the potential-energy surface (see Figure 1), the electron transfer may take place only along a path on the energy surface on which the energies of the initial and final state are equal. The embodiment of this condition in the Marcus and Hush treatments is found in the thermodynamic analyses which lead to the free energy of activation. In the Levich–Dogonadze theory, energy conservation is ensured by the presence of the Dirac delta function.

It was shown in Part I that, aside from differences in the manner of mathematical analysis involved, the Marcus and Levich–Dogonadze theories are essentially the same. The parameters in the activation energy quantity determined by both of these treatments are the same. Specifically, both theories predict that the rate-constant expression for a simple outer-sphere electron-transfer reaction has the form of equation (3), where E_r is the solvent repolarization energy, defined by equation (4).

$$k \propto \exp[-(\Delta G_0 + E_r)^2/4E_r k_B T] \quad (3)$$

$$E_r = \Delta e^2 (1/\varepsilon_{op} - 1/\varepsilon_s)(1/2a_1 + 1/2a_2 - 1/R) \quad (4)$$

Here R is the inter-reactant separation, a_1 and a_2 are the solvated ionic radii of the donor and acceptor, Δe is the charge transferred, and ε_{op} and ε_s are the optical and static solvent dielectric constants. ΔG_0 is the difference in the energy levels for the transferred electron in the initial and final states in solution. This last quantity, ΔG_0, depends upon the vacuum energies of the electron associated respectively with the donor and acceptor species, together with the free energy of solvation of these species [equation (5)], and e_i is the vacuum energy of the electron on species i, and $E_{s(i)}$ is the solvation free energy of that species in the state i.

$$\Delta G_0 = e_i - e_j + E_{s(i)} - E_{s(j)} \quad (5)$$

It is a characteristic of the simplest outer-sphere electron transfer that no internal molecular degrees of freedom undergo changes of state during the course of the electron transfer. This aspect of the reaction system is implicit in the definition of the *simple* outer-sphere reaction. However, this definition can be enlarged. It is still possible to define an outer-sphere reaction for systems within which there is a degree of change in some of the internal degrees of freedom of the molecular reactant. In particular, a reaction can still be considered outer-sphere if there is alteration of the equilibrium bond lengths and only slight variations in molecular geometry. The extent of the variations in bond length and position allowed, however, is limited. It is limited by the definition of an outer-sphere reaction. The definition itself is limited in part by considerations of mathematical tractability. Nevertheless, within the bounds of these limitations, mathematical and descriptive, modifications and improvements of the theory are possible. In 1964 Marcus[5] reported a model-independent, general account of the outer-sphere reaction. In his treatment, contributions to the activation energy due to solvent repolarization as well as contributions due to adjustments of bond lengths and position are included. In principle, his arguments are sufficiently general to consider significant changes in these quantities. For example, the Morse potential can be used to describe the states of interatomic motion within the molecule. Consequently, account can be given of a number of changes which depend upon anharmonic contributions within the reactants. In practice, however, owing to the complexity of the molecular structures of the reactants, one often is limited to the consideration of low-order harmonic-limit changes. An account of the general intramolecular contributions to the activation energy has not been given until recently.[7,9]

Modifications, improvements, and a general extension of the range of applicability of the original Levich–Dogonadze theory are also possible. The modifications are very close to those of Marcus. In the past six years, Donogadze, Kuznetsov, and Vorotyntsev[19,20] have investigated the application of general methods in the radiationless transition theory to the electron-transfer system in the non-adiabatic limit. Specifically, they have examined the application of the Kubo–Toyozawa theory.[21] Many of their results extend the Kubo–Toyozawa analysis, and furthermore, go beyond the Marcus treatment[5] of the intramolecular contributions to the activation energy. In particular, Dogonadze, et al.,[19,20] have determined that highly endo- and exo-thermic electron-transfer reactions can be barrierless or activationless, respectively, and these states of reactivity are not handled adequately in Marcus's semi-classical statistical-mechanical formulation.[5] Experimental investigations have substantiated this view.[22]

The essential distinction between the Marcus and Levich–Dogonadze simple outer-sphere reaction theories, aside from the differences in mathematical manipulation, is that the quantal theory of Levich and Dogonadze is manifestly non-adiabatic. The classical statistical-mechanical limit of Marcus's treatment is manifestly adiabatic. Semi-classically, Marcus's theory admits both adiabatic limits. An account of non-adiabaticity in the Marcus theory is introduced with the use of a velocity-weighted transition probability, κ, and adiabaticity in the Levich–Dogonadze theory is allowed for by replacing the pre-exponential factor in the non-adiabatic limit by the adiabatic quantity. It is arguable, but generally held, that the manner of the introduction of the velocity-weighted transition probability in the semi-classical statistical-mechanical theory of Marcus is more acceptable in terms of physical and mathematical rigour than the initial, and less formal, approach by means of the apparent caveat of Levich and Dogonadze. This view is overly harsh, and derives from the prejudices established by the long-standing and widely accepted use of the semi-classical approach to all types of reaction system. To the degree that both approaches are *ad hoc* and 'boot-strap' in nature, both methods of approach to the specification of the adiabatic and non-adiabatic limits are nearly equivalent in terms of rigour.

There is an ancillary argument, due to Dogonadze,[23,24] which supports their approach to the determination of the adiabatic limit, and furthermore, illustrates

[19] R. R. Dogonadze, A. M. Kuznetsov, and M. A. Vorotyntsev, *Phys. Status Solidi* (b), 1973, **54**, 125.
[20] R. R. Dogonadze, A. M. Kuznetsov, and M. A. Vorotyntsev, *Phys. Status Solidi* (b), 1973, **54**, 425.
[21] R. Kubo and Y. Toyozawa, *Progr. Theor. Phys.*, 1955, **13**, 160.
[22] R. P. Van Duyne and S. F. Fischer, *Chem. Phys.*, 1974, **5**, 183.
[23] R. R. Dogonadze, *Doklady Akad. Nauk S.S.S.R.*, 1962, **142**, 1108.
[24] R. R. Dogonadze and A. M. Kuznetsov, 'Physical Chemical Kinetics', Tome 3, VINITI, Moscow, 1973 (in Russian).

an additional level of correspondence between the Levich–Dogonadze and Marcus theories, at least for the simple outer-sphere reaction. Dogonadze[23] has carried out the semi-classical Landau–Zener[25,26] type of treatment of the electron-transfer reaction, based on the use of the Levich–Dogonadze analysis of the polar solvent modes. In this manner Dogonadze demonstrated that the non-adiabatic limit obtained from the Landau–Zener treatment is the same as Levich and Dogonadze obtained using quantum-mechanical arguments. The Landau–Zener treatment also provides an adiabatic limit. A comparison of the two limits thereby indicates the manner by which it is possible to *suggest* the adiabatic limit, provided one has on hand only the non-adiabatic-limit expression for the transition probability or the rate constant. To this extent, it is justified certainly, in the simple Levich–Dogonadze theory,[6] to argue that in the adiabatic limit the pre-exponential factor in the transition-probability expression of equation (6), where L_{fi} is the electronic matrix element of the Coulomb operator, can be replaced by $\omega_0/2\pi$ [equation (7)], where ω_0 is an average frequency associated with the librational excitation modes of the polar solvent medium.

$$w_{n.a.} = |L_{fi}|^2 (\pi/\hbar^2 k_B T E_r)^{\frac{1}{2}} \exp[-(\Delta G_0 + E_r)^2/4E_r k_B T] \quad (6)$$

$$w_{a.} = (\omega_0/2\pi)\exp[-(\Delta G_0 + E_r)^2/4E_r k_B T] \quad (7)$$

Marcus' theory,[5,16] on the other hand, assumes the necessity of a Landau–Zener or equivalent analysis in order to determine the factor κ, the velocity-weighted transition probability. To the extent that the Landau–Zener-type treatment generally admits the non-adiabatic limit from a semi-classical treatment, Marcus's theory admits the non-adiabatic limit to the same extent. As we show later, for barrierless and activationless transitions, the Marcus semi-classical approach is not adequate.[19,20,22]

The Linear-response Theory of Chemical Reactivity.—Attempts to formulate general theories of electron-transfer reactions require the consideration of two factors. First, given any adequate description of the stationary and non-stationary states of the reaction system, it is necessary to find an expression which provides a rate constant that is sufficiently general to account for any degree of strength of interaction between the states of the system. Furthermore, there should be no restrictions as to the magnitudes of coupling constants associated with various interactions between the individual subsystems of the entire system. In other words, we need to find a general form of rate-constant expression, which, when evaluated for a specific system, yields expressions that are valid over all reaction limits from adiabatic to non-adiabatic. Of course, for a specific system there should apply one expression for the rate constant which automatically provides the correct limit. In principle, the rate-constant expression obtained with the use of the non-equilibrium linear response theory[1] satisfies this need. In practice, however, other (mathematical) limitations so far have restricted the generality of this approach.

The second factor is equally important. It is simply that we must find adequate means to describe the stationary and non-stationary states of the system. Certain very simple, ideal model systems can be found (*viz.*, the two-level system interacting

[25] L. D. Landau, *Phys. Z. Sowjet.*, 1932, **1**, 89; 1932, **2**, 96.
[26] C. A. Zener, *Proc. Roy. Soc.*, 1933, **A137**, 696; 1933, **A140**, 660.

with a continuum of environmental oscillators) which, under an additional set of mathematical restrictions, provide solutions that are valid in the adiabatic and non-adiabatic limits.[24] The general situation, which is posed by the complexity of real systems and the complexity of their actual quantum-mechanical analysis, has not yielded a method of analysis adequate to resolve the component degrees of freedom, their mutual interactions, and their contributions to the electron-transfer rate constant. In the past, and in the foreseeable future, use has been made of the Born–Oppenheimer[27,28] scheme of separation. Few alternatives exist, and these alternatives have so far been associated only with extremely ideal systems. It is possible that suitable analyses can be found within the basic Born–Oppenheimer analysis, through its mathematical evolution. This is a matter of great concern for the electron-transfer theory, as well as for general chemical reactivity. The consideration of this problem, however, is outside the scope of this article.

The linear response theory provides a general form of expression for the rate constant. This general expression can be transformed into a specific rate constant that is applicable to an electron transfer once a suitable model system has been specified. The limitations inherent in the existing electron-transfer-rate theory are shown shortly to be limitations that are in part contained within the representations of the model system and in part mathematical.

The general approach to the consideration of the electron-transfer reaction, therefore, requires first that we determine the most general expression for the rate constant. Second, it is necessary to find suitable descriptions, *i.e.* models, of the reaction system. The remainder of this section is devoted to the first of these considerations. Later sections address the problem of the search for good representations of the various states of the system.

We begin this discussion of the general reaction rate theory with an account of the derivation of the Yamamoto expression[1] for the rate constant. The derivation presented follows Fischer's[29] method, which generalizes the original treatment given by Yamamoto.

Consider a general reaction system[29] which consists of a set of reactants A_i and products A_i' together with the associated stoicheiometric coefficients v_i. The reaction can be written as shown in equation (8). The forward reaction is associated with a

$$\sum v_i A_i \underset{k_2}{\overset{k_1}{\rightleftarrows}} \sum v_i' A_i' \tag{8}$$

rate constant k_1, the reverse with k_2. Let N_i and μ_i, respectively, be the number of molecules and the chemical potential associated with species A_i. If the reaction system is not in an equilibrium state, all the N_i differ from their equilibrium values N_i^0. In this case the chemical force (the affinity) is non-vanishing; it is the driving force associated with the chemical reaction flux. The quantities ΔN_i represent the deviations of the reactants from their equilibrium values. Moreover, the ratios of the individual deviations ΔN_i to the stoicheiometric coefficients, *viz.* $\Delta N_i/v_i$, are independent of the index i. Therefore equation (9) is valid, and the affinity is defined by equation (10).

---

[27] M. Born and J. R. Oppenheimer, *Ann. Physik*, 1927, **84**, 457.
[28] M. Born, *Nachr. Akad. Wiss. Gott.*, 1951, No. 6.
[29] S. F. Fischer, *J. Chem. Phys.*, 1970, **53**, 3195.

$$a\Delta N = \sum_i \mu_i \Delta N_i + \sum_i v_i' \Delta N_i' \tag{9}$$

$$a = \sum_i v_i' \mu_i' - \sum_i v_i \mu_i \tag{10}$$

We assume that the system in a non-equilibrium state has associated with it a total Hamiltonian operator H_T of the form shown in equation (11), where H contains the kinetic and potential-energy operators for all the species in the system, and H_F is given by equation (12). This term can be regarded as the potential-energy

$$H_T = H + H_F \tag{11}$$

$$H_F = aN \tag{12}$$

function from which derives the chemical force, the affinity. The variable N, as is indicated later, is a number operator which acts on the population states of the various reactants and products in the system.

The chemical flux can be expressed in terms of a linear dependence on the affinity.[1,29] In particular, in terms of the time variable, it is expressed as the convolution of the affinity with a quantity $G(t)$: *viz.* equation (13). The energy- (or

$$d\Delta N/dt \equiv \mathcal{T} = G \otimes a \tag{13}$$

frequency-)space Fourier transform, which is needed shortly, is simply given by equation (14). Thus, it can be seen that $G(\varepsilon)$ is effectively the chemical admittance function.

$$\mathcal{T}(\varepsilon) = G(\varepsilon)a(\varepsilon) \tag{14}$$

The chemical flux also can be expressed as the ensemble average of the flux operator $\dot{N}$ [equation (15)], where the trace operator, $\text{tr}[\rho(t)\text{Op}]$, is defined by equation (16a) for any operator Op. The quantity $\rho(t)$ is the system density operator. It is

$$\mathcal{T} = \text{tr}[\rho(t)\dot{N}] \tag{15}$$

$$\text{tr}[\rho(t)\text{Op}] = \sum_i \exp[-\hbar\omega_i(t)\beta]\langle i|\text{Op}|i\rangle \tag{16a}$$

$$\beta = 1/k_B T \tag{16b}$$

a quantity which is important in the subsequent analysis. In order to establish a relationship between the chemical flux, as expressed by equations (15) and (13), and ultimately to express a relationship between the rate constant k and G, it is necessary to examine the equation of motion for the density operator. In the Heisenberg representation [*cf.* Part I, equation (b4c)], the operator equation of motion for the density operator is equation (17), where $[A,B]$ is the quantum-mechanical commutator bracket: $[A,B] = AB - BA$.

$$i\hbar\dot{\rho}(t) = [H_T, \rho(t)] \tag{17}$$

Assume that for $t = 0$, *i.e.* at the beginning of the experiment, the system is in a state of chemical equilibrium. Thus, the equilibrium density operator is defined by equation (18). The complete density operator applies to the system in a non-equilibrium state in the presence of a force, the affinity. A measure of the distance of the system from equilibrium is afforded by the relationship (19). If the system is

$$\boldsymbol{\rho}_0 = \exp(-\beta H_0)/\text{tr}[\exp(-\beta H_0)] \tag{18}$$

$$\boldsymbol{\rho}_1(t) = \boldsymbol{\rho}(t) - \boldsymbol{\rho}_0 \tag{19}$$

$$i\hbar\dot{\boldsymbol{\rho}}_1(t) \doteq [H,\boldsymbol{\rho}_1(t)] + [H_F,\boldsymbol{\rho}_0] \tag{20}$$

near to the equilibrium state, it is sufficient for the purpose of determining the rate of reaction only to consider terms that are linear in the affinity. Therefore, if we assume that $\boldsymbol{\rho}_1(t)$ depends linearly on a, then the equation of motion can be written as equation (20). This equation of motion has the following formal solution:[30] Define the new density operator quantity $\boldsymbol{\rho}'(t)$ by equation (21). By means of a direct differentiation, we find that equations (22) are valid, where equation (20)

$$\boldsymbol{\rho}'(t) = \exp(iHt/\hbar)\boldsymbol{\rho}_1\exp(-iHt/\hbar) \tag{21}$$

$$i\hbar\dot{\boldsymbol{\rho}}'(t) = -[H,\boldsymbol{\rho}_1(t)] + i\hbar\exp(iHt/\hbar)\dot{\boldsymbol{\rho}}_1(t)\exp(-iHt/\hbar) \tag{22a}$$

$$= \exp(iHt/\hbar)[H_F,\boldsymbol{\rho}_0]\exp(-iHt/\hbar) \tag{22b}$$

has been used in the transition from the first to the second line. This expression now is integrated directly, to yield equation (23). Finally, with the use of the definition of

$$\boldsymbol{\rho}'(t) = -\frac{i}{\hbar}\int_0^t d\tau \, \exp(iH\tau/\hbar)[H_F,\boldsymbol{\rho}_0]\exp(-iH\tau/\hbar) \tag{23}$$

equation (21), equation (23) can be re-expressed in terms of the density operator $\boldsymbol{\rho}_1(t)$, which depends on the affinity only to the first order. This re-expression is equation (24).

$$\boldsymbol{\rho}_1(t) = -\frac{i}{\hbar}\int_0^t d\tau \, \exp[-iH(t-\tau)/\hbar][H_F,\boldsymbol{\rho}_0]\exp[iH(t-\tau)/\hbar] \tag{24}$$

It is now possible to obtain an explicit expression for the chemical flux. In order to simplify the notation, as well as the subsequent manipulations, it is convenient to introduce the Kubo identity[2] in equation (24), to produce equation (25). Thus, $\boldsymbol{\rho}_1(t)$

$$\frac{i}{\hbar}[N,\boldsymbol{\rho}_0] = \boldsymbol{\rho}_0\int_0^\beta d\lambda \, \exp(\lambda H)\dot{N}\exp(-\lambda H) \tag{25}$$

assumes the form of equation (24a), and the flux, defined in terms of the trace of the flux operator $\dot{N}$, $cf.$ equation (15), is given by equation (26). Since this expression is

$$\boldsymbol{\rho}_1(t) = -\int_0^t d\tau \int_0^\beta d\lambda \, \exp[-iH(t-\tau)/\hbar]\boldsymbol{\rho}_0\dot{N}(i-\hbar\lambda)a \, \exp[iH(t-\tau)/\hbar] \tag{24a}$$

$$\mathcal{T} = -\int_0^t d\tau \int_0^\beta d\lambda \, \text{tr}\{\boldsymbol{\rho}_0 \exp(iH\tau/\hbar)\dot{N}(-t-i\hbar\lambda) \exp(-iH\tau/\hbar)\dot{N}(0)\}a(t-\tau) \tag{26}$$

[30] D. ter Haar, in 'Fluctuation, Relaxation and Resonance in Magnetic Systems', ed. D. ter Haar, Oliver and Boyd, Edinburgh, 1962, pp. 109–117.

the convolution of the affinity and another factor, we can now identify that factor as $G(\varepsilon)$ [equation (27)], where the brackets $\langle \ldots \rangle$ imply the trace operation.

$$G(\varepsilon) = \int_0^\infty dt\, \exp(-i\varepsilon t) \int_0^\beta d\lambda \langle \dot{\mathbf{N}}(-t-i\hbar\lambda)\dot{\mathbf{N}}(0)\rangle \tag{27}$$

It now remains to determine the relationship between G and the rate constant k. Equation (13), and equivalently equation (15), expresses the flux, or rate, of chemical reaction in the presence of an applied force, the affinity. The rate of reaction can also be expressed in terms of the deviation $\Delta N(t)$ of the system from equilibrium; this is the familiar form of the chemical rate, *viz.* equation (28). This

$$d\Delta N(t)/dt = -k\Delta N(t) \tag{28}$$

expression specifically gives the rate of chemical relaxation to a state of equilibrium. The initial conditions for the two different flux expressions differ. The initial condition for equation (13) is chemical equilibrium. The presence of the generalized field, the affinity, induces a system response away from chemical equilibrium. On the other hand, the initial condition for equation (28) is a non-equilibrium one.

If one assumes that all the system degrees of freedom, except the affinity, are in equilibrium states, then an expression for the density operator for this situation reads as equation (29). This expression can be linearized with respect to the affinity and the difference ΔN to give equation (30). It is possible to identify a density operator component $\boldsymbol{\rho}_F$, as defined by equation (31), and now $\boldsymbol{\rho}_F$ appears as the cause of

$$\boldsymbol{\rho}_1 = \exp\{-\beta[\mathbf{H}+a(t)\mathbf{N}]\}/\mathrm{Tr}(\exp\{-\beta[\mathbf{H}+a(t)\mathbf{N}]\}) \tag{29}$$

$$\boldsymbol{\rho}_1 = \boldsymbol{\rho}_0 + \boldsymbol{\rho}_0 a(t)\int_0^\beta d\lambda\, \exp(\lambda\mathbf{H})[\mathbf{N}-\langle\mathbf{N}\rangle]\exp(-\lambda\mathbf{H}) \tag{30}$$

$$\boldsymbol{\rho}_F = \boldsymbol{\rho}_0 a(t)\int_0^\beta d\lambda\, \exp(\lambda\mathbf{H})[\mathbf{N}-\langle\mathbf{N}\rangle]\exp(-\lambda\mathbf{H}) \tag{31}$$

the chemical disturbance, whereas before it was $\mathbf{H}_F$. This density quantity $\boldsymbol{\rho}_F$ does not carry a chemical flux. It does, however, determine the quantity ΔN. Consequently, it is possible to write equation (32) or (33). A comparison of the Fourier transform of this equation with the transform of equation (13) allows one to write equation (34), which is the relationship between k and G that was sought.

$$\Delta N(t) = \mathrm{tr}\{\boldsymbol{\rho}_F\mathbf{N}\} = \int_0^\beta d\lambda[\langle\mathbf{N}(-i\hbar\lambda)\mathbf{N}(0)\rangle - \langle\mathbf{N}\rangle^2] \tag{32}$$

$$d\Delta N/dt = -k\int_0^\beta d\lambda[\langle\mathbf{N}(-i\hbar\lambda)\mathbf{N}(0)\rangle - \langle\mathbf{N}\rangle^2] \tag{33}$$

$$k = G\left[\int_0^\beta d\lambda\{\langle\mathbf{N}(-i\hbar\lambda)\mathbf{N}(0)\rangle - \langle\mathbf{N}^2\rangle\}\right]^{-1} \tag{34}$$

Equation (34) is reduced to a final useful form by means of the following observations. With particular reference to the electrochemical systems, the quantity $\mathbf{N}$ is a measure of the number of transfer systems in a prescribed initial state. In other words, at the start of the experiment the electron-donor and -acceptor species are specified. The relative numbers of these species are fixed, and the experiment is allowed to begin. The number operator N commutes with the Hamiltonian operator to the lowest order. Therefore, the integration in equation (34) can be carried out, with the result that one obtains equation (35). [As is shortly to be discussed, the

$$k = G/[\beta \langle N_1 \rangle (1 - \langle N_1 \rangle)] \tag{35}$$

occupation-number operator N is idempotent, *i.e.* $N^2 = N$, for the system as defined. When NN operates on the initial state vector, the result is the same as N operating on the same state. This behaviour of the operator N is a consequence of the fermion characteristics of the state vectors associated with the transfer electron.]

The rate constant k, however, is the sum of the forward and reverse rate constants: $k = k_1 + k_2$. At equilibrium, equation (36) is valid. Moreover, if $\langle N_1 \rangle$ and $\langle N_2 \rangle$ are the mole fractions of electrons in the initial (donor) and final (acceptor) states, then equation (37) holds. As a result, the final form of the rate constant for

$$k_1 \langle N_1 \rangle = k_2 \langle N_2 \rangle \tag{36}$$

$$\langle N_1 \rangle + \langle N_2 \rangle = 1 \tag{37}$$

the forward reaction, k_1, can be written as equation (38). This expression is used in the remaining discussions of the electron-transfer reaction in this section.

$$k_1 = G(\varepsilon = 0)/\beta \langle N_1 \rangle \tag{38}$$

Activated-complex Theory.—It is possible to show,[3] starting with suitable assumptions and definitions of the reacting system, that the activated-complex theory due to Marcus[5] is contained as a limit in the general formulation presented in the last subsection. It is also possible to obtain the quantum-mechanical Levich–Dogonadze theory as another limit.[4] This demonstration is considered in the next subsection.

Equation (38) is fundamental to the analysis. In order to obtain explicit expressions for the rate constant, it is necessary to make a number of specifications of the physical and chemical characteristics of the model system. In particular, it is necessary to specify the system Hamiltonian operator for the initial and final states, as well as the appropriate operators which couple these two states.

Kestner, Logan, and Jortner[31] have presented a rigorous and general analysis of the Born–Oppenheimer separation, based on Holstein's methods,[32] of the electron-transfer system. Their analysis is considered in the next section. For our purposes, in this section it is sufficient to assume that such an analysis of the system Hamiltonian operator and associated eigenstates has been made. Therefore, it is possible (i) to specify the reaction system, and (ii) to specify the effective Hamiltonian operator that is applicable to that system and sufficient for the subsequent mathematical manipulation.

The chemical reaction can be specified as:

$$D_A{}^e + A_B = A_A + D_B{}^e$$

where the symbols D and A mean donor and acceptor respectively, and the

[31] N. R. Kestner, J. Logan, and J. Jortner, *J. Phys. Chem.*, 1974, **78**, 2148.
[32] T. Holstein, *Ann. Phys.* (*N.Y.*), 1959, **8**, 325, 343.

subscripts A and B indicate the distinct molecular species. Thus, for example, for the reaction:

$$Fe^{III}* + Fe^{II} = Fe^{II}* + Fe^{III}$$

where Fe* implies isotope labelling, $Fe^{III}*$ and Fe^{III} on the left- and right-hand sides are the acceptor species for the forward and reverse reactions. We can specify a general Hamiltonian operator of the form shown in equation (39), in which $\mathbf{p}_i$ is

$$H = \sum_i \mathbf{p}_i^2/2m_e + \sum_n \mathbf{P}_n^2/2M_n + U(r,R) \qquad (39)$$

the momentum operator for the electron i and $\mathbf{P}_n$ is the momentum operator for the atomic particle n. The summations include, in principle, all electrons and atoms. $U(r,R)$ is the total potential-energy operator. This last quantity U must, of necessity, contain all the pertinent interactions between the electrons and their atomic and molecular cores, between the electrons and other atoms and molecules, and between ionic and molecular species in solution. At this point, it is sufficient to claim that, by means of the 'appropriate' analyses (*e.g.* the Born–Oppenheimer separation or an equivalent treatment), the complete, formal Hamiltonian operator of equation (39) can be transformed into an expression of the form (40).

$$H = (V_a^{(o)} + J_a\mathbf{a}^\dagger\mathbf{a}) + (V_b^{(o)} + J_b\mathbf{b}^\dagger\mathbf{b}) + \sum_n \mathbf{P}_n^2/2M_n + C(R)\mathbf{a}^\dagger\mathbf{b} + C^\dagger(R)\mathbf{a}\mathbf{b}^\dagger \qquad (40)$$

The various terms in this expression have the following meanings: the operators $\mathbf{a}$ and $\mathbf{a}^\dagger$ are annihilation and creation operators which operate on a state function $\Psi_A = |A,N\rangle$ associated with the species at site A. N is the state of the species at A; if $N = 1$, the A-species is an electron donor; if $N = 0$, it is an acceptor. The occupation-number operator $N = \mathbf{a}^\dagger\mathbf{a}$ has the following effect on $|A,N\rangle$: for a donor species, $N|A,1\rangle = 1|A,1\rangle$; for an acceptor, $N|A,0\rangle = 0|A,0\rangle$. A similar operator $\mathbf{b}^\dagger\mathbf{b}$ operates on the donor and acceptor states of the B-species.

It should be noted at this point that the effective operator of equation (40) is not renormalized, or diagonalized, with respect to the environmental degrees of freedom. An example in which renormalization is specifically included is considered for the Levich–Dogonadze theory in the next subsection. The fact that this effective Hamiltonian operator is used, so to speak, 'as is' has implications for the treatment of adiabatic reactions. This subject is considered shortly.

The quantities $V^{(o)}$ and $V^{(r)}$ are the complete potential-energy functions for the species in the oxidized and reduced states. Thus, the generalized heavy-centre kinetic-energy operators $\mathbf{P}_n^2/2M_n$, together with the remaining terms in the various V's, specify the degrees of freedom of the first co-ordination or solvation sphere, as well as the polarization states of the solvent. In this general form of representation, however, the specification of the response of the solvent system to the solute species does not necessarily have to be regarded as a polarization response. This general formulation, in principle, can be regarded as a complete many-body representation with which it is possible to derive a polarization representation and response. In practice, of course, the polarization response is currently the only effective means by which practical expressions for the rate constant can be derived.

The quantities J in equation (40) are defined by equations (41). The energies e_i are the vacuum energies of the transfer electron in a donor state of the i-species. The definition of these J quantities can be seen simply as a device to associate the appropriate system potential-energy function V with the state selected by the occupation-number operators $\mathbf{a}^\dagger\mathbf{a}$ and $\mathbf{b}^\dagger\mathbf{b}$.

$$J_a = e_a + V_a^{(r)} - V_a^{(o)} \tag{41a}$$

$$J_b = e_b + V_b^{(r)} - V_b^{(o)} \tag{41b}$$

Finally, the operator $\mathbf{C}(R)\mathbf{a}^\dagger\mathbf{b}$, and its conjugate $\mathbf{C}^\dagger(R)\mathbf{a}\mathbf{b}^\dagger$, represent the terms which can couple the initial and final states of the system. The specific form of the quantity $\mathbf{C}(R)$, which depends on the electronic matrix elements of the electrostatic operators, can be determined with the use of the detailed terms in the Hamiltonian operator together with the Born–Oppenheimer separation. This matter was considered in Part I, and is taken up again in a later section of this chapter.

An alternative, equivalent form of the rate-constant expression is given by equation (42), and it is this expression that we use to determine the specific form of the rate constant for the activated-complex theory.

$$k = \int_{-\infty}^{\infty} dt \langle \dot{\mathbf{N}}(0)\dot{\mathbf{N}}(t) \rangle \tag{42}$$

By concentrating our attention only on one donor species, and ignoring all but the nearest-neighbour electron transfers, the occupation-number operator needed to evaluate the rate-constant expression is that given in equation (43). The equation of motion for this operator [equation (44)] is found with the use of equation (40). This quantity is used in equation (42).

$$\mathbf{N} = \mathbf{a}^\dagger\mathbf{a} \tag{43}$$

$$i\hbar\dot{\mathbf{N}} = \mathbf{C}(R)\mathbf{a}^\dagger\mathbf{b} + \mathbf{C}^\dagger(R)\mathbf{a}\mathbf{b}^\dagger \tag{44}$$

At this point, by recalling the details of the derivation of the Yamamoto[1] rate-constant expression in the last subsection, it is possible to make partial contact with the Marcus treatment of the generalized activated state for the electron-transfer reaction. In the discussion of equation (28), and its comparison with equation (13), it was noted that the typical chemical rate expression represents the relaxation of the system to an equilibrium state, governed by the fact that only the number of molecules in the system deviates from equilibrium, and all other degrees of freedom are in their equilibrium states. Marcus, in his formulation of the generalized activated-state reaction-rate theory, stipulates the following.[5] In the region of the intersection of the initial and final potential-energy surfaces, the distribution of systems is an equilibrium one. The only aspect of the activated-state theory which admits a non-equilibrium distribution is again that aspect related to the distribution of the number of molecules. That is, as with the Yamamoto theory, and with the Marcus theory as well, only the number of molecules in the system is a non-equilibrium number. All other degrees of freedom are in their equilibrium states.

The derivation of the Marcus activated-complex expression for the rate constant in a non-adiabatic limit is now given. Following this discussion, the general case is stated, but not considered in detail. This form of presentation succinctly illustrates the problems associated with any attempt to formulate a general rate theory that is capable of simultaneously expressing the adiabatic or non-adiabatic limits, as well as any limit in between, as they apply to a given situation.

Some of the analysis of the following paragraphs depends upon the use of quantum-mechanical operator algebra for which the ordering of operators is important. The techniques are discussed in a number of texts; *cf.* ref. 85.

The rate constant is written as defined in equation (45), where H is the complete Hamiltonian operator. With the use of the operator $\dot{N}$, as given by equation (44), one can derive equation (46). To the lowest order in H′, the replacement of H by H_0

$$k_1 = \int_{-\infty}^{\infty} dt \langle \dot{N}(0) \exp(iHt/\hbar) \dot{N}(0) \exp(-iHt/\hbar) \rangle \tag{45}$$

$$k_1 = \frac{1}{\hbar^2} \int_{-\infty}^{\infty} dt \langle C(R) a^\dagger b \exp(iHt/\hbar) C^\dagger(R) a b^\dagger \exp(-iHt/\hbar) \rangle \tag{46}$$

in the exponential operator function leads to equation (47). A further reduction of

$$\mathrm{tr}\{\rho_0 [C(R) a^\dagger b \exp(iH_0 t/\hbar) C^\dagger(R) a b^\dagger \exp(-iH_0 t/\hbar)\}$$
$$= \sum_i \langle i | \rho_0 C(R) a^\dagger b \exp(iH_0 t/\hbar) C^\dagger(R) a b^\dagger \exp(-iH_0 t/\hbar) | i \rangle \tag{47}$$

this expression depends on the use of the (assumed) completeness relation (48).

$$\sum_f | f \rangle \langle f | = 1 \tag{48}$$

Thus, one can derive equation (49), where now the initial- and final-state environ-

$$\sum_{i,f} \langle i | \rho_0 C(R) a^\dagger b \exp(iH_0 t/\hbar) | f \rangle \langle f | C^\dagger(R) a b^\dagger \exp(-iH_0 t/\hbar) | i \rangle$$
$$= \mathrm{tr}\{\rho_0 [C(R) \exp(iH_0^f t/\hbar) C^\dagger(R) \exp(-iH_0^i t/\hbar)]\} \exp[i(e_b - e_a)t/\hbar] \tag{49}$$

mental mode and co-ordination-shell Hamiltonian operators are defined by equations (50), and e_a and e_b have the same definition as given earlier; *cf.* equation

$$H_0^i = \sum P_n^2 / 2M_n + V_a^{(r)} + V_b^{(o)} \tag{50a}$$

$$H_0^f = \sum P_n^2 / 2M_n + V_a^{(o)} + V_b^{(r)} \tag{50b}$$

(41). The steps in the sequence of operations which lead to equation (49) depend upon the use of the operator relations (51) and the relationships (52). These last relationships follow as a consequence of the properties of the Fermi operators $a^\dagger$ and b. The same result is obtained for $\langle f | ab^\dagger | i \rangle$.

$$\exp(iH_0 t/\hbar) | i \rangle = \exp(iE_0^i t/\hbar) | i \rangle \tag{51a}$$

$$H_0 | i \rangle = E_0^i | i \rangle \tag{51b}$$

$$\langle i, a^\dagger b | f \rangle = \langle 1_a, 0_b | a^\dagger b | 0_a, 1_b \rangle \tag{52a}$$

$$= \langle 1_a, 0_b | b | 1_a, 1_b \rangle \tag{52b}$$

$$= \langle 1_a, 0_b | 1_a, 0_b \rangle = 1 \tag{52c}$$

At this point, we re-define the initial- and final-state Hamiltonian operators as shown in the equations (53), where the potential-energy components are defined by equations (54). The definition of the above potential-energy functions brings this treatment into a form consistent with Marcus's theory.[5]

$$H_i = \sum P_n^2/2M_n + U_r \tag{53a}$$

$$H_f = \sum P_n^2/2M_n + U_p \tag{53b}$$

$$U_r = e_a + V_a^{(r)} + V_b^{(o)} \tag{54a}$$

$$U_p = e_b + V_a^{(o)} + V_b^{(r)} \tag{54b}$$

The rate-constant expression now has the form of equation (55). At this point,

$$k_1 = \hbar^{-2}\int_{-\infty}^{\infty} dt \langle \mathbf{C}(R)\exp(i H_f t/\hbar)\mathbf{C}^\dagger(R)\exp(-i H_i t/\hbar)\rangle \tag{55}$$

the classical limit is taken. Thus, in effect, the commutator $[H_i, H_f]$ can be ignored, $\mathbf{CC}^\dagger = \mathbf{CC}^* = |\mathbf{C}|^2$ [we write simply $\mathbf{C}^2(R)$], and classical averaging techniques can be used. Thus, one may derive equation (56), in which Z is the partition function, which is given by equation (57), and $d(P)d(R)$ is the volume element in phase space.

$$k_1 = \frac{1}{Z\hbar^2}\int_{-\infty}^{\infty} dt \,\langle \exp[it(U_p - U_r)/\hbar]\mathbf{C}^2(R)\rangle_{cl} \tag{56}$$

$$Z = \iint d(P)d(R)\exp(-\beta H_i) \tag{57}$$

Note that in equation (56) the time integration involving the exponential function defines the Dirac delta function, with the argument $U_p - U_r$. Therefore, we can write equation (58). The potential-energy functions contained as arguments of the

$$k_1 = \frac{2\pi}{Z\hbar}\iint d(P)d(R)\mathbf{C}^2(R)\delta(U_p - U_r)\exp(-\beta H_i) \tag{58}$$

delta function depend upon the co-ordinate variables R. As a result, it is possible to use a well-known property of the delta function to write the rate constant in the form of equation (59). The dependence of the delta function is now explicit in R.

$$k_1 = \frac{2\pi}{Z\hbar}\iint d(P)d(R)\exp(-\beta H_i)\mathbf{C}^2(R)[\partial(U_p - U_r)/\partial R]^{-1}\delta(R_p - R_r) \tag{59}$$

The integration involving the delta function can be carried out directly. The result of the integration restricts the rate constant to an expression which gives the probability that the system crosses the reaction hypersurface S on which $U_p = U_r$ everywhere. Specifically, equation (60) is valid, where dS is an element of area on

$$k_1 = \frac{2\pi}{Z\hbar}\iint d(P)dS\,\mathbf{C}^2(R')[\partial(U_p - U_r)/\partial R']\exp[-\beta H_i(P, R')] \tag{60}$$

the hypersurface. This result can be expressed specifically in terms consistent with Marcus's analysis. In particular, with the observation that the generalized co-ordinate q^r is that co-ordinate which is constant on the reaction hypersurface, it is zero there, the rate-constant equation (60) assumes the form of equation (60a), in which ρ is now the equilibrium phase-space density, defined in equation (61). The quantity $\dot{q}^r$ [defined in equation (62)] can be seen to be an effective velocity. If it is

$$k = \iint d(P)dS\rho\{2\pi \mathbf{C}^2(R)/\hbar[\partial(U_p - U_r)/\partial R]_S\} \tag{60a}$$

$$\rho = Z^{-1}\exp(-\beta H_i) \tag{61}$$

$$2\pi \mathbf{C}^2(R)/\hbar[\partial(U_p - U_r)/\partial R]_S \doteq \dot{q}^r \tag{62}$$

possible, which generally it is not, to assume that the quantity (62) depends linearly on $\dot{q}^r$, then the subsequent analysis required to obtain the explicit form of the rate-constant expression is the same as that done by Marcus.[5]

If we assume that, in general, the effective velocity term defined in equation (62) is independent of any direct dependence on q^r, then it is possible to carry out the remainder of the analysis, leading to the final form of the rate constant. The method of analysis is identical to that of Marcus. We write the rate constant as shown in equation (63), where Q is the configurational integral for the system in the initial equilibrium state, and where the left-hand term in equation (64) defines a net force

$$k_1 = \frac{2\pi}{\hbar Q}\int dR'[\mathbf{C}^2(R)/(F_p - F_r)]\exp(-\beta U) \tag{63}$$

$$[\partial(U_p - U_r)/\partial R]_S = F_p - F_r \tag{64}$$

acting on the system in the region of the intersection of the initial and final system state potential-energy surfaces. Marcus has discussed the method whereby the element of surface, dS, on the reaction hypersurface is factored into internal and external activated-complex components.[5] The external co-ordinate components are those associated with translation and rotation of the activated complex. The integral defining the rate constant, in equation (63), therefore is the one shown in equation (65). The integral involving dR is related to the external contributions; the integral

$$k_1 = \frac{2}{\hbar Q}\int_R dR\ R^\alpha \int_{S'} dS'\exp(-\beta U)(m^\neq)^{-\frac{1}{2}} \tag{65}$$

involving dS' deals with the internal contributions. The factor α is 0 or 2, depending upon whether the rate constant applies to unimolecular (0), heterogeneous (0), or bimolecular (2) reactions. $m^\neq$ is an effective mass, which generally is dependent on S. The further evaluation of equation (65) depends on the use of a relationship which can be defined between the integral (66) and the volume integral (67) over V', the volume for the internal co-ordinates.

$$\int_{S'} dS'(m^\neq)^{-\frac{1}{2}}\exp(-\beta U) \tag{66}$$

$$\int_{V'} dV'\exp(-\beta U^*) \tag{67}$$

The potential-energy function U^* is defined by equation (68), where m is a Lagrange multiplier, which must be determined for a particular system of interest. The origin

$$U^* = U_r + m(U_r - U_p) \tag{68}$$

of m is in the consideration of the distribution of non-equilibrium states in the region of the hypersurface. The determination of this term is analogous to the determination of a similar term in Marcus's treatment of the electron transfer in a continuum polar solvent; *cf.* ref. 16 and Part I of this survey.

In order to evaluate the volume integral it is necessary to introduce an auxiliary function defined by the expression (69), where $\mathbf{a}^{rr}$ is the tensor conjugate to an element a_{rr} in the line element of the many-dimensional configuration space. The volume element dV' itself is related to dS' and dq^r by equation (70). Therefore, we can write the relationship (71), and this relation holds because $U^* = U_r$ on S. If

$$\exp[-I(q^r,R)] = \int dS'(\mathbf{a}^{rr})^{-\frac{1}{2}} \exp(-\beta U^*) \tag{69}$$

$$dV' = (\mathbf{a}^{rr})^{-\frac{1}{2}} dS' dq^r \tag{70}$$

$$\int_{V'} dV' \exp(-\beta U^*) = \int dq^r \exp[-I(q^r,R)] \tag{71}$$

the function $I(q^r,R)$ is expanded in a Taylor series to terms that are second order in q^r, the above integral involving q^r can be evaluated, with the results shown in equation (72) [the details of the substantiation of this step are given by Marcus[5]]. $\mathbf{g}^{rr}$ is

$$\int_{V'} dV' \exp(-\beta U^*) = \langle (\mathbf{g}^{rr})^{\frac{1}{2}} \rangle \frac{\sqrt{\pi}}{[2I''(0,R)]^{\frac{1}{2}}} \int_{S'} dS' \exp(-\beta U) \tag{72}$$

conjugate to g_{rr}, which is analogous to a_{rr} for the mass-weighted velocity space. The substitution of the integral (66) in (72) is possible because, at the point $q^r = 0$, $U^* = U_r$ on S' as indicated.

The volume integral involving U^* is just the configurational free energy of a system with a potential U^*, as defined by equation (73). Hence, the surface integral

$$\int_{V'} dV' \exp(-\beta U^*) = \exp[-\beta F^*(R)] \tag{73}$$

$$\int_{S'} dS' \exp(-\beta U) = \left(\frac{2I''(0,R)}{\pi}\right)^{\frac{1}{2}} \exp[-\beta F^*(R)] \langle (\mathbf{g}^{rr})^{\frac{1}{2}} \rangle \tag{74}$$

(66) has the value shown in equation (74). The factor $I''(0,R)$, *i.e.* the second derivative of I with respect to q^r, evaluated at $q^r = 0$, is related to the mean-square deviation of the perpendicular distance s from the hypersurface, $\langle (\delta s)^2 \rangle$. As a result, it is possible to transform equation (74) into (75). With the use of the Laplace method, the rate constant is obtained by replacing the integrand over R by the expression (76), and in this expression R is now the optimum value for which the (assumed Gaussian) integral is a maximum. The ratio ρ is defined by equation (77), and it has an order of magnitude of unity.

$$\{(m^*)^{-\frac{1}{2}} / [\langle (\delta s)^2 \rangle]^{\frac{1}{2}}\} \exp[-\beta F^*(R)] \tag{75}$$

$$k_1 = \frac{2\pi}{\hbar Q} \frac{\mathbf{C}^2(R)}{F_p - F_r} \rho R^\alpha (m^*)^{-\frac{1}{2}} \exp[-\beta F^*(R)] \tag{76}$$

$$\rho = [\langle(\delta R)^2\rangle/\langle(\delta s)^2\rangle]^{\frac{1}{2}} \quad (77)$$

This rate-constant quantity can be re-expressed in terms of the free energy of activation, $\Delta F^{\neq}(R)$ [see equation (78)]; $F^r(R)$ is the configurational free energy of the reactants as defined in equation (79). The work required to bring the reactants

$$\Delta F^{\neq}(R) = F^*(R) - F^r(R) \quad (78)$$

$$F^r(R) = \beta^{-1}\ln Q \quad (79)$$

from infinity to R is $w^r = F^r(R) - F^r$. Therefore, k_1 can be defined as shown in equation (80). It should be recalled that F_p and F_r are forces defined in terms of the potential-energy functions U_p and U_r, and are not free-energy quantities.

$$k_1 = \frac{2^{3/2}}{\sqrt{\pi}\hbar} \frac{\mathbf{C}^2(R)}{F_r - F_p} \rho R^\alpha \exp(-\beta w^r)\exp[-\beta\Delta F^{\neq}] \quad (80)$$

The final form of the rate constant, given by equation (80), is an expression in the non-adiabatic limit. It differs considerably from Marcus's non-adiabatic expressions,[5] and it is necessary to examine this treatment and compare it with his to see the nature of the various sets of assumptions and approximations made.

Before presenting the details of this comparison, however, it is worth noting the physical implications of the several separate operations performed in the mathematical analysis leading to equation (80).

The separation of variables that is used to isolate the internal degrees of freedom of the activated complex from the translations and rotations of the activated complex as a whole, is familiar to all the treatments of the electron-transfer reaction. Marcus's approach[5] is very general. That this separation marks an extension of the simple outer-sphere electron-transfer reaction theory is easily seen. If this treatment is specialized to the consideration of the simple outer-sphere electron transfer, as in the Marcus and the original Levich–Dogonadze treatments, then the only 'internal' degrees of freedom which participate in the activated complex are those associated with the polar modes of the solvent. The activated complex, therefore, consists of the transfer electron, in a suitably defined state, together with the states of the solvent which respond to changes of state of the transfer electron. This facet of the activated complex is assumed implicitly in the Born–Oppenheimer method of analysis, as applied to the system of the transfer electron and its solvation environment.

By relaxing the restrictions of the simple outer-sphere model with the use of a much more general, and inclusive, method of analysis, it is possible also to consider intramolecular degrees of freedom in the activated complex. That is, by considering as 'internal' co-ordinates all those co-ordinates not directly related to the translations and rotations of the activated complex, one automatically has included a provision to account for the contributions due to intramolecular vibrations, *etc.* This treatment, therefore, presents the problem of accounting for changes in the intramolecular degrees of freedom in terms of the evaluation of their contribution to the free energy of activation. In principle, the evaluation of the free energy of activation can be carried out with the use of classical statistical mechanics. If, for example, the polar-solvent model is used, then the familiar activation-energy contributions which

The Theory of Electron-transfer Reactions in Polar Media

arise from the solvent repolarization appear, as well as new contributions associated with the deformations of the molecular topography during the course of the electron transfer.

There are several steps in the Marcus treatment[5] which appear to restrict the generality of it. As indicated above, if one notes that the non-adiabatic quantity (62) depends linearly on $\dot{q}^r$, then the substitution shown in equation (81) can be made.

$$\frac{2\pi}{\hbar}\frac{C^2(R)}{\partial(U_p - U_r)/\partial R} = \kappa \dot{q}^r \tag{81}$$

The integration over $\dot{q}^r$, related to p^r, followed by the configuration-space integrations identical to those shown above, leads to Marcus's results,[5] provided one additional assumption is made. That assumption is that the quantity κ is represented as an average [equation (82)], which generally is only dependent on R. The

$$\kappa = \frac{2\pi}{\hbar}\langle C^2(R)/\partial(U_p - U_r)/\partial R \rangle_{\text{velocity}} \tag{82}$$

average indicated by equation (82) involves the same hypersurface co-ordinates as the subsequent analysis, which leads to the final form of the rate constant. The assumption specifically is that it is acceptable to assume that the internal co-ordinate-dependence of the transmission coefficient is effectively vanishing. This assumption was made in the analysis presented previously in this section. There is no reason to assume, however, that this assumption is generally valid, although, on the basis of many investigations of vibronic interactions in many molecular systems, this seems to be reasonably justified. It is, in fact, a direct manifestation of the Condon approximation. To the degree that this approximation is valid, this particular aspect of the Marcus theory[5] or the analysis presented above is general.

Another aspect of the Marcus treatment seems more limiting. If one compares the analysis presented above with Marcus' treatment, it is seen that the Marcus treatment does not seem to admit a legitimate adiabatic limit. The treatment presented here is manifestly non-adiabatic. If one assumes an explicit non-adiabatic-limit form for κ in Marcus's treatment, a similar result to ours is found. That result is an acceptable non-adiabatic limit. Conversely, if one takes the adiabatic limit for κ, i.e. $\kappa = 1$, it appears that no provision has been made to account for the fact that, in the adiabatic limit, the free energy of activation depends sensitively on the degree of separation of the two adiabatic potential-energy surfaces. That is, in the avoided-crossing region, the initial and final reaction potential-energy surfaces split. The magnitude of the splitting depends upon the interaction quantity $C(R)$. Marcus[5] has not indicated an explicit dependence of $\Delta F^{\neq}(R)$ on the surface-splitting in the avoided-crossing region. However, a measure of the splitting is implicit in his analysis. It is assumed in this type of analysis that the reaction potential-energy surface is known in considerable degree. In particular, a Born–Oppenheimer analysis defines such an adiabatic reaction surface. The maximum height of the surface in the region associated with the activated complex is measured with respect to the ground state of the reactants. This height reflects the degree of splitting encountered. Thus, in principle, one can subtract a quantity f^* from the formally defined intersection of the potential-energy surfaces, and this result equals the activation free energy. Christov[33] has formulated a quantum-mechanical theory of reaction rates in which this true barrier height is specifically displayed.

In a number of instances, certainly for the gaseous phase, we are aware of the shapes of a number of potential-energy surfaces for relatively simple reacting systems. By extension, we may assume that the activation free energy for an electrochemical or electron-transfer reaction should be lowered by an energy quantity that is roughly the order of magnitude of the interaction term $C(R)$ evaluated at the surface col. It must be borne in mind, however, that, apart from the relatively few analyses carried out on very restricted, simple systems,[24] no general, rigorous prescription exists for the proper account of the effect of splitting on the activation energy, as determined entirely within the framework of a statistical-mechanical rate theory.

As indicated, Christov,[33] in an elaborate (but nevertheless *ad hoc*) treatment of kinetics in general and electron transfer in particular,[34] has introduced specific account of the height of the barrier in the intersection region. His analyses, which are generally useful for approaching an answer to the questions concerning the extent of tunnelling, non-adiabaticity, *etc.*, are based on the use of a general energy quantity which is equal to the degree of lowering, due to splitting, of the adiabatic potential-energy surface with respect to the energy defined by the surface crossing. However, it is still not possible rigorously to define this energy quantity (for an electron-transfer system) and to assign a magnitude to it.

The manner by which Marcus[5] and Christov[33] introduce account of non-adiabaticity, through κ, into the general statistical-mechanical treatment of reaction rates is again adequate, but not rigorous. Both Marcus and Christov define probability functions for the reaction system in a given energy state, with a given total energy, and in a definite volume element in phase space. This probability function does not explicitly or implicitly contain any information or factor expressing transmission. This factor must be appended as a suitably defined and evaluated transmission coefficient, κ.[35] On the other hand, it is clear from the linear-response theory, even as handled here in the non-adiabatic limit, that this theory (in principle) contains sufficient inherent information to obtain both the degree of splitting and the proper adiabatic limit. To see that this is the case, it is sufficient to note equation (46). The total Hamiltonian operator H may be partitioned into a zero'th-order contribution H_0 and an interaction term H'. The exponential operator-dependent functions in equation (46) may be 'disentangled' with the use of the U-matrix [equation(83)],[36,37] where $U(t_1, t_2)$ is defined as shown in equation (84), and where T is the time-ordering operator. H'(s) denotes the interaction representation shown in equation (85). The time-ordered exponential operator in equation (84) is in fact a short-hand notation for a complete, iterated series representation of the solution to equation (86). As a result, it represents a complete, but nevertheless formal, solution to the complete perturbation series. For certain systems it has proven possible to sum the

[33] S. G. Christov, *Ber. Bunsengesellschaft phys. Chem.*, 1972, **76**, 507; 1974, **78**, 537.
[34] S. G. Christov, *Ber Bunsengesellschaft phys. Chem.*, 1975, **79**, 357.
[35] R. A. Marcus, *J. Chem. Phys.*, 1964, **41**, 2614.
[36] R. P. Feynman, *Phys. Rev.*, 1951, **84**, 108.
[37] P. Roman, 'Advanced Quantum Theory', Addison-Wesley, Reading, Massachusetts, 1965.

$$\exp[-i(H_0 + H')t/\hbar] = \exp(-iH_0 t/\hbar)U(t) \tag{83}$$

$$U(t_1, t_2) = U(t_1)U^\dagger(t_2) = T \exp\left[-i \int_{t_2}^{t_1} ds H'(s)\right] \tag{84}$$

$$H'(s) = \exp(iH_0 s/\hbar)H' \exp(-isH_0/\hbar) \tag{85}$$

$$\frac{\hbar}{i}\frac{\partial \Psi(t)}{\partial t} + H'\Psi = 0 \tag{86}$$

complete series for certain classes of perturbation interaction.[38] However, except for the model system considered by Dogonadze and Kuznetsov,[24] this has not been done in general for the electron-transfer system.

In concluding this subsection on the general treatment of chemical reactions on the basis of the linear-response theory, it is necessary to point out that the linear-response theory is not the only approach to a complete theory which may be used. In fact, proper statistical averaging of the complete transition probability derived from the quantum-mechanical S-matrix yields identical, and identically rigorous, results. Thus, for example, Dogonadze and Kuznetsov[24] start with the general expression of equation (87) to obtain a form for the transition probability expressed

$$w_{if} = \frac{2\pi}{\hbar} \sum_{\alpha,\alpha'} \exp[\beta(F_i - E_{i\alpha})] \left| \langle f\alpha' | \sum_{l=0}^{\infty} [V_f G_i(E_{i\alpha}) V_i G_f(E_{i\alpha})]^l \right.$$

$$\left. \times \{V_i + V_f G_i(E_{i\alpha}) V_i\} | i\alpha \rangle \right|^2 \delta(E_{i\alpha} - E_{f\alpha'}) \tag{87}$$

as equation (88), where α in equation (88) is related to the slope of the potential-energy surface of the initial state in the region of the intersection of the initial and final states. $G(E)$ in (87) is a Green function, defined by equation (89), where $H_{i(f)}$ is the Hamiltonian operator of the medium and δ is a small convergence factor.

$$w_{if}(R) = \omega_{eff} \exp\{-\beta E^* - 2\beta |V_{if}|[\alpha(1-\alpha)]^{\frac{1}{2}}\}/2\pi \tag{88}$$

$$G_{(if)}(E_{i\alpha}) = -\beta \int_0^{i\infty} d\tau \exp[\beta\tau(E_{i\alpha} - H_{i(f)} + i\delta)] \tag{89}$$

The Linear-response Analysis of the Levich–Dogonadze Theory.—In the last subsection it was shown that the Marcus[5] theory may be regarded as being contained within the general linear-response theory of chemical reaction rates. In this subsection it is shown that the Levich–Dogonadze theory[6] also is contained in this general representation.[39] In effecting this analysis it is possible further to refine the essentially mathematical differences between the Marcus and Levich–Dogonadze approaches. This treatment also serves to provide a basis for the demonstration of the limitation of the Marcus theory to handle highly exothermic and endothermic reaction limits.

The discussion presented in the following paragraphs builds on the discussion of the Levich–Dogonadze theory given in Part I, Section 7.

---

[38] A. L. Fetter and J. D. Walecka, 'Quantum Theory of Many-Particle Systems', McGraw-Hill Book Co., New York, 1971.
[39] P. P. Schmidt, *J. Phys. Chem.*, 1974, **78**, 1684.

The Hamiltonian operator used for the analysis of the Levich–Dogonadze theory is essentially the same as that used in the examination of the Marcus theory; *cf.* equation (40). The difference now is that the specific continuum polarization response model of the solvent modes is used in place of the non-specific, but nevertheless general, solvent medium representation of equation (40).

With reference to Part I, Section 7, the specification of the Born–Oppenheimer adiabatic Hamiltonian operator for the solvent modes develops as follows. Considering two reacting species A^{M+} and B^{N+}, and the reaction:

$$A^{M+} + B^{N+} = A^{(M+1)+} + B^{(N-1)+}$$

it is clear that each charged reactant and product species makes a contribution to the polarization response in the initial and final electron-transfer states. That is, associated with each charged species is an induction field D for the ion in solution. In some instances this induction field for the complete charge distribution can be related to the field G associated with the same distribution in vacuum.[40] In terms of the simple polar solution model[18] and in terms of the refined treatment of the polar continuum model which includes account of spatial and temporal dispersion,[40—44] it is possible to associate a generalized Boson co-ordinate variable with the state of polarization of the solvent, *viz.* q_k. The equilibrium values for the co-ordinates $\bar{q}_k$ are related to the charge distributions of the ions in the initial and final states.[40—44] As a consequence, it is possible to write the model system Hamiltonian operator in the form of equation (90). The quantities ε_a and ε_b are the energies of the electron in

$$H = \varepsilon_a a^\dagger a + \varepsilon_b b^\dagger b + \frac{\hbar\omega}{2}\sum_k [p_k^2 + q_k^2](a^\dagger a + b^\dagger b)$$
$$+ \hbar\omega \sum_k [a^\dagger a(\bar{q}_{ak}^{(r)} + \bar{q}_{bk}^{(0)}) + b^\dagger b(\bar{q}_{ak}^{(0)} + \bar{q}_{bk}^{(r)})]q_k$$
$$+ C(R)a^\dagger b + C^\dagger(R)ab^\dagger \qquad (90)$$

the donor state of species A and B respectively; the acceptor state, *i.e.* an unoccupied orbital, is specified as the zero of energy for each species. As before, the quantities $C(R)a^\dagger b$ and $C^\dagger(R)ab^\dagger$ specify the interaction operators responsible for the electron transfer which takes the system from the initial to the final state.

The Hamiltonian operator (90) is simply rearranged to equation (91), where $\bar{\varepsilon}_a$ is defined by equation (92), and there is a similar term for $\bar{\varepsilon}_b$. The quantity E_s

$$H = \bar{\varepsilon}_a a^\dagger a + \bar{\varepsilon}_b b^\dagger b + (\hbar\omega/2)\sum_k [p_k^2 + (q_k + \bar{q}_{ak}^{(r)} + \bar{q}_{bk}^{(0)})^2]$$
$$+ (\hbar\omega/2)\sum_k [p_k^2 + (q_k + \bar{q}_{ak}^{(0)} + \bar{q}_{bk}^{(r)})^2] + C(R)a^\dagger b + C^\dagger(R)ab^\dagger \quad (91)$$

[equation (93)] is simply the polarization self energy of the charge distribution in

---

[40] Yu. I Kharkats, A. A. Kornyshev, and M. A. Vorotyntsev, *J.C.S. Faraday II*, 1976, **72**, 361.
[41] R. R. Dogonadze, A. M. Kuznetsov, and V. G. Levich, *Doklady Akad. Nauk S.S.S.R.*, 1969, **188**, 383.
[42] R. R. Dogonadze and A. M. Kuznetsov, *Elektrokhimiya*, 1971, **7**, 763.
[43] R. R. Dogonadze and A. A. Kornyshev, *Phys. Status Solidi (b)*, 1972, **53**, 439; (correction) 1973, **55**, 843.
[44] R. R. Dogonadze and A. A. Kornyshev, *J.C.S. Faraday II*, 1974, **70**, 1121.

$$\bar{\varepsilon}_a = \varepsilon_a - (\hbar\omega/2)\sum_k (\bar{q}_{ak}^{(r)} + \bar{q}_{bk}^{(0)})^2 \tag{92}$$

$$-(\hbar\omega/2)\sum_k (\bar{q}_{ak}^{(r)} + \bar{q}_{bk}^{(0)})^2 = E_s \tag{93}$$

the initial state. In practice, this quantity generally is referred to as the free energy of solvation; for the charges represented as charged metallic spheres, *i.e.* the Born model,[45] the above quantity reduces to three terms, shown in equation (94), where

$$E_s = -\frac{Z_A^2 e^2}{2r_A}(1-1/\varepsilon_s) - \frac{Z_B^2 e^2}{2r_B}(1-1/\varepsilon_s) - \frac{Z_A Z_B e^2}{R}(1-1/\varepsilon_s) \tag{94}$$

ε_s is the static dielectric constant for the polar solvent and R is the distance separating the species A and B. The quantities ε_a and ε_b, therefore, represent the energies of the solvated transfer electron that is bound, respectively, to the species A and B, together with the free energy of solvation due to the acceptor species and the interaction energy for the two species.

A further simplification of the Hamiltonian operator (91) is possible, but by no means essential. It is possible to consider the unitary transformation of equation (95), which leaves the effect of the Hamiltonian operator unaltered, but which does

$$\mathcal{H} = \exp(iS) H \exp(-iS) \tag{95}$$

eliminate the linear solvent mode displacement terms in equation (90), or equivalently, equation (91). The operator S is given by equation (96). The effect of the

$$\mathbf{S} = -\mathbf{a}^\dagger \mathbf{a} \sum_k (\bar{q}_{ak}^{(r)} + \bar{q}_{bk}^{(0)}) p_k - \mathbf{b}^\dagger \mathbf{b} \sum_k (\bar{q}_{ak}^{(0)} + \bar{q}_{bk}^{(r)}) p_k \tag{96}$$

unitary transformation (95) is to give the operator $\mathcal{H}$ [equation (97)], where $\mathbf{B}(i)$

$$\mathcal{H} = \bar{\varepsilon}_a \mathbf{a}^\dagger \mathbf{a} + \bar{\varepsilon}_b \mathbf{b}^\dagger \mathbf{b} + (\hbar\omega/2)(\mathbf{a}^\dagger \mathbf{a} + \mathbf{b}^\dagger \mathbf{b}) \sum_k [p_k^2 + q_k^2]$$

$$+ \mathbf{C}(R)\mathbf{a}^\dagger \mathbf{b} \mathbf{B}^\dagger(i)\mathbf{B}(f) + \mathbf{C}^\dagger(R)\mathbf{a} \mathbf{b}^\dagger \mathbf{B}(i)\mathbf{B}^\dagger(f) \tag{97}$$

and $\mathbf{B}(f)$ are defined by equations (98) and (99).

$$\mathbf{B}(i) = \exp\left[-i\sum_k (\bar{q}_{ak}^{(r)} + \bar{q}_{bk}^{(0)}) p_k\right] \tag{98}$$

$$\mathbf{B}(f) = \exp\left[-i\sum_k (\bar{q}_{ak}^{(0)} + \bar{q}_{bk}^{(r)}) p_k\right] \tag{99}$$

At this point, we determine the equation of motion for the N operator analogous to equation (44), obtaining equation (100). Consequently, it is now possible to

$$i\hbar \dot{\mathbf{N}} = \mathbf{C}(R)\mathbf{a}^\dagger \mathbf{b} \mathbf{B}(i)^\dagger \mathbf{B}(f) + \mathbf{C}^\dagger(R)\mathbf{a}\mathbf{b}^\dagger \mathbf{B}(i)\mathbf{B}(f)^\dagger \tag{100}$$

evaluate the expression for the rate constant [equation (45)]. The result can be expressed in the form of equation (101), the quantity ζ_k being defined as shown in

---

[45] M. Born, *Z. Physik*, 1920, **1**, 45.

$$k = \int_{-\infty}^{\infty} dt \, \langle \mathbf{C}(R)\mathbf{a}^{\dagger}\mathbf{b} \, \exp[i\sum_{k}\zeta_k p_k]\exp[i\mathcal{H}t/\hbar]\mathbf{C}^{\dagger}(R)\mathbf{a}\mathbf{b}^{\dagger}$$
$$\times \exp[-i\sum_{k}\zeta_k p_k]\exp[-i\mathcal{H}t/\hbar]\rangle \tag{101}$$

equation (102). The operations involving the Fermion operators, **a**, **b**, *etc.*, are not easily carried out in equation (101) without further simplification. It should be

$$\zeta_k = \bar{q}_{ak}^{(0)} - \bar{q}_{ak}^{(r)} + \bar{q}_{bk}^{(r)} - \bar{q}_{bk}^{(0)} \tag{102}$$

noted that equation (101) is exact for this model system representation in the same sense as equation (46) is exact for the Marcus-type activated-complex-theory treatment. In effect, equation (101) represents the complete summation of all terms of the total perturbation series expansion of the interaction-dependent transition probability. Here, as before, it is possible (in principle) to disentangle the operator-dependent arguments of the exponential functions in equation (101), with the result that the explicit form of the perturbation series representation can be displayed. At this time, however, even for this model system representation, this cannot be carried out satisfactorily except with the use of additional restrictions and assumptions. The model treatment due to Dogonadze and Kuznetsov,[24] previously mentioned, is an example of the complete summation of the series for a simple model system.

The progress which can be made with equation (101) depends upon the use of the same assumptions used previously in our consideration of the non-adiabatic limit of the activated-complex theory. That is, to the lowest order in the interaction, the Hamiltonian operator may be decomposed into two terms [see equation (103)], where H_s is simply defined as shown in equation (104). The interaction term is

$$\mathcal{H} = \mathbf{a}^{\dagger}\mathbf{a}\bar{\varepsilon}_a + \mathbf{b}^{\dagger}\mathbf{b}\bar{\varepsilon}_b + (\mathbf{a}^{\dagger}\mathbf{a} + \mathbf{b}^{\dagger}\mathbf{b})H_s + H_{\text{int}} \tag{103}$$

$$H_s = (\hbar\omega/2)\sum_{k}[p_k^2 + q_k^2] \tag{104}$$

easily identified. Consequently, if H_{int} in equation (103) is ignored in the exponential parts of equation (101), the rate-constant expression can be expressed as shown in

$$k = \int_{-\infty}^{\infty} dt \, \langle \mathbf{C}(R) \exp[i\sum_{k}\zeta_k p_k(0)]\mathbf{C}^{\dagger}(R) \exp[-i\sum_{k}\zeta_k p_k(t)]\rangle$$
$$\times \exp[i(\bar{\varepsilon}_b - \bar{\varepsilon}_a)t/\hbar] \tag{105}$$

equation (105), where the dependence of $p_k(t)$ upon time is expressed in the Heisenberg representation as shown in equation (106).

$$p_k(t) = \exp[iH_s t/\hbar]p_k(0)\exp[-iH_s t/\hbar] \tag{106}$$

There is now virtually no difference between the expression for the rate constant (105) and the Levich–Dogonadze expression;[6] *cf.* equation (234) in Part I. We have demonstrated, therefore, that, just as the non-adiabatic limit for the activated-complex theory is embedded in the general linear-response expression for the rate constant, the same is true for the non-adiabatic limit of the Levich–Dogonadze theory. We found the non-adiabatic limit for the Marcus activated-complex theory by using assumptions equivalent to those made by Marcus.[5] Consequently, the result we obtained is general for the non-adiabatic limit to the same degree as Marcus's original treatment is general. In the preceding linear-response analysis of the simple Levich–Dogonadze theory the generality of the final expression for the rate constant is limited to the same extent that the original Levich–Dogonadze theory was limited. It is essential to note, however, that the analysis outlined above is in no way limited to the simple model system representation propounded originally by Levich and Dogonadze.[6] Recent work by Dogonadze, especially, has generalized the original approach of Levich and Dogonadze to a state of fairly great usefulness. (Much of this recent work by Dogonadze and his colleagues is considered in the next section.) It is possible to effect the above analysis for any of the recent, refined treatments, and hence there is in fact no essential restriction implied or imposed. The analysis presented above used the original Levich–Dogonadze model-system Hamiltonian operator merely for illustrative purposes.

It would appear that the conclusions reached in this section contribute little which is new and enlightening or which may be useful. Certainly, it has been amply demonstrated that by beginning with a sophisticated and general approach involving linear-response theory we can obtain limits already found by other, less tedious and involved means. This view, however, is harsh, and moreover obscures the aim of treatments of this type. We seek some form of general analysis capable of applying to and yielding results (*i.e.* rate constants) for any given system subject to a collection of constraints of any order of magnitude. That it is not yet possible to realize our goal of obtaining some form of omnibus rate constant should not deter us from the quest. We have pointed out that there are at least two independent and equally valid approaches to the determination of general forms of rate-constant expressions for any given system: the linear response theory, as formulated for reaction systems by Yamamoto,[1] and the S-matrix theory of quantum scattering, suitably statistically averaged, as has been illustrated by Dogonadze and Kuznetsov.[24] We have ascertained that these general approaches yield the familiar non-adiabatic limits. Moreover, we have determined that there are some limitations on the generality of the statistical-mechanical activated-complex theory; limitations implicit with the *ad hoc* accounting for several of the contributions to the magnitude of the rate constant, *viz.* the velocity-weighted transition probability. It is evident from Dogonadze and Kuznetsov's treatment of a model two-level system[24] that, when it does become possible to handle more general representations of the electron-transfer system, these analyses, based either on the S-matrix methods or the linear-response analysis, will automatically yield expressions for the activation energy and the pre-exponential factors which adequately account for such factors as the degree of adiabaticity and the degree of splitting of the adiabatic reaction potential-energy surfaces. At present, many of these quantities can be estimated with the use of separate treatments. However, our level of confidence in, for example, a Landau–Zener treatment of the transmission coefficient must be tempered by the observation that the analysis applies itself to an ideal model system. Progress has been made in recent attempts to generalize the Landau–Zener theory.[46] Nevertheless, it remains essentially suggestive.

[46] M. S. Child, 'Molecular Collision Theory,' Academic Press, London, 1974.

On the other hand, Levich,[47] in his review article, and Kestner, Logan, and Jortner[31] have argued that the non-adiabatic limit may apply more generally to more reaction systems than hitherto has been suspected. If this should be the case, we are in a very fortunate position. As the remaining sections in this survey illustrate, the substantive recent advances in the theory of electron transfer and electrochemical reactions have been found to lie entirely within the framework of the non-adiabatic limit as expressed by the first-order time-dependent perturbation-theoretic transition probability. The search for an adequate representation of the rate constant for any limit of adiabaticity, however, cannot be abandoned. Very fast reactions, *e.g.* proton- and electron-transfer reactions, may operate in the adiabatic limit. Moreover, many of these reactions are highly endo- or exo-thermic. As we shall show, the existing framework of the equilibrium statistical-mechanical treatment of reaction rates, *e.g.* the activated-complex theory, incorrectly expresses the activation energy through an incorrect choice of saddle point in the complex time integration as expressed by equation (45). The quantum-mechanically based non-adiabatic theories, on the other hand, yield accurate expressions in these regions of excess reaction energy. It is clear that the accurate analysis of any reaction system must be based at least on the assumption of the most general mathematical representation of the rate and rate constant.

3 Generalizations of the Polar Continuum Theory of Outer-sphere Reactions: the Consideration of the First Solvation or Co-ordination Shell

Refinements in the Model Representation and Associated Hamiltonian Operator.—In the last section it was indicated that Marcus, in his general treatment of the activated-complex theory,[5] published in 1965, considered the contribution to the activation free energy of the degrees of freedom associated with the inner solvation or co-ordination species. That treatment was general, and for the time, vague. It was clear then, as it is now, that if an adequate normal-mode analysis exists for the vibrations of the species of the inner solvation shell, then it is possible to evaluate their contribution to the activation free energy. This can be done certainly in the harmonic-oscillator limit for the molecular vibrations. It can be done as well for more accurate potential-energy functions, as, for example, the Morse potential. In spite of the assumed generality implicit in the 1965 Marcus treatment, it is now clear that, for purposes of actual calculation, it is necessary to specify very carefully the physical and chemical 'nature' of the system, and carefully to effect the decomposition of the degrees of freedom of the system into accurate normal-mode contributions. Ultimately, however, it is more important to be able to assess the magnitudes of the various contributions to the activation process in terms of their relative energies. The reason for the importance of this assessment process lies entirely with the problem of choosing the proper saddle point in the time integration leading to the final form of the transition probability or rate constant.

In many respects, *i.e.* for many reaction systems and limits, Marcus's treatment, vague and unfulfillable in 1964, is now adequate and applicable. The advent of the systematic use of high-speed digital computers over the last twelve years has muted

---

[47] V. G. Levich, *Adv. Electrochem. Electrochem. Engineering*, 1966, **4**, 249.

the difficulties originally apparent in many formal treatments. As a consequence, we are in a position now to be able to test many more theoretical expressions.

Recent examinations of the electron-transfer reaction have sought numerical estimation of the activation energy and the rate constant.[7,22,31] Attempts have been made to compare theoretical values with experimental ones.[7,31] Although the match is not yet complete, agreement between theory and experiment is becoming closer. There is good reason for confidence in our eventual ability to be able to dissect the electron-transfer components and to understand their various interactions and operations.

In this first subsection of this general section, we examine the refinements of the simple outer-sphere model of the electron-transfer system needed to proceed to the consideration of more complicated processes. The refinements which have been made, and which have been generally successful, continue to employ the polar-continuum representation of the dielectric solvent in which the transfer species are dissolved. The polar continuum model itself has undergone a considerable refinement in the past six years,[40—44] and a brief account of these refinements is given in the next subsection. It is anticipated that a detailed account of the polar solvent will be given by Dogonadze and Kornyshev in another chapter of these Reports.

Kestner, Logan, and Jortner[31] have presented an analysis of the realistic electron-transfer system based on the application of Holsteins' analysis of the small polaron system.[32] It is useful, therefore, to summarize the Kestner, Logan, and Jortner (KLJ) discussion as a basis for further consideration.

Let the complete Hamiltonian operator for the system (which includes the donor and acceptor centres, the transfer electron, the solvent molecules, the co-ordination shell, and all remaining electrons) be given by equation (107), where H_{ea} and H_{eb}

$$\mathscr{H} = H_{ea} + V_{eb} + T_n = H_{eb} + V_{ea} + T_n \tag{107}$$

$$H_{ea} = T_e + V_{ea} + H_a + H_b + V_{ab} + H_s + H_c + V_{int}^s + V_{int}^c \tag{108}$$

$$H_{eb} = T_e + V_{eb} + H_a + H_b + V_{ab} + H_s + H_c + V_{int}^s + V_{int}^c \tag{109}$$

are defined by equations (108) and (109), respectively. The various terms have the following meanings:

T_e = kinetic-energy operator for the transfer electron
T_n = kinetic-energy operator for all atoms, *i.e.* heavy centres
H_a, H_b = Hamiltonians for the non-transfer electrons in the bare oxidized ions $A^{(N+1)+}$ and B^{M+}
V_{ea}, V_{eb} = interaction energies operating between the transfer electron and the donor cores
V_{ab} = electrostatic interaction operating between the ion cores
H_s = Hamiltonian operator for the solvent molecules
H_c = Hamiltonian operator for the molecules of the first co-ordination shell
V_{int}^s = ionic (total) interaction energy operating between the ion and transfer electron and the solvent
V_{int}^c = ionic (total) interaction energy operating between the ion and transfer electron and the molecules of the first co-ordination shell.

The two equivalent complete expressions for the total Hamiltonian operator given by equation (107) enable one to focus attention on states for which the transfer electron is localized respectively on the centre A or B.

The representation given by equation (107), together with all the defined terms, provides a formal, and in principle complete, basis for the subsequent Born–Oppenheimer-type analytical separation of the degrees of freedom of the transfer electron from the remaining degrees of freedom associated with the solvent and the first co-ordination shell. In addition, the analysis is basically sufficiently complete to provide an account of radiationless transitions within the donor species to excited vibronic states without the passage of the transfer electron to the acceptor state. Transitions from excited donor states to the acceptor can also be considered. Hence, generally, the Hamiltonian operator (107), together with the remaining analysis, to be illustrated in the following paragraphs, provides a complete and flexible method of approach to the consideration of any electron-transfer transition.

We shall see, however, in the subsequent consideration of this general treatment, that once the general expression for the electron-transfer transition-probability expression has been obtained, its specific evaluation requires care. In particular, it is necessary to effect a further separation of the degrees of freedom of the first co-ordination shell from the degrees of freedom of the solvent. This separation is required in order accurately to determine the various system contributions to the activation energy, *etc.* In a number of treatments this part of the analysis has been inadequately carried out. We shall return to this matter shortly.

It should be noted that, in the method of factoring the complete Hamiltonian operator for the system, certain classes of contributions have been grouped together for convenience in the subsequent analysis. Thus, for example, the terms H_s and H_c, the operators for molecules of the solvent and co-ordination shell, respectively, contain potential-energy operator terms which, through the medium of the Born–Oppenheimer analysis, eventually yield specific interaction quantities operating between the molecules of the co-ordination shell and the remainder of the solvent. Thus, in this initial part of the analysis, the vibrational degrees of freedom of the co-ordination shell and the solvent are found to be coupled. This coupling is important, and it must be dealt with before it is possible accurately to determine the final form of the transition-probability expression for the transfer.

Let r be the set of all system electrons, $r = \{r_i, i = \text{all electrons}\}$. Let Q be the set of all heavy-centre (atomic) co-ordinates: $Q = \{Q_n, n = \text{all system atoms}\}$. Then, we can define the eigenvalue equations (110) (following the conventional methods of the Born–Oppenheimer analysis).[31] The wavefunction Ψ_{ai} is an element of the

$$H_{ea}\Psi_{ai}(r,Q) = \varepsilon_{ai}(Q)\Psi_{ai}(r,Q) \tag{110a}$$

$$H_{eb}\Psi_{bf}(r,Q) = \varepsilon_{bf}(Q)\Psi_{bf}(r,Q) \tag{110b}$$

complete orthonormal set $\{\Psi_{ai}\}$. The elements of this set represent all the electronic states of the total system for the case in which the transfer electron is localized on centre A; specifically, the reaction state $A^{N+} + B^{M+}$. Similarly, the wavefunction Ψ_{bf} is an element of the complete set of functions for the state $A^{(N-1)+} + B^{(M+1)+}$. As Kestner, Logan, and Jortner point out,[31] the elements of either one of these complete sets of basis functions are sufficient to form the basis of a Born–Oppenheimer expansion of the total system wavefunction. However, the usual approach is to use elements of both sets (which are not mutually orthogonal). This approach is chemically intuitive, in that it recognizes the local character of the initial and final reaction states. It has the advantage of requiring a smaller number of elements of either of the two complete sets $\{\Psi_{ai}\}$ and $\{\Psi_{bf}\}$; that is, the use only of $\{\Psi_{ai}\}$, for example, requires the consideration of a large number of basis functions (including states of the continuum) in order to describe the final state in which the electron is localized on the B centre. As a consequence of these considerations, therefore, if the complete wavefunction is given by equation (111), where the index α spans both ai and bf, equation

$$\overline{\Psi}(r,Q,t) = \sum_\alpha \chi_\alpha(Q,t)\Psi_\alpha(r,Q) \qquad (111)$$

(107) can be used to derive an equation for the expansion coefficients $\chi_a(Q,t)$. The wavefunction (111) satisfies the complete, time-dependent Schrödinger equation (112). Consequently, by rewriting equation (107) as equation (113), where α,β = ai, bf, we find the relationship (114). Multiplying this equation on the left by $\Psi_\beta^*(r,Q_n)$,

$$\mathcal{H}\overline{\Psi}(r,Q,t) - i\hbar\partial\overline{\Psi}/\partial t = 0 \qquad (112)$$

$$\mathcal{H} = \mathbf{T}_n + \mathbf{H}_{e\alpha} + \mathbf{V}_{e\beta} \qquad (113)$$

$$\sum_\alpha [(\mathbf{T}_n + \mathbf{H}_{e\alpha} + \mathbf{V}_{e\beta})\chi_\alpha(Q,t)\Psi_\alpha(r,Q) - i\hbar\Psi_\alpha\partial\chi_\alpha/\partial t] = 0 \qquad (114)$$

and integrating over all system electronic co-ordinates, gives equation (115). The

$$\sum_\alpha S_{\beta\alpha}[\mathbf{T}_n + \varepsilon_\alpha(Q) - i\hbar\partial/\partial t]\chi_\alpha(Q,t) + \sum_\alpha \bigg(\langle\Psi_\beta|\mathbf{V}_{e\beta}|\Psi_\alpha\rangle$$

$$- \hbar^2\langle\Psi_\beta|2\partial/\partial Q|\Psi_\alpha\rangle\frac{\partial\chi_\alpha}{\partial Q} - \hbar^2\langle\Psi_\beta|\partial^2/\partial Q^2|\Psi_\alpha\rangle\chi_\alpha \bigg) = 0 \qquad (115)$$

last two terms represent the Born–Oppenheimer 'break-down' operators which express the coupling between the electronic and nuclear degrees of freedom. For the general electron transfer, for which the dependence of the electronic degrees of freedom on the solvent modes as well as the vibrational modes of the first co-ordination shell is weak, generally it is justifiable to ignore the last two terms in equation (115).[6] On the other hand, it is possible, and perhaps likely more often than generally suspected, that for some electron-transfer transitions which can take place in reasonably well defined, fairly tightly bound electron-transfer encounter complexes, these non-adiabatic operators may play an important role in the rate of the reaction.[48] When one is justified in ignoring the non-adiabatic operators [*i.e.* the last two terms in equation (115)], the remaining non-adiabatic contribution depends on the electronic matrix elements of the electrostatic operator $\mathbf{V}_{e\alpha}$. These matrix elements generally have been assumed to take the entire responsibility for the determination of the magnitude of the interaction between the initial and final states.[6,31]

In equation (115), $S_{\beta\alpha}$ is an element of the overlap matrix. Following Holstein,[32] KLJ[31] define the inverse element $S_{\beta\alpha}^{-1}$ such that equation (116) is valid. With the use

$$\sum_\gamma S_{\beta\gamma}^{-1} S_{\gamma\alpha} = \delta_{\beta\alpha} \qquad (116)$$

of this relationship, it is possible to write equation (115) in the form of equation

[48] P. P. Schmidt, *Z. Naturforsch.*, 1974, **29a**, 880.

(117), in which the diagonal elements have been separated from the non-diagonal

$$[\mathbf{T}_n + \varepsilon_\gamma(\mathbf{Q}) + \langle\Psi_\gamma|\mathbf{V}_{e\gamma}|\Psi_\gamma\rangle + \sum_\beta S_{\gamma\beta}^{-1}\langle\Psi_\beta|\mathbf{V}_{e\beta}|\Psi_\gamma\rangle - i\hbar\partial/\partial t]\chi_\gamma$$
$$= -(\sum_{\substack{\alpha \\ \alpha \neq \gamma}} \langle\Psi_\gamma|\mathbf{V}_{e\alpha}|\Psi_\alpha\rangle + \sum_{\alpha \neq \gamma}\sum_{\beta \neq \gamma} S_{\gamma\beta}^{-1}\langle\Psi_\beta|\mathbf{V}_{e\alpha}|\Psi_\alpha\rangle)\chi_\alpha \qquad (117)$$

elements and grouped together on the left-hand side of the equation. The equation (117) is complete in the sense that all electronic transitions are included. That is, in addition to the electron-transfer transition which takes an electron from the donor to the acceptor centre, all intramolecular electronic transitions are included. Thus, for example, it is possible to use equation (117) to describe an electron-transfer reaction in which the transfer electron occupies an excited vibronic state of the donor species. For the moment, however, we shall not consider this possibility, although it is of interest. Instead, we concentrate on the development of an accurate representation of the simple electron transfer which involves the electronic ground states of the donor and acceptor species.

For the simpler case of the transfer of an electron between electronic ground states, it is necessary only to consider a representative electronic two-level system. Thus, the further specialization of equation (117) leads to the expression (118). The

$$[\mathbf{T}_n + \varepsilon_a(\mathbf{Q}) + \langle\Psi_a|\mathbf{V}_{eb}|\Psi_a\rangle + S_{ab}^{-1}\langle\Psi_b|\mathbf{V}_{eb}|\Psi_a\rangle - i\hbar\partial/\partial t]\chi_a$$
$$= -(\langle\Psi_a|\mathbf{V}_{eb}|\Psi_b\rangle + S_{ab}^{-1}\langle\Psi_b|\mathbf{V}_{eb}|\Psi_b\rangle)\chi_b \qquad (118)$$

adiabatic vibrational wavefunctions $\chi_a(\mathbf{Q})$ are found as solutions to equation (119) and a similar equation for χ_{bv}.

$$[\mathbf{T}_n + \varepsilon_a(\mathbf{Q}) + \langle\Psi_a|\mathbf{V}_{eb}|\Psi_a\rangle + S_{ab}^{-1}\langle\Psi_b|\mathbf{V}_{eb}|\Psi_a\rangle - E_{av}]\chi_{av} = 0 \qquad (119)$$

The vibrational wavefunctions expressed as shown in equation (120) can be used to obtain the expression for the transition probability by means of the usual perturbation-theory methods used to determine the quantities $C_{\alpha v}(t)$ to the various orders in the relevant interaction.[31]

$$\chi_\alpha(\mathbf{Q},t) = \sum_v \mathbf{C}_{\alpha v}(t)\chi_{\alpha v}(\mathbf{Q})\exp(-iE_{\alpha v}t/\hbar) \qquad (120)$$

The Schrödinger equation for the vibrational wavefunctions in the adiabatic limit, (119), is investigated in more detail in the following sections. The analysis presented above, due to Kestner, Logan, and Jortner,[31] presents a general equation for the determination of the various solvent and transfer electron contributions to the rate constant for electron transfer, expressed through the non-adiabatic transition probability. In the next subsection we examine the sophisticated treatment of the polar solvent due to Dogonadze, Kornyshev, and Kuznetsov,[40—44] together with some of the recent results of the Reporter,[49—54] to determine an accurate representation of the solvate-co-ordinated electron-transfer system.

[49] P. P. Schmidt and J. M. McKinley, *J.C.S. Faraday II*, 1976, **72**, 143.
[50] P. P. Schmidt, *J.C.S. Faraday II*, 1976, **72**, 171.
[51] P. P. Schmidt, *J.C.S. Faraday II*, 1976, **72**, 1048.
[52] P. P. Schmidt, *J.C.S. Faraday II*, 1976, **72**, 1061.
[53] P. P. Schmidt, *J.C.S. Faraday II*, 1977, **73**, in the press.
[54] P. P. Schmidt, *J.C.S. Faraday II*, 1977, **73**, in the press.

The Polar-solvent Model and Models of the Source Charge Densities.—The first part of the following discussion deals with the recent contributions to the theory of the polar solvent due to Dogonadze and his colleagues.[40—44] The second part deals with recent attempts[49—54] to formulate realistic, accurate, and computationally feasible model representations of the ionic and molecular charge densities associated with the electroactive species that are involved in electron-transfer reactions. In the next subsection, the application of the general treatments of the solvent system and the source charge densities to the calculation of the various contributions to the activation energy is considered.

The simple model of the polar solvent has been described extensively in the literature dealing with the simple electron transfer.[5,6] A summary of the simple theory is given in Part I. This simple polar-solvent model is characterized, usually, by a single frequency ω_0 associated primarily with the librational modes of the molecules in the medium. Furthermore, and perhaps more importantly, the same representation gives no consideration to spatial dispersion in the solvent medium. Specifically, the spatially non-disperse representation implies only the operation of instantaneous local interactions between external charges and the induced polarization charge density. The situation may be characterized in terms of the Coulomb interaction V between two charged species A and B, with charges $Z_A e$ and $Z_B e$, in a medium with a static dielectric constant ε_s, as shown in equation (121) Similarly, the self energy, or free energy of solvation, of either of the charges individually is given by the Born formula (122),[45] where a is the radius of a metallic sphere taken as a representation (model) of the ion A.

$$V = Z_A Z_B e^2 / \varepsilon_s R_{AB} \tag{121}$$

$$w_s = -Z_A^2 e^2 (1 - 1/\varepsilon_s)/2a \tag{122}$$

The local representation of the polar medium is summarized also in terms of the interaction energy operating between an external charge density and the induced polarization charge of the medium. For a point charge located at R, we have equation (123), where the value of $4\pi P(r)$ is defined as shown in equation (124). Thus, the interaction energy depends upon the value of the potential associated with the polarization P at R.

$$-\int d^3r\, D(r,R) P(r) = \frac{1}{4\pi} \int d^3r\, \nabla\left(\frac{e}{r-R}\right) \nabla \Phi(r) = e\Phi(R) \tag{123}$$

$$4\pi P(r) = \nabla \Phi(r) \tag{124}$$

Real solvent systems, on the other hand, are characterized by a non-local character of the response of the medium to an external charge distribution. That is, if $\varepsilon_{\alpha\beta}^{-1}(r,r')$ is an element of the medium permittivity tensor (where α and β label spatial co-ordinates), then the electrostatic field at a point r is related to the induction field $D(r')$ associated with the external charge distribution by equation (125).[40] If the medium is local, then one may write equation (126).[40] If, moreover, the medium is uniform, with a constant dielectric constant, then equation (127) is valid. All of these conditions apply to the simple representation of the polar medium.

$$E_\alpha(r) = \sum_\beta \int d^3r' \varepsilon_{\alpha\beta}^{-1}(r,r') D_\beta(r') \qquad (125)$$

$$\varepsilon_{\alpha\beta}^{-1}(r,r') = \delta_{\alpha\beta}\delta(r-r')/\varepsilon(r) \qquad (126)$$

$$E(r) = D(r)/\varepsilon \qquad (127)$$

The actual polar medium, however, is spatially disperse, *i.e.* $\varepsilon_{\alpha\beta}^{-1}(r,r')$ depends on $(r-r')$. This implies a non-local response of the medium to an external charge density. More in particular, it implies that the response of the medium to the external charge density is distributed in space. This induced charge density generally decays with distance from the charge density, and the degree of decay with distance is characterized by a correlation length. Thus, the real response of the medium to any externally imposed charge density is complicated, and it can be seen that the simple representation characterized by a local response function ignores this potentially important behaviour of the medium.

The situation is even more complicated than the above discussion implies. The general solvent system is characterized by a number of responses to time-dependent electromagnetic perturbations. In particular, the solvent system exhibits a number of absorption bands associated with the absorption of light of various wavelengths. As is known from the London theory of the van der Waals forces,[55] the R^{-6} interaction between atomic and molecular systems depends on the instantaneous polarizabilities (related to the optical indices of refraction) of the individual species. Associated with the vibrational degrees of freedom in the medium is another class of polarization response, and finally, the librational modes contribute a long-wavelength, long-correlation-length response. The complete response of a polar medium to any external charge distribution, therefore, is a summation of the individual responses which can be associated with classes of degrees of freedom of the medium. The correlation lengths associated with the optical response modes of the polar medium, for example, are small, and of the order of the Bohr radius.[44] Correlation lengths associated with the i.r. bands of the solvent are of the order of an Ångstrom; the librational mode correlation lengths are of the order of 11 Å.[44] Thus, the various types of motion of the medium play important parts in the determination of the magnitude of the electrostatic interaction, depending upon the magnitude of the separation between charges. For example, Dogonadze and Kornyshev have shown[56] that for small separations between ions in a polar solvent, the form of the electrostatic interaction depends strongly upon the instantaneous, electronic modes of the medium. The interaction, therefore, is characterized by the optical dielectric constant together with the short-wavelength correlation length associated with the optical excitations of the solvent. At the opposite extreme, for large separations between charges, the optical modes, as well as the vibrational excitations of the solvent, play a much reduced role in the total interaction. In fact, for sufficiently large values of R, the classical interaction energy is recovered[55] [equation (121)].

The recent efforts of Dogonadze and his colleagues, in particular Kornyshev and Kuznetsov, to place the polar-solvent model on a firm, accurate, and useful footing have involved the specific consideration of temporal and spatial dispersion in the medium.[40—44] The results, so far, are impressive. It has been possible to obtain

[55] E. A. Power, 'Introductory Quantum Electrodynamics', Longmans, London, 1964.
[56] R. R. Dogonadze and A. A. Kornyshev, *Doklady Akad. Nauk S.S.S.R.*, 1972, **207**, 896.

marked improvements in the determination of the ionic solvation free energies both with the use of the Born ion model[44] (*i.e.* the representation of an ion as a hard, charged metallic sphere of radius a), and with the use of continuous charge densities modelled after accurate quantal charge densities.[49—54] In particular, it has proved possible to calculate solvation free energies with the use of reasonable ionic radii (*i.e.* the Goldschmidt[57] or Gourary–Adrian[58] radii) without the introduction of the familiar radial extensions characteristic of the earlier treatments.[59,60] Very recently, Dogonadze's methods have been used together with the quantum-mechanically defined charge densities for the species H_2^+, H_2, and H_2^- in an attempt to determine a quantum-mechanical basis which can enable one to relate solvation free energies, *etc.*, to atomic and molecular charge and bond densities.[53,54] The results indicate that it should be possible to establish rules of solvation based upon orbital populations and bond order. Ionic and molecular solvation free energies appear to depend upon individual bond-additivity principles much as the free energy of formation of a molecule depends upon bond additivity.

It is apparent that if these general methods prove successful for the consideration of real systems, many of the difficulties previously encountered in the general treatments of ionic and molecular solvation and solution-phase reactivity should disappear.

We now turn to the consideration of the form of the polar solvent theory due to Dogonadze *et al.*[40—44,56] This discussion is abbreviated to the extent that only those aspects of the general polar solvent theory needed to consider the electron-transfer reaction will be considered. Dogonadze and Kornyshev provide a more detailed and incisive survey in another chapter in this series.

In Part I it was shown that the simple, non-disperse theory of the polar solvent in the continuum representation yields an expression for the solvent-system Hamiltonian operator of the form (128), where the Hamiltonian density $\mathbf{H}_s(\mathbf{R})$ is given by equation (129), C is defined as shown in equation (130), n is the index of refraction,

$$\mathbf{H}_s = \int d^3R\, \mathbf{H}_s(\mathbf{R}) \tag{128}$$

$$\mathbf{H}_s = (2\pi/C)[\mathbf{P}_{ir}^2 + \omega_s^{-2}\dot{\mathbf{P}}_{ir}^2] \tag{129}$$

$$C = 1/n^2 - 1/\varepsilon_s \tag{130}$$

and ε_s the static dielectric constant. This representation of the polar solvent is a strictly local one. The expansion of the polarization vectors $\mathbf{P}$ and $\dot{\mathbf{P}}$ as Fourier series followed by the application of Bose quantization conditions leads to the (harmonic oscillator) representation given by equation (131). In this expression, as well as the previous one, ω_s is the single frequency associated with the long-wavelength librational modes of the solvent system.

$$\mathbf{H}_s = \tfrac{1}{2}\hbar\omega_s \sum_k [p_k^2 + q_k^2] \tag{131}$$

On the other hand, it is clear that the real solvent system must be characterized by several frequencies, associated with absorption bands, and must be non-local. In the absence of a detailed theory of the polar solvent, based on molecular and statistical

[57] V. M. Goldschmidt, *Skrifter Norske Videnskaps-Akad. Oslo Mat. Naturv. Kl.*, 1926, **8**, 1.
[58] B. S. Gourary and F. J. Adrian, *Solid State Physics*, 1960, **10**, 127.
[59] E. Glueckauf, *Trans. Faraday Soc.*, 1964, **60**, 572.
[60] J. S. Muirhead-Gould and K. J. Laidler, *Trans. Faraday Soc.*, 1967, **63**, 944.

mechanics, and in view of the success of the simple continuum solvent model in providing closed, complete expressions for the repolarization and solvation energies which enter the electron-transfer activation energy, Dogonadze, Kuznetsov, and Levich[41] proposed a new phenomenological treatment of the solvent. Their break with the simple continuum representation was made possible with the use of a new classical Hamiltonian operator for the medium. This operator was constructed in a manner analogous to Pitaevskii's[61] which Pitaevskii applied to the system of ⁴He.

Thus, in particular, Dogonadze, Kuznetsov, and Levich[41] assumed that the polarization state of the solvent medium is described by a polarization vector $P(r)$ together with a conjugate quantity $G(r)$ related to $P(r)$. A classical Hamiltonian operator which takes into account the non-local character of the polar medium can be written as equation (132),[41—43] where $\alpha, \beta = x, y, z$, and ν is a summation

$$H_s = \tfrac{1}{2} \sum_\nu \sum_{\alpha,\beta} \iint d^3 r d^3 r' \left[F_\nu^{\alpha\beta}(r,r') G_\nu^\alpha(r) G_\nu^\beta(r') + \Phi_\nu^{\alpha\beta}(r,r') P_\nu^\alpha(r) P_\nu^\beta(r') \right] \quad (132)$$

index which labels the elementary excitations of the solvent. The total polarization, expressed in terms of the components of the polarization vector, is given by equation (133).[43] It can be seen from the form of equation (133) that $P_\nu^\alpha(r)$ is a component

$$P^\alpha(r) = \sum_{\nu,\beta} \int d^3 r' \, b_\nu^{\alpha\beta}(r,r') P_\nu^\beta(r') \quad (133)$$

of the vector field due to an elementary excitation in the medium. The factors $b_\nu^{\alpha\beta}(r,r')$ are in effect statistical weights which are related to the oscillator strengths for the elementary excitations. The magnitude, and spatial extent, of the contribution of a particular elementary excitation of the medium to the total polarization field is expressed in the b-quantities. In equation (132) the factors F and Φ are related to the dielectric properties of the medium. Specifically, the Fourier transform of Φ, $\Phi_p(k)$, is essentially the inverse of the solvent structure factor $S(k)$; *cf.* refs. 41—43.

As with the simple continuum treatment of the polar solvent, it is useful here to perform the Fourier transformation, through the use of the Fourier series decomposition of the various quantities, of the Hamiltonian operator for the solvent medium. The result can be expressed as equation (134), with $\omega_{\nu f}^2$ defined as shown in equation (135) and $f = (k, \alpha)$.

$$H_s = \tfrac{1}{2} \sum_{\nu,f} (Q_{\nu f}^2 + \omega_{\nu f}^2 \Pi_{\nu f}^2) \quad (134)$$

$$\omega_{\nu f}^2 = \Phi_{\nu f} F_{\nu f} \quad (135)$$

Although the Hamiltonian operator (132), and its Bose harmonic-oscillator form (134), remain phenomenological, this refinement in the treatment of the polar solvent is definitely an improvement. In particular, the formulation does account explicitly for the important spatial and temporal dispersion of the solvent medium. The Hamiltonian operator (133) does not account for these properties specifically in terms of statistically weighted molecular parameters, although, in principle, this detailed type of approach could be taken. It is known, for example, that the Fourier transform of the space–time correlation function for the polarization states is related to the structure factor for the solvent medium.[41—43] The theory of the liquid

[61] L. P. Pitaevskii, *Soviet Phys. J.E.T.P.*, 1956, **31**, 536.

state does attempt to relate molecular and fluid structures to the structure factor $S(k)$.[62—64] On the other hand, as we shall see, in keeping with the phenomenological approach, it is generally sufficient to provide a reasonably accurate, physical approximation to the structure factor. Once this factor has been specified, it is possible to determine a number of important properties of the medium such as the ionic solvation free energy and the repolarization energy for an electron-transfer transition.[43]

In the absence of any external electrostatic field, the expectation value of the polarization field vanishes, viz. $\langle P(r) \rangle = 0$. However, this is not the case when foreign charge distributions, e.g. ions, are introduced into the medium. The interaction between the polarization field of the medium and the electrostatic field associated with the external charge density is given by equation (136). The use of the

$$V_{\text{int}} = -\int d^3 r\, P(r) \cdot E(r) \tag{136}$$

Fourier decomposition of the polarization vector and the electrostatic field due to the external charges allows one to determine the Hamiltonian operator for the solvent system in this case as shown in equations (137).[43] The form of the operator

$$H_s = \tfrac{1}{2} \sum_\xi [Q_\xi^2 + \omega_\xi^2 (\Pi_\xi - \bar{\Pi}_\xi)^2] - E_p \tag{137a}$$

$$\xi = (v, f) \tag{137b}$$

(137a) follows when one performs a straightforward diagonalization; cf. the last section. The energy quantity E_p is in fact the self energy of the charge distribution with respect to the polarization field. It is conveniently given in terms of a Fourier transformation representation as shown in equation (138),[43,49] where $D(k)$ is the

$$E_p = \frac{1}{4(2\pi)^4} \int d^3 k\, C_N(k)\, D^*(k) \cdot D(k) \tag{138}$$

Fourier transform of the induction field associated with the external, ionic charge distribution. It is a simple matter to show that, in terms of the Fourier transform of the external charge-density function $\rho(k)$, D is given by equation (139),[49] where $\hat{k}$ is

$$D(k) = -4\pi i (\hat{k}/k) \rho(k) \tag{139}$$

the unit radial vector in k-space (i.e. momentum space). In general, given the coordinate space charge distribution function $\rho(r)$, the Fourier transform is defined as equation (140). The Fourier transform of the induction field $D(r)$ is given by a similar expression.

$$\rho(k) = \int d^3 r\, \rho(r) \exp(i k \cdot r) \tag{140}$$

The quantity $C_N(k)$ in equation (138) has been determined by Dogonadze, Kuznetsov, and Kornyshev to have the form of equation (141),[42,43] where D_n is related to the dielectric constants associated with various of the elementary excitations of the

---

[62] P. Eglstaff, 'An Introduction to the Liquid State', Academic Press, New York, 1967.
[63] C. A. Croxton, 'Liquid State Physics – A Statistical Mechanical Introduction' Cambridge University Press, London, 1974.
[64] B. J. Berne and G. D. Harp, *Adv. Chem. Phys.*, 1970, **17**, 63.

medium [equation (142)] and $f_n(k)$ is a function of the structure factor of the solvent

$$C_N(k) = \sum_n D_n f_n(k) \qquad (141)$$

$$D_n = 1/\varepsilon_{n+1} - 1/\varepsilon_n \qquad (142)$$

medium. In particular, $f_n(k)$ is a spectral function which is related to the structure factor by equation (143), where S_0 is defined by equation (144).[43] Thus, $f_n(k)$ is a

$$f_n(k) = S(k)/S_0 \qquad (143)$$

$$S_0 = S(k=0) = \frac{4\pi}{3} \sum_{\alpha,\beta} \int dR\, R^2 S_{\alpha\beta}(R) \qquad (144)$$

dimensionless, normalized quantity. It must have the properties shown in equations (145). It is important to note at this point, and to bear in mind for the remaining

$$\lim_{k \to 0} f_n(k) = 1 \qquad (145a)$$

$$\lim_{k \to \infty} f_n(k) = 0 \qquad (145b)$$

discussion, that the function $f_n(k)$ carries the essential information about the spatial dispersion of the polarization field associated with a particular elementary excitation. The index n now labels the transparency bands which separate the various absorption bands associated with the collective excitations of the medium. Thus, if we consider the optical, vibrational, and librational degrees of freedom of the solvent, we can identify three transparency bands associated with the regions between the vacuum and optical, the optical and vibrational, and finally the vibrational and librational bands. Furthermore, associated with each of these transparency regions is a characteristic correlation length.

The particular application of the Dogonadze–Kornyshev formula for the solvation free energy,[43] equation (138), involves the following identification of parameters. The summation index in equation (138) runs from zero to two. The dielectric constants are identified as shown in equations (146), and here n is the index of refraction

$$\varepsilon_0 = \varepsilon_{\text{vac}} = 1 \qquad (146a)$$

$$\varepsilon_1 = \varepsilon_{\text{op}} = n^2 \qquad (146b)$$

$$\varepsilon_2 = \varepsilon_{\text{vib}} \qquad (146c)$$

$$\varepsilon_3 = \varepsilon_{\text{lib}} = \varepsilon_{\text{st}} \qquad (146d)$$

of the solvent. The spectral function $f_n(k)$ has a specific dependence on the correlation length, which can be expressed by writing equation (147), where λ_n is a correla-

$$f_n(k) \equiv f_n(\lambda_n k) \qquad (147)$$

tion length. Several forms of the spectral function $f_n(\lambda_n k)$ have been used in particular applications. It is not difficult to see that if the spectral function is approximated as a constant (unity), which implies a delta-functional form for the structure factor

$S(r)$, then the solvation free energy reduces to the form of equation (148). Since

$$w_s = -E_p = \frac{-1}{4(2\pi)^4}(1-1/\varepsilon_{st})\int d^3k \, D^*(k) \cdot D(k) \tag{148}$$

this expression is Parseval's theorem,[65] as also is equation (138), it is easy to see that equation (149) may be written. If the Born model of the ion is used, then equation (147) immediately gives equation (150),[44,49] where a is the ionic radius.

$$w_s = -(1-1/\varepsilon_{st})\frac{1}{8\pi}\int d^3r \, D(r) \cdot D(r) \tag{149}$$

$$w_s(\text{Born}) = -(1-1/\varepsilon_{st})Z^2e^2/2a \tag{150}$$

Forms of the spectral function which provide accounts of the degree of spatial dispersion associated with the various modes of the solvent system are shown in equation (151),[44,49] which is related to a structure factor of Gaussian form, $\exp(-r^2/\lambda_n^2)$ [see equation (152)], which is related to a simple exponential representa-

$$f_n(\lambda_n k) = \exp(-\lambda_n^2 k^2) \tag{151}$$

$$f_n(\lambda_n k) = 1/(1+\lambda_n^2 k^2)^2 \tag{152}$$

tion $\exp(-r/\lambda_n)$. A step-function approximation also has been used.[44] In general, it has been found that for most applications the Gaussian form is the most useful in that it is the most easily applied.[49—53] The use of (152) in many instances leads to complicated integrations which must be performed using most of the machinery available from molecular quantum mechanics.[49,66] The Gaussian forms, on the other hand, can be treated with great generality, and the results are generally simple to implement for purposes of computation; see, in particular, refs. 49 and 51.

The solvent-system Hamiltonian operator [equation (137)], which describes the state of the solvent in the presence of a charge density, is similar in form to the expressions obtained in Part I on the basis of the use of the simple continuum solvent representation. In particular, the polarization mode shift quantities, here written as $\overline{\Pi}_\xi$, enter into the expression for the repolarization energy for the electron transfer.[43] The specific form of the repolarization energy depends upon the differences of the induction fields of the various species in their initial and final states. In its general form, the repolarization energy is essentially the same as found in the simpler theory.

The ionic self energy with respect to the polarization field, E_p, is a far more important quantity in the general electron-transfer theory than has been recognized previously. We have established a relationship between E_p and the simple Born ionic solvation model in equation (149). In the remaining paragraphs of this subsection, we consider the necessary extensions of the model of ionic solvation. These extensions involve the use of more realistic charge-distribution functions,[49—53] and ultimately the use of molecular quantal charge densities defined in terms of molecu-

---

[65] G. Arfken, Mathematical Methods for Physicists', Academic Press, New York, 1970, 2nd edn.
[66] H. Silverstone, in 'Physical Chemistry. An Advanced Treatise', ed. H. Eyring, D. Henderson, and W. Jost, Academic Press, New York, 1975, vol. XIA.

lar and atomic orbitals. This analysis definitely will show that the vibrational states of the molecules in the inner solvation or co-ordination shell can be influenced strongly by the presence of and interaction with the solvent.

The obvious extension of the Born ion model, still on a phenomenological scale, is to assume a continuous charge-density representation which is modelled upon the accurate quantum-mechanical charge densities. Thus, for example, if we consider a simple solvated alkali-metal cation, such as lithium or sodium, in a polar solvent, then it is reasonable to approximate the ionic charge distribution by a function of the form (153),[49] where the relationship of the quantity p to an 'effective Bohr radius' a is shown by equation (154). In equation (153), Γ is the usual gamma function.[65]

$$\rho_{ns}(r) = \frac{Zep^{2n+3}}{4\pi\Gamma(2n+3)} r^{2n} \exp(-pr) \tag{153}$$

$$p = 2(n+1)/a \tag{154}$$

The index $n = 0, 1, 2 \ldots$ Thus, for $n = 0$, the continuous, or soft, charge distribution has the form of a Slater $1s$ charge distribution.[67] The effective Bohr radius, therefore, reflects the region, measured radially from the origin of the charge density, where there is the greatest concentration of charge. That this single, continuous charge-density function is a reasonable representation of a cation is easily argued. In fact, it is the electronic charge of the subvalence electrons which is continuous and delocalized. However, this charge screens the uncompensated core charge. The net effect, therefore, is an effective distribution of positive charge. A single, continuous charge-distribution representation for a simple anion seems immediately physically reasonable.

A second general form of continuous, spherical charge distribution which has been investigated is Gaussian, and is represented by equation (155).[49] In this case the effective Bohr radius a is related to p by equation (156).

$$\rho_{G(ns)}(r) = \frac{Ze2^n p^{2n+3}}{(2n+1)!!\pi^{3/2}} r^{2n} \exp(-p^2 r^2) \tag{155}$$

$$p = (n+1)^{1/2}/a \tag{156}$$

The two charge densities, equations (153) and (155), are normalized to the total charge on the ion [see equation (157)].

$$Ze = \int d^3r\, \rho(r) \tag{157}$$

The analysis, which proceeds from either of the above expressions for the charge density to the ultimate expression for the solvation free energy, is as follows. In most instances, and in particular for non-simply spherical charge distributions, it is extremely easy to carry out all aspects of the analysis leading to w_s in the momentum-space representation. Thus, we work specifically with equation (138). It is necessary, first, to obtain the expression for the Fourier transform of the ionic charge density. A number of examples have been considered, and have been reported in the literature.[49—52] For the purpose of illustration, we consider here only the Gaussian form,

---

[67] J. C. Slater, *Phys. Rev.*, 1930, **36**, 57.

equation (155). The Gaussian forms yield the simplest analytical expressions. They are accurate, and they have been the most extensively investigated.[49–52]

The Fourier transform of the charge density is given in general terms by equation (140). In order to implement the transformation, use is made of the Rayleigh expansion[65] [equation (158)], where Y_{lm} is the spherical harmonic function and

$$\exp(i\mathbf{k} \cdot \mathbf{r}) = 4\pi \sum_{l,m} i^l Y_{lm}(\hat{k}) Y_{lm}^*(\hat{r}) j_l(kr) \tag{158}$$

$j_l(kr)$ is the spherical Bessel function of the first kind. In order to evaluate equation (140) with equations (155) and (158) it is necessary to express equation (155) in terms of its angular functional component; this is done simply by writing equation (155a). The zero'th-order spherical harmonic function has the simple form $(1/4\pi)^{\frac{1}{2}}$.

$$\rho_{G(ns)}(r) = \frac{Ze 2^{n+1} p^{2n+3}}{(2n+1)!!\pi} Y_{00}(\hat{r}) r^{2n} \exp(-p^2 r^2) \tag{155a}$$

The orthogonality of the spherical harmonic functions can be used to obtain the integral shown in equations (159), where $H_n(x)$ is the Hermite polynomial.[64] Limiting our considerations to the lowest-order charge-density function, for which $n = 0$, we have equation (160).

$$\rho_{G(ns)}(k) = \frac{2^{n+3}}{(2n+1)!!}(p/k)Y_{00}(\hat{k})\int_0^\infty du\, u^{2n+1}\exp(-u^2)\sin(ku/p) \tag{159a}$$

$$= (-1)^n \frac{\pi^{\frac{1}{2}}(p/k)}{2^{n-1}(2n+1)!!} Y_{00}(\hat{k})\exp(-k^2/4p^2)H_{2n+1}(k/2p) \tag{159b}$$

$$\rho(k) = 2\sqrt{\pi} Y_{00}(\hat{k})\exp(-k^2/4p^2) = \exp(-k^2/4p^2) \tag{160}$$

With the use of equation (139), which relates the Fourier transform of the induction field $D(k)$ to the charge density $\rho(k)$, we can write equation (138) in the form (138a). [It is important to note that this expression (138a) can be used only if

$$w_s = -\frac{1}{4\pi^2}\int d^3k\, k^{-2} C_N(k)\rho^*(k)\rho(k) \tag{138a}$$

$C_N(k)$ is a zero'th-order tensor. If angle-dependent anisotropic properties of the solvent medium are considered, and accounted for in $C_N(k)$, then it is necessary to work directly with equation (138). This is especially true if the charge-density functions are also complicated functions of the angular variables.]

With the use of equation (160) in (138a), we find directly that the single ionic solvation free energy for the 1s Gaussian-type charge distribution is given by equation (161),[49] where the quantity A_n is given by equation (162). Written explicitly, equation (161) is expanded to equation (161a). If all the correlation length quantities

$$w_s = -\frac{Z^2 e^2}{2a}\sum_{n=0}^{2} D_n(2/\pi)^{\frac{1}{2}} A_n \tag{161}$$

$$A_n = (1 + \lambda_n^2 p^2)^{-\frac{1}{2}} \tag{162}$$

$$w_s = -\frac{Z^2e^2}{2a}(2/\pi)^{\frac{1}{2}}[(1-1/n^2)A_0 + (1/n^2 - 1/\varepsilon_{vib})A_1 + (1/\varepsilon_{vib} - 1/\varepsilon_{lib})A_2]$$

(161a)

vanish, an expression for the solvation free energy similar to the Born formula results, as is shown in equation (163). The factor $(2/\pi)^{\frac{1}{2}}$ follows only because of the continuous nature of the model charge distribution.

$$w_s(\lambda = 0) = -\frac{Z^2e^2}{2a}(1 - 1/\varepsilon_{lib})(2/\pi)^{\frac{1}{2}} \qquad (163)$$

Apart from the fact that the more realistic, and accurate, phenomenological treatment of the polar solvent medium is expected to yield improvements for all energy quantities calculated, it is possible to see that in the non-spatially-disperse limit [equation (163)] the model alone is also an improvement. For cations especially, the Born formula is known to yield solvation free energies which are in excess of the experimental quantities.[68] When these energy quantities are estimated with the use of the Goldschmidt, (or some other suitable) radii, agreement with experiment is found to be poor. Generally, radial adjustments have been made,[69] or some other device has been used[70] to modify the dielectric constant of the solvent in the so-called 'dielectrically saturated' region in the vicinity of the ion. Even without the consideration of spatial and temporal dispersion, equation (163) yields values for the free energy of solution that are 80% of the Born value for the same value of the radial quantity. By taking spatial dispersion into account, further reductions are found. In fact, it is found that, when spatial and temporal dispersion are considered, the major contribution to the complete solvation self energy arises from the short-wavelength optical mode contributions. That is, the first term in equation (161a) generally dominates.

With respect to the various treatments of the solvation free energy, the following has been done. Dogonadze and Kornyshev have applied their phenomenological solvation theory to the calculation of the solvation free energy of the Born ion model.[44] They considered a number of simple cations and anions in solution in water. The values for the ionic radii used by them were those determined by Gourary and Adrian.[58] They represented the optical- and vibrational-mode spectral function contributions in terms of the Fourier transform of the Heaviside step function. The long-wavelength librational-mode spectral function was taken to be that given by equation (152). Dielectric constants and correlation lengths were determined by means of the analyses of a number of experimental data. Their results represent a marked improvement upon the simple Born theory of ionic solvation. Still within the framework of the Born representation, Kharkats[71] has examined several complicated charge distributions. Schmidt, in a series of papers (refs. 49—52), has examined the application of the continuous charge-distribution representations to a number of problems. In particular, single- and double-charge-distribution repre-

---

[68] B. Case, in 'Reactions of Molecules at Electrodes', ed. N. S. Hush, J. Wiley, New York, 1971, pp. 45—134.
[69] E. Glueckauf, *cf.* ref. 59.
[70] F. Booth, *J. Chem. Phys.*, 1951, **19**, 391, 1327, 1615.
[71] Yu. I. Kharkats, *Elektrokhimiya*, 1974, **7**, 1137.

sentations have been used to model the simple anions and cations.[49] The double charge distribution,[49] which is a device to account for the different degrees of charge delocalization due to the core and electronic charges, has been found to be reasonably successful and accurate. The continuous-charge-distribution representation has been applied to ring[50] and elliptic[52] charge distributions in an attempt to provide an accurate representation of such species as aromatic radical anionic and cationic species of the benzenoid and naphthalenic varieties. The simple spherical-charge-distribution representations have been modified through the introduction of tetragonal and octahedral distortion parameters.[51] The object of this particular analysis was to determine whether the delocalization of ionic charge in the neighbouring solvent molecules has an effect on solvation and the free-energy quantities. It was found[51] that a successful, simple representation of the solvation free energy could be given in terms of a Gaussian charge distribution together with a degree of tetragonal distortion. Moreover, it was determined that, within one group of ions (*e.g.* the alkali metals, alkaline earths, *etc*.), the degree of distortion required to obtain the experimental self energies was related to the ionic radius by a simple relationship. This regularity of behaviour was not found for the octahedral case.

The extent of the investigations of ionic solvation that have been carried out so far with the use of Dogonadze's polar solvent model and the continuous, phenomenological charge distributions is encouraging. Useable results are relatively easy to obtain, and one can anticipate fairly good agreement with experiment. Thus, this combined approach seems to offer some hope of being useful on a predictive basis.

The connection between the polar solvent theory, in its refined form (due to Dogonadze *et al.*), and the molecular theory of the fluid state is reasonably evident. The problems in implementing the connection are nearly insurmountable for all but the simplest fluid systems. On the other hand, the continuous, phenomenological charge distributions can be replaced by accurate quantum-mechanical charge-density functions which are constructed from molecular and atomic orbitals. An initial treatment has recently been completed which attempts to provide a direct formulation of molecular (neutral) and ionic solvation in terms of atomic and bond properties.[53,54] We consider some of the aspects of this general formulation in the remainder of this subsection. The discussion to follow has a very important bearing on the problem of the resolution of the molecular vibrations of the inner solvation or co-ordination shell from the degrees of freedom of the remaining solvent.

The determination of the ionic solvation energy with the use of molecular quantum-mechanical charge-density functions depends upon the specification of the total molecular electronic wavefunction. The complete molecular charge-density function is constructed with the use of the molecular wavefunction. The resultant electronic density, together with the atomic nuclear charges, provides, therefore, a representation that is adequate to express the solvation free energy.

We consider a single-determinant wavefunction [see equation (164)] for the complete closed-shell molecular ionic system and a similar expression for an open-shell system.[72] In equation (164), Φ_r is the rth molecular orbital, which can be expressed

$$\Psi = \frac{1}{(2n)!} \left| \Phi_1 \alpha(1) \ldots \Phi_n \beta(2n) \right| \tag{164}$$

---

[72] J. A. Pople and R. K. Nesbet, *J. Chem. Phys.*, 1954, **22**, 571.

as a linear combination of atomic orbitals [equation (165)]. The spin wavefunctions

$$\Phi_r = \sum_{N,n} c_{r,Nn} \phi_{Nn}(r_N) \tag{165}$$

are given by α and β. The index N labels the Nth atom in the molecule; n represents a set of indices, n, l, and m, associated with atomic wavefunctions, *viz.* the quantum numbers associated with $1s$, $2s$, $2p_x$, $2p_y$, $2p_z$, *etc.* atomic orbitals.

For a closed-shell molecule, the bond-order matrix $p(Nn \mid N'n')$ is given by equation (166),[73] and for an open-shell molecule it is (167).[72] The molecular

$$p(Nn \mid N'n') = 2 \sum_{r=1}^{n} c_{r,Nn} c_{r,N'n'} \tag{166}$$

$$p(Nn \mid N'n') = \sum_{r=1}^{n} c_{r,Nn} c_{r,N'n'} + \sum_{r=1}^{n} c_{r,Nn} c_{r,N'n'} \tag{167}$$

electronic charge density is expressed simply in terms of the total molecular wavefunction as $\Psi^*\Psi$. In terms of the atomic orbitals which enter the LCAO–MO approximation, we find the value for $\rho(r)$ shown in equation (168). Thus, the com-

$$\rho(r) = \sum_{Nn,N'n'} p(Nn \mid N'n') \phi^*_{Nn}(r_N) \phi_{N'n'}(r_{N'}) \tag{168}$$

plete charge-density function consists, in part, of individual atomic contributions. Because of the nature of equation (168), the charge density due to the electrons consists of individual atomic charge densities as well as contributions from delocalized electrons.

The total charge density is the sum of the electronic and core charge densities [equation (169)]. In terms of a laboratory-based co-ordinate system, the core charge-density function can be expressed as shown in equation (170). The delta-function

$$\rho(r) = \rho_{\text{core}} - \rho_{\text{el}}(r) \tag{169}$$

$$\rho_{\text{core}}(r) = \sum_{N} Z_N e \delta(r - R_N) \tag{170}$$

approximation of the core charges is typical in molecular quantum mechanics. The use of this function for vacuum phase calculations leads to infinite self energies; however, these infinite, but constant, energies are ignored universally. On the other hand, in solution, all the atomic self energies contribute to the complete molecular self energy. For a spatially and temporally disperse polar solvent, the assumption of point core charges for the atomic nuclei does not lead to infinite self energies. The point-charge model is, of course, only an approximation. However, it has been shown that when nuclei of finite radii are considered, because of the smallness of the nuclear radii compared to the solvent correlation lengths, there is only a small alteration of the solvation free energy due to the finite size of the ions.[53]

In terms of the charge-density function, equation (169), the solvation free energy can be written as equation (171).[53] In terms of the decompositions of the charge

$$w_s = -\frac{1}{4\pi^2} \int d^3k \; k^{-2} \{ \rho^*_{\text{core}}(k) \rho_{\text{core}}(k) + \rho^*_{\text{el}}(k) \rho_{\text{el}}(k)$$

$$- [\rho^*_{\text{core}}(k) \rho_{\text{el}}(k) + \rho_{\text{core}}(k) \rho^*_{\text{el}}(k)] \} C_N(k) \tag{171}$$

[73] K. Ruedenberg, *Rev. Mod. Phys.*, 1962, **44**, 326.

densities into the core and electronic atomic and molecular orbital contributions [equations (168) and (169)], it is possible to relate the solvation free energy to individual molecular atomic and bond contributions. Thus, it is possible to relate ionic and neutral molecular solvation to principles of additivity of bond and atom energies.[53] For our immediate objective, which is accurately to determine the nature of the molecular vibrations in the electroactive species, we concentrate attention on the individual charge terms. These terms are used in the next subsection in the construction of the solvated molecular vibrational potential-energy function.

We can express the quantal electronic charge-density terms in the form of equations (172). The last term in equations (172) is general; it reduces to the single-

$$\left. \begin{array}{l} d(nN|nN) = \phi_n^*(r_N)\phi_n(r_N) \\ d(nN|n'N) = \phi_n^*(r_N)\phi_{n'}(r_N) \end{array} \right\} \text{ same centre} \qquad (172a)$$

$$d(nN|n'N') = \phi_n^*(r_N)\phi_{n'}(r_{N'}) \qquad \text{two centres} \qquad (172b)$$

centre representation when $n = n'$ and $N = N'$. Equations (172), however, do display the various types of individual charge density associated with the electrons in the molecular system which contribute to the free energy of solvation. In particular, the decomposition into equations (172) illustrates that all atomic and molecular electrons contribute (in principle) to the free energy. However, in actual calculations, which have not yet been extensively carried out, it is expected that the subvalence electrons will contribute only marginally to the total energy.

The solvation free energy w_s is expressed by equation (171) as an integral in momentum space. Consequently, it is necessary to consider the Fourier transformations of the electronic charge densities represented in terms of the products of atomic orbitals. It is known, from the theory of the Fourier transform,[74] that a function of R can be expressed in terms of the Fourier transform of the same function, evaluated in terms of a co-ordinate r: *viz*. equation (173), where f(k) is defined by equation (174). Thus, in general, it is easy to see that the Fourier trans-

$$\text{F.t.}\{F(r-R)\} = f(k)\exp(i k \cdot R) \qquad (173)$$

$$f(k) = \int d^3 r \, F(r) \exp(i k \cdot r) \qquad (174)$$

form of the single-centre atomic charge-density function $d(nN \mid nN)$ is given by equation (175), where R_{0N} is the vector directed from the origin of the r-co-ordinate

$$d(nnR_N|k) = d(nn0|k)\exp(i k \cdot R_{0N}) \qquad (175)$$

system to N. The term $d(nn0 \mid k)$ is the Fourier transform of the wavefunction product with respect to an arbitrary r-co-ordinate. It can be shown also that the two-centre charge density can be handled in a similar manner.[75—77] The analyses needed for the two-centre case are complicated, and this is especially the case when Slater

---

[74] P. Dennery and A. Krzywicki, 'Mathematics for Physicists', Harper and Row, New York, 1967.
[75] H. Silverstone, *J. Chem. Phys.*, 1967, **47**, 537.
[76] P. P. Schmidt, *J. Phys.* (*B*), submitted for publication.
[77] P. P. Schmidt, *J. Phys.* (*B*), submitted for publication.

orbitals are used.[75] In general, therefore, we see that the density transform $d(nn'R_N R_{N'} \mid k)$ implies a dependence on the separation between atomic centres N and N' of the form $R_{NN'} = R_N - R_{N'}$. For a single-centre contribution, $R_{NN'} = 0$.

As a consequence of the preceding considerations, it is now possible to express the molecular solvation free energy in the form shown in equation (176). The

$$w_s = -\frac{1}{4\pi^2} \int d^3k \; k^{-2} C_N(k) \Bigg[\sum_{nN \in \{nN\}} p(N_i n_i \mid N_j n_j) p(N_k n_k \mid N_l n_l)$$
$$\times d(n_i n_j R_i R_j \mid k) d(n_k n_l R_k R_l \mid k)$$
$$+ \sum_{N,N'} Z_N Z_{N'} e^2 \exp[i\mathbf{k} \cdot (R_N - R_{N'})]$$
$$- 2 \sum_{nN \in \{nN\}} p(Nn \mid N'n') d(nn'R_N R_{N'} \mid k) \exp(i\mathbf{k} \cdot R_{N''}) \Bigg] \quad (176)$$

expansion of the first, electronic, contribution in equation (176) reveals that the free energy w_s depends upon one-, two-, three-, and four-centre atomic electronic-charge-distribution terms. These terms are identical in form [apart from the mediation due to the solvent expressed in $C_N(k)$] to the electrostatic potential-energy matrix elements obtained in the usual quantum-chemical analyses.[53] Ruedenberg,[73] in fact, has pointed out that part of the state energy of a molecule is merely the self energy of the total charge distribution. It can be seen in equation (176) that the second set of terms are similar in form to the nuclear penetration integrals found in quantum-chemical calculations. The last set of terms consists of the individual core self energies as well as all the core–core interactions.

The free energy of solvation does not depend upon the contributions of the matrix elements involving the total molecular momentum operator. On the other hand, the enthalpy of solvation does depend on these elements.[53]

The analysis summarized above illustrates that it is possible to determine solvation free energies by means of the direct use of the charge distributions defined in terms of the molecular and atomic orbitals which enter into the analysis leading to the state energies of any particular solvated species. The analysis described above has been applied to the simple species H_2^+, H_2, and H_2^-. The effect of the disperse, polar solvent on the bonding characteristics of molecules is taken up in the next subsection.

In concluding this discussion on solvation, several remarks are necessary. Kharkats et al.[40] recently examined the electrostatic bases for various models of ions in solution. It is well known, for example, that when the Born ion model is used to represent solvated charges, severe problems of dielectric saturation are generally encountered.[68] The use of the Born model or any extension of it requires the formation of a cavity in the dielectric medium. The ion itself is represented as a charged metallic sphere. As a consequence, it is necessary carefully to consider the boundary conditions which apply at the ion–solvent interface. For simple, totally spherical ions, there is no difficulty in relating the field due to the charge distribution in vacuum, $G(r)$, to the induction field of the ion in solution, $D(r)$. There exists, moreover, a well-defined relationship between $D(r)$ and $E(r)$, the field due to the charge in the medium.[40] The equipotential surfaces are all closed and continuous.

On the other hand, when one attempts to represent a dipolar charge distribution, for example, as two point charges separated by a distance s, and located in a cavity in the dielectric, the relationship between $G(r)$ and $D(r)$ does not hold.[40] If one uses the vacuum charge distribution, and its associated field G, for the dipole in the general expression for the solvation free energy, an error of as yet undetermined magnitude results.[40]

Conversely, if the external charge distribution in the polar medium is continuous, and if no cavity is formed in the polar solvent to accomodate the charge density, then the difficulties mentioned in the last paragraph do not result.[40,78—80] The situation with respect to the continuous charge distributions[49—54] is the same as for the electron gas or the plasma.[78—80] The continuous charge distributions are vanishing at the co-ordinate origin. Consequently, infinite self energies, even in vacuum, do not arise. The distributions are well-behaved functions, and as a result, it is found that the self-energy expressions associated with these charge-distribution models are well-behaved also.[49—54]

Problems of dielectric saturation do not appear, or are muted, with the use of the model of a spatially disperse polar medium together with the continuous external charge density. It can be seen that, to a degree, the use of the individual polar medium elementary excitation modes to represent different levels of interaction (or response) between the solvent and the external charge accounts for dielectric-saturation effects. The optical modes, for example, express the short-range interactions which arise from the interactions between the external charges and induced dipole moments (molecular optical polarizations) of the solvent molecules.

The following restrictions apply to the continuous-charge-distribution model of an ion. Although the electronic and core charges can be represented in terms of continuous charge densities which penetrate the dielectric, no consideration is given to the effect of electronic exchange interactions between the ions and the solvent molecules. That is, the ionic charge distribution interacts with the solvent disperse polar modes in an entirely classical manner. This aspect of the treatment is true both of the phenomenological and quantum-chemical approaches. In the quantum-chemical analysis, exchange interactions take place only between the atoms in the charge distribution specifically included in the molecular system. If the quantum-chemical analysis is expanded to consider, for example, a central ion together with its complement of solvate molecules in the first co-ordination layer, then exchange interactions within that defined system are included. Exchange interactions between the ion–solvent complex and the remainder of the polar medium are not considered.

It must be noted that the synthesis of the continuous-charge-distribution model with the Dogonadze polar-solvent theory specifically represents the system as one consisting of two interpenetrating fluids. This concept, although perhaps alien to chemists, is familiar to a number of areas of many-body physics. The two-fluid model of the nucleus is well known.[15] Equally well known are the two-fluid representations for the superfluid and superconducting states.[81] One may argue the validity of such models. However, for practical calculations, under a number

[78] J. Henrix, *Phys. Rev. (B)*, 1973, **8**, 1346.
[79] J. Linhard, *Kgl. Dansk Vid. Sil. Mat.-Fys. Medd.*, 1954, **28**, 8.
[80] V. P. Silin and A. A. Rukhadze, 'Electromagnetic Properties of Plasma,' Atomizdat, Moscow, 1961.
[81] Z. M. Glasiewicz, 'Superconductivity and Quantum Fluids', Pergamon Press, Oxford, 1970.

of circumstances, the approach proves useful and accurate. Indeed, it is possible to argue the existence of real connections between the two-fluid representations and the detailed many-body treatments. This has been done for the nuclear problem[82] and for superfluid helium[81] and other systems as well.

Having paved the way to a more detailed and accurate general representation of the solvated ionic system, in the next subsection we turn to the consideration of the consequences of the above theory to molecular vibrations of the inner co-ordination or solvation sphere.

Molecular Vibrations of the Inner Solvation or Co-ordination Shell.—We return to the consideration of the Born–Oppenheimer separation of the various degrees of freedom in the solvated electroactive species. Thus, now we complete the discussion begun in the first part of this section.

In the analysis to follow, use is made both of the treatment by Kestner, Logan, and Jortner[31] as well as the polar-solvent theory of Dogonadze.[40—44] However, as has been shown recently,[83] it is necessary to modify the KLJ treatment slightly in order properly to extract the correct form of the inner-sphere vibrational potential-energy function. This departure from the KLJ treatment is outlined below.

The reason for considering any departure from the KLJ analysis lies not with any inaccuracy in that treatment. Rather, it has been found[83] that, although the analysis is formally completely correct, and in principle applicable to the process of determining the inner- and outer-sphere contributions to the rate, it is cumbersome to use. Specifically, the analysis tends to over-generalize the role of the solvent in modifying the interactions and chemical bonds within the solvated or complexed electroactive species. The remedy to this shortcoming, given below, is to partition the total system Hamiltonian operator in such a manner as to display the specific effect of the solvent terms on the inner-sphere vibrational modes.

The result of this analysis not only provides an accurate, solvent-modified inner-sphere vibrational potential-energy function, but also it indicates clearly the proper definition of thermodynamic quantities associated with the electron-transfer system.[83] A characteristic of most treatments of the electron-transfer reaction system in the non-adiabatic limit has been the neglect of work terms associated with the formation of the electron-transfer aggregate state. These work terms definitely are included in the adiabatic theories.[5,16,17] It is also true that the usual work terms are included in the non-adiabatic treatment.[83] The careful application of the Born–Oppenheimer–Hostlein (BOH)[32] separation to the electron-transfer system displays these quantities.[83]

We consider a general form of electron-transfer reaction which schematically can be represented as equation (177). An aggregate state for the electron transfer from

$$A^{M+} + B^{N+} \rightleftarrows A^{(M+1)+} + B^{(N-1)+} \qquad (177)$$

A to B (*i.e.* the reaction from left to right) is defined in terms of a vector separation between the reactants referred to the centres of mass of the species. The aggregate state is essentially an encounter configuration; it is not necessarily a transition-state configuration. This aggregate state defines a configuration of reactants (and pro-

---
[82] W. Wild, *Sitz. ber. Math. Naturw. Kl. Bayer. Akad. Wiss.*, München, 1955, p. 371.
[83] P. P. Schmidt, *J. Electroanalyt. Chem.*, 1977, in the press.

ducts) which enables us further to define energy states of the entire system in terms of energy states of the reactants. In this respect, the BOH method of analysis of the electron-transfer system (177) has many features in common with the quantum-chemical methods of atoms- and molecules-in-molecules first considered by Moffitt.[84] Moreover, as we shall see, with the use of an actual molecules-in-molecules approach as a measure of the strong-coupling limit, it is possible to place more stringent limits on the definition of a non-adiabatic electron-transfer reaction.

We consider an electron transfer to take place in the system in a defined state of aggregation. In view of the Franck–Condon principle, the electron transfer itself takes place in the system with the heavy-centre co-ordinates essentially immobile. Thus, the geometric arrangements of atoms and molecules in the initial aggregate state apply to the final state. Consequently, the initial and final states for the electron-transfer process can be viewed as interacting configurations which contribute to the state of binding in a supermolecular aggregate. In the strong-coupling limit, this aggregate state is a genuine molecular bound state. In the weak-coupling, non-adiabatic limit, the association is a loose one.

Instead of using the Hamiltonian operators (107)—(109) in constructing a practical analysis of the electron-transfer system, we prefer to use the total system operator without the partitioning introduced by KLJ. In formal terms, the interaction between the atomic charges and the solvent subsystem is expressed in terms of point-charge distributions for the atoms. The diagonalization of the operator which yields a solvent term of the form of equation (137) can be carried out.[83] It must be borne in mind, however, that now the renormalized solvent-subsystem Hamiltonian operator is an operator with respect to the atoms in the molecular species which constitute the electron-transfer system. The details of the diagonalization process are contained in ref. 83. The result is shown in equation (178). In

$$\mathcal{H} = \mathbf{T}_N + \mathbf{T}_n + e^2 \sum_{N \leqslant N'} \{1 - \delta_{N,N'} - C_N(\mathbf{R}_N, \mathbf{R}_{N'})\} \frac{Z_N Z_{N'}}{|\mathbf{R}_N - \mathbf{R}_{N'}|}$$

$$+ e^2 \sum_{n \leqslant n'} \{1 - \delta_{n,n'} - C_N(\mathbf{r}_n, \mathbf{r}_{n'})\} \frac{1}{|\mathbf{r}_n - \mathbf{r}_{n'}|}$$

$$- e^2 \sum_{N,n} \{1 - C_N(\mathbf{R}_N, \mathbf{r}_n)\} \frac{Z_N}{|\mathbf{R}_N - \mathbf{r}_n|} + \mathbf{H}_s\{(\mathbf{R}, \mathbf{r})\} \qquad (178)$$

this operator expression the index N refers to atomic centres and n to any electron associated with any atom N. The expectation value of this operator with respect to molecular electronic wavefunctions yields the specific forms for the solvent shift quantities $\overline{\Pi}_s$. The third through fifth terms include all the pertinent electrostatic quantities. Pure solvation self energies emerge only for $N = N'$ and $n = n'$. [In the vacuum molecular quantum mechanics, the infinite single-charge self energies are ignored.] The solvent-modified interaction potential-energy operator can be given an explicit form in co-ordinate space. With the use of equations (141) and (152), we find that it is possible to write equation (179), where $R_>$ ($R_<$) is the greater (lesser) of

[84] V. F. Weisskopf and E. Wigner, *Z. Physik*, 1930, **63**, 54.

$$\{1-C_N(R_1,R_2)\}\frac{1}{|R_1-R_2|} = 4\pi \sum_{L,M} Y_{LM}^*(\hat{R}_1) Y_{LM}(\hat{R}_2)$$

$$\times \left(\frac{1}{\varepsilon_{st}(2L+1)}(R_<^L/R_>^{L+1}) - \sum_\nu D_\nu(1/2\lambda)\{(2L+3)i_L(x_\nu)k_L(y_\nu) \right.$$

$$\left. - x_\nu i_{L-1}(x_\nu)k_L(y_\nu) + y_\nu i_L(x_\nu)k_{L-1}(y_\nu)\} \right) \quad (179)$$

(R^1, R^2). The quantities x_ν and y_ν are defined by $x_\nu = R_</\lambda_\nu$ and $y_\nu = R_</\lambda_\nu$. The functions i_L and k_L are the modified spherical Bessel functions.[65] As a consequence of the potential function (179), it is still possible to carry out quantum-chemical calculations for solvated electroactive systems in the usual co-ordinate representation; in view of the complicated nature of the potential-energy operator, however, it is obvious that these calculations are much more complicated than is the case for the typical vacuum systems.

By means of the analogy between the BOH analysis and the molecules-in-molecules methods of quantum chemistry, we can define solvent-modified Hamiltonian operators for the species A and B. The discrete states of the species, donor, acceptor, and excited states, are defined in terms of the wavefunctions and energy eigenvalues. Thus, we can write two solvent-modified wave equations as (180a) and (180b), where α and δ respectively signify the acceptor and donor states, and $R_{a(b)}$

$$H_{e(a)}\psi_{a(\alpha \text{ or } \delta)} = E_a(R_a)\psi_{a(\alpha \text{ or } \delta)} \quad (180a)$$

$$H_{e(b)}\psi_{b(\alpha \text{ or } \delta)} = E_b(R_b)\psi_{b(\alpha \text{ or } \delta)} \quad (180b)$$

stands for the complete set of atomic and molecular co-ordinates associated with the system in its aggregate state. The Hamiltonian operator $H_{e(a)}$ is given by equation (181);[83] a similar expression can be written for $H_{e(b)}$. The difference $H - H_{e(a)} - H_{e(b)}$ defines an electrostatic interaction operator $V_{int(s)}$ [where the designation

$$H_{e(a)} = -\frac{\hbar}{2m_e}\sum_a \nabla_a^2 - e^2 \sum_{A,a}\{1-C_N(R_A,r_a)\}\frac{Z_A}{|R_A-r_a|}$$

$$+ e^2 \sum_{A<A'}\{1-C_N(R_A,R_{A'})\}\frac{Z_A Z_{A'}}{|R_A-R_{A'}|}$$

$$- e^2 \sum_{a<a'}\{1-C_N(r_a,r_{a'})\}\frac{1}{|r_a-r_{a'}|} \quad (181)$$

int(s) reminds us that the operator is solvent-modified] which operates between the species A and B [equation (182)]. With the use of these expressions, it is possible to

$$V_{int(s)} = -e^2 \sum_{A,b}\{1-C_N(R_A,r_b)\}\frac{Z_A}{|R_A-r_b|}$$

$$- e^2 \sum_{B,a}\{1-C_N(R_B,r_a)\}\frac{Z_B}{|R_B-r_a|} + e^2 \sum_{A<B}\{1-C_N(R_A,R_b)\}$$

$$\times \frac{Z_A Z_B}{|R_A-R_B|} + e^2 \sum_{a<b}\{1-C_N(r_a,r_b)\}\frac{1}{|r_a-r_b|} \quad (182)$$

proceed to develop a general form of BOH separation for the electron-transfer system. The proper definition of the electronic basis functions in terms of the solvent-modified Hamiltonian operators clearly is essential if the solvent-dependence of the inner-sphere modes is to be accurately accounted for.

We use the molecules-in-molecules approach to examine the aggregate system in order to determine which types of individual molecular basis states should enter into the total system wavefunction. The quantum-mechanical composite molecular theories are based on the use of the total system electronic Hamiltonian operator, expressed as in equation (183). The wavefunction for the system is constructed in

$$H = H_a + H_b + V_{int} \qquad (183)$$

terms of the wavefunctions of the isolated molecular fragments as basis functions. The coefficients in the expansion of the total system wavefunction are determined in the usual fashion by means of variational calculations. Taking into consideration all possible states of the electron-transfer aggregate state, we can write a system wavefunction as (184). The summations involving the indices η and ν cover the

$$\Psi = \sum_{\eta,\nu} C^{\delta}_{ab(\eta,\nu)} \psi_{a(\delta,\eta)} \psi_{b(\alpha,\nu)} + \sum_{\eta,\nu} C^{\alpha}_{ab(\eta,\nu)} \psi_{a(\alpha,\eta)} \psi_{b(\delta,\nu)} \qquad (184)$$

ground and excited states of the species A and B in an aggregate state. These indices exclude the electron-transfer state. The first term is that for which the species A formally is the donor and B the acceptor; the second term expresses the reverse situation. Thus, in addition to recognizing the fact that the binding of the aggregate state depends upon the interaction of the ground and excited states of the species, we have also specifically displayed the electron-transfer contribution. The implication that this wavefunction has for the general analysis is simply that it is possible to consider not only electron-transfer transitions between the donor and acceptor species ground states, but also transitions to excited states with and without electron transfer. The energy of the ground state of the aggregate state, as well as of the excited states, can be found by means of the variational calculation of equation (185).

$$\delta\{\langle\Psi|H|\Psi\rangle - E\langle\Psi|\Psi\rangle\} = 0 \qquad (185)$$

The basic form of the composite-molecule analysis carries over to the BOH treatment if one makes the replacements $C^{\delta}_{ab(\eta,\nu)}$ to $X^{\delta}_{ab(\eta,\nu)}$ and $C^{\alpha}_{ab(\eta,\nu)}$ to $X^{\alpha}_{ab(\eta,\nu)}$, with the factors X defined as shown in equation (186), where R is the set of inner-

$$X^{\alpha,\delta}_{ab(\eta,\nu)} = X^{\alpha,\delta}_{ab(\eta,\nu)}(R,Q,t) \qquad (186)$$

sphere co-ordinates and Q the set of generalized solvent-system co-ordinates. The coefficients X now are time-dependent. The BOH analysis yields the equation of motion (187) for the coefficients X;[83] the operator $V_{int(s)}$ is defined by equation (182).

$$\left(T_N + H_{s(\gamma)} - E_{p(\gamma)} + E_\gamma + \langle\gamma|V_{int(s)}|\gamma\rangle + \sum_\beta S^{-1}_{\gamma\beta}\langle\beta|V_{int(s)}|\gamma\rangle - i\hbar\partial/\partial t\right) X_{ab}$$

$$= -\Big(\sum_{\alpha\neq\gamma}\langle\gamma|V_{int(s)}|\alpha\rangle + \sum_{\alpha\neq\gamma}\sum_{\beta\neq\gamma}S^{-1}_{\gamma\beta}\langle\beta|V_{int(s)}|\alpha\rangle\Big) X^{\alpha}_{ab} \qquad (187)$$

The simplest composite system wavefunction that one can assume merely consists of two terms. For the electron-transfer system in its aggregate state, we can consider these two basis functions constructed from the ground-state wavefunctions for A as a donor and B as an acceptor, and the converse. As a consequence, the BOH analysis yields the simpler equation of motion (188) for the system vibrational

$$\left(\mathbf{T}_N + \mathbf{H}_s(\mathbf{R}_\delta) - E_{p(\delta)} + E_\delta + \langle\delta|\mathbf{V}_{\text{int(s)}}|\delta\rangle + S_{\delta\alpha}^{-1}\langle\alpha|\mathbf{V}_{\text{int(s)}}|\delta\rangle - i\hbar\partial/\partial t\right) X_{\text{ab}}^\delta$$
$$= -\left(\langle\delta|\mathbf{V}_{\text{int(s)}}|\alpha\rangle + S_{\delta\alpha}^{-1}\langle\alpha|\mathbf{V}_{\text{int(s)}}|\alpha\rangle\right) X_{\text{ab}}^\alpha \quad (188)$$

functions X. The adiabatic ansatz of the BOH theory applied to equation (188) gives equation (189) for the determination of the system vibrational wavefunctions and energy levels.

$$\left[\mathbf{T}_N + \mathbf{H}_s(\mathbf{R}_\delta) - E_{p(\delta)} + E_\delta + \langle\delta|\mathbf{V}_{\text{int(s)}}|\delta\rangle + S_{\delta\alpha}^{-1}\langle\alpha|\mathbf{V}_{\text{int(s)}}|\delta\rangle - E_{\delta n}\right] X_n^\delta = 0$$
$$(189)$$

Equation (189) will be used to obtain the system vibrational energy levels and wavefunctions. Before considering the problem of the separation of the solvent-system modes from the inner-sphere vibrations, and the subsequent identification of the inner-sphere vibrational potential-energy function, it is useful to compare the results obtained with the composite molecular analysis further. The comparison yields insight into the limits of applicability of the non-adiabatic treatment of the electron-transfer reaction.[83]

Let a complete configuration interaction wavefunction for a molecular system be written as shown in equation (190) for fixed $\mathbf{R}$ and $\mathbf{Q}$. Then, the energy expression $\langle\Psi|\mathbf{H}-E|\Psi\rangle = 0$ yields equation (191). The variation of the coefficients yields

$$\Psi(r,\mathbf{R},\mathbf{Q}) = \sum_\alpha C_\alpha \psi_\alpha(r,\mathbf{R},\mathbf{Q}) \quad (190)$$

$$\sum_\alpha C_\alpha^*[S_{\alpha\beta}(E_\beta - E)C_\beta + \sum_\beta C_\beta \langle\alpha|\mathbf{V}|\beta\rangle] = 0 \quad (191)$$

$$\sum_\beta S_{\alpha\beta}(E_\beta - E)C_\beta + \sum_\beta \langle\alpha|\mathbf{V}|\beta\rangle C_\beta = 0 \quad (192)$$

equation (192), which can be rearranged to give equation (193). It can be seen that

$$\{E_\gamma + \langle\gamma|\mathbf{V}|\gamma\rangle + \sum_\beta S_{\gamma\beta}^{-1}\langle\beta|\mathbf{V}|\gamma\rangle - E\}C_\gamma$$
$$= -\Big(\sum_{\alpha\neq\gamma}\langle\gamma|\mathbf{V}|\alpha\rangle + \sum_{\alpha,\beta\neq\gamma}S_{\gamma\beta}^{-1}\langle\beta|\mathbf{V}|\alpha\rangle\Big)C_\alpha \quad (193)$$

the form of this equation is very close to the BOH expression (188). If we apply the equivalent of the BOH adiabatic ansatz to equation (193), then the condition for a non-trivial solution to the remaining equations is that (194) be valid for all γ.

$$E = E_\gamma + \langle\gamma|\mathbf{V}|\gamma\rangle + \sum_\beta S_{\gamma\beta}^{-1}\langle\beta|\mathbf{V}|\gamma\rangle \quad (194)$$

Clearly, the energies (194) cannot be regarded as accurate or even valid for the case that the molecular species form a strongly bonded supermolecular complex. As a consequence, there must be a distance and configuration of closest approach.

As generally the energies (194) reduce to the appropriate infinite-separation limit, distances and configurations greater than the minimum value may be well represented by (194); although in view of the terms omitted in (193), this need not be the case. One must conclude, therefore, that if the energies (194) are to be accurate in any analysis, then the number of aggregate-state configurations available to the reactants in the electron-transfer reaction in particular is limited. This conclusion must apply to the BOH analysis because of the similarity of approximations.

If, in fact, an actual composite molecule quantum-mechanical analysis of a particular system indicates the formation of a strongly bound state, then one must conclude that the non-adiabatic treatment of the electron-transfer reaction for that system is not valid. It is possible that in systems in which bound aggregate states are formed one can treat the electron-transfer transition within the aggregate as a radiationless transition between molecular states.[48] The transition involved would resemble the familiar spectroscopic charge-transfer transition.[48]

We continue with the analysis leading to the definition of the inner-sphere vibrational potential-energy function.

It is clear that the solvent subsystem Hamiltonian operator $H_{s(\gamma)}$ in equation (187) contains a residual dependence on atomic position co-ordinates within the species of the aggregate state. This operator is the expectation value of the original operator used in equation (178) [see equation (195)]. In order to be able to determine the form of the inner-sphere vibrational potential-energy operator, it is necessary to separate this residual dependence on atomic co-ordinates in H_s from the solvent-mode dependence.

$$H_{s(\gamma)} = \langle \psi_{a(\gamma)}\psi_{b(\gamma)} | H_s(\{R,r\}) | \psi_{a(\gamma)}\psi_{b(\gamma)} \rangle \qquad (195)$$

In principle, it is possible to formulate the general electron-transfer problem, embodied in its Hamiltonian operator, in terms of Green functions*. Subsequent manipulations with the system Green functions can lead to a resolved form. An alternative approach is to apply the Weisskopf–Wigner strong coupling perturbation theory[84] to the system. In view of the usual approximations required to make the Green-function analysis manageable, the Weisskopf–Wigner method is effectively equivalent to the Green-functional analytic approach in the form of the results. The Weisskopf–Wigner analysis, although phenomenological to a substantial extent, is easy to understand, and it is possible readily to extract the physically and chemically meaningful limits once the analysis is complete. In view of the state of our partial knowledge about the inner spheres, in particular, it makes sense at this time certainly only to consider this more suggestive, less formal, Weisskopf–Wigner analysis.[83]

We consider the form of equation (137) (ignoring E_p, which already has been used elsewhere in the BOH analysis) together with its expectation value [equation (195)]. If we assume that the major effect of the residual atomic co-ordinate dependence in H_s can be expressed in terms of the first-order term in a Taylor expansion, then it is possible to write equation (196).[83] It is implied in this expression that $(\nabla_{R_n}\bar{\Pi}_\zeta)$

---

* It has been learned recently by the Reporter, from a conversation with Dogonadze, that he has in fact been able to carry out a Green function analysis as indicated above. Although the Dogonadze analysis is clearly mathematically more rigorous than the analysis to be outlined, the conclusions of each are the same. Dogonadze's work is to be published.

$$H_{s(\gamma)} = H_s(\{R_\gamma\}) \cong H_s(\{R_\gamma^0\}) + \sum_{n \in N, \xi} \omega_\xi^2 (\nabla_{R_n} \Pi_\xi) \cdot (R_n - R_n^0)(\Pi_\xi - \bar{\Pi}_\xi) \quad (196)$$

is evaluated in terms of the set of equilibrium co-ordinates R_n^0 after the differentiation operation has been carried out. The zero'th-order term $H_s(\{R_\gamma^0\})$ no longer is an operator in the co-ordinate space of the atoms in the aggregate-state molecules. It is, however, still an operator with respect to the solvent subsystem modes. The bi-linear term in equation (196) now contains the specific dependence on the co-ordinate operators of the solvent- and inner-sphere-systems that is needed.

The vibrational Hamiltonian operator for the solvent modes and inner-sphere modes now can be expressed as equation (197).[83] The inner-sphere co-ordinates

$$H_{vib(\gamma)} = H_s(\{R_\gamma^0\}) + T_N + \tfrac{1}{2}\Sigma(R_n - R_n^0) \cdot W \cdot (R_{n'} - R_{n'}^0)$$
$$+ \Sigma \omega_\xi^2 (\nabla R_n \bar{\Pi}_\xi) \cdot (R_n - R_n^0)(\Pi_\xi - \bar{\Pi}_\xi) \quad (197)$$

can be expressed in normal-mode form. The quantity W in equation (197) is a frequency–mass tensor which applies to the inner-sphere vibrational modes. For an individual inner-sphere vibrational normal mode, the result of the application of the Weisskopf–Wigner analysis is equation (198),[83] where $K(\xi, v)$ is a coupling constant

$$q_v(t) = q_v^0(t)\exp(-\gamma_v t/2) + \sum_\xi \frac{K(\xi, v)}{\Delta\omega^2 + \gamma_v^2/4}[1 - \exp(-i\Delta\omega t - \gamma_v t/2)]$$
$$\times \left[\Delta\omega(\omega_\xi/\omega_v)^{\frac{1}{2}} \Pi_\xi^0 - \frac{\gamma_v}{2\sqrt{\omega_\xi \omega_v}} Q_\xi^0\right] \quad (198)$$

which depends upon $(\nabla_{R_n} \bar{\Pi}_\xi)$ and elements of the frequency–mass tensor, $\Delta\omega = \omega_\xi - \omega_v - \omega_s$, where ω_s is defined by equation (199),[83,85] which represents the frequency

$$\omega_s = -P \int_{-\infty}^{\infty} d\omega_\xi |K(\xi, v)|^2 \rho(\omega_\xi)(\omega_\xi - \omega)^{-1} \quad (199)$$

shift of the inner-sphere normal mode due to the interaction with the solvent [$\rho(\omega_\xi)$ is a lineshape factor for the solvent modes which act as a loss field]. The damping factor for the normal mode v, γ_v, is given by equation (200).[83,85] This

$$\gamma_v = 2\pi |K(\xi, v)|^2 \rho(\omega) \quad (200)$$

analysis indicates that the residual coupling between the inner-sphere and solvent modes, when treated in this approximation limit, leads to a damping of the inner-sphere modes. The form of the coupling responsible for this damping can be seen from equation (200) to be due to dipolar and higher multipole interactions between the charge distribution of the inner-sphere complex and the polar solvent molecules.[83]

The introduction of the damping of the inner-sphere (and solvent) vibrational normal modes presents a complication with respect to the subsequent evaluation of the electron-transfer rate constant. Recently, however, mathematical methods have been devised to handle the non-analytic aspects, and as a result there is no longer any practical prohibition to the consideration of damping effects.[86-88]

[85] W. Louisell, 'Radiation and Noise in Quantum Electronics', McGraw-Hill, New York, 1964.
[86] P. P. Schmidt, *J. Phys. (B)*, 1976, **13**, 2331.
[87] P. P. Schmidt, *J.C.S. Faraday II*, 1973, **69**, 1122.
[88] P. P. Schmidt, *J.C.S. Faraday II*, 1976, **72**, 1736.

The Weisskopf–Wigner expression for the damping constant can be evaluated in principle. Given the explicit form of the coupling constants $K(\xi,\nu)$, together with a reasonable expression for the solvent-system shape factor $\rho(\omega)$, the evaluation of the damping terms can be carried out along lines similar to those used in the general theory of radiationless transitions. At present, such calculations for the electron-transfer system have not yet been carried out, although there is a clear need for calculations to be performed. In the absence of any numerical estimates of the magnitudes of these damping factors, therefore, we must be content merely with qualitative estimates: in essence, guesses. It is a common assumption in almost all current treatments of the electron-transfer problem that it is justifiable to regard the inner-sphere vibrational degrees of freedom as independent of any significant influence by the solvent. This is a working assumption in that it allows one relatively easily to obtain explicit expressions for the electron-transfer rate constants. It is an assumption which is, however, by no means generally defensible. Indeed, for molecular radiationless transitions Nitzan and Jortner,[89] in particular, have pointed out that strong vibrational relaxation can have a substantial effect upon a radiationless vibronic transition. Considering the electron-transfer reaction, it is reasonable to suspect that, for the weakly bound solvation or co-ordination shells about an electroactive species, the coupling to the solvent-system degrees of freedom can be strong. Certainly, in terms of the analysis we have developed here, damping of the inner-sphere vibrational modes will be pronounced.

In the remaining discussion of this section, we assume that the damping factors are small, and may be ignored. As a consequence, it is clear that only tightly bound solvated or co-ordinated species may be considered. In the next section we investigate a model of the inner sphere which allows a relaxation of this restriction.

When it is allowable to ignore the inner-sphere normal-mode damping which results from the coupling with the continuum of solvent modes, then the inner-sphere–solvent subsystem Hamiltonian operator reduces to the form shown in equation (201). In principle, the specification of the inner-sphere vibrational

$$H_{vib(\gamma)} = H_s(\{R_\gamma^0\}) + T_N + \tfrac{1}{2}\Sigma(R_n - R_n^0)\cdot W \cdot (R_{n'} - R_{n'}^0) \qquad (201)$$

problem now is complete. It is complete in the sense that the complete harmonic potential-energy operator for the aggregate state of the electron-transfer system has been specified; specifically, it is specified in the third term of equation (201). For a given configuration of this aggregate state, it is possible to carry out the usual normal-mode analysis for the complete system. Such an approach is useful, in general, for the evaluation of the Franck–Condon integrals which arise in the expression for the electron-transfer transition probability, and hence the rate constant. This complete approach, on the other hand, tends to mask or even bury the contributions of the vibrations of the individual species to the activation process. A remedy for this obscuring generality is to consider either (i) the states of the complete system in terms of basis states and functions of the individual species, or (ii) to diagonalize the complete Hamiltonian operator with respect to the individual species and the interaction terms. Either route is acceptable. The first, which has been explored in general terms,[83] leads to expressions for the ground- and excited-

---

[89] A. Nitzan and J. Jortner, *Theor. Chim. Acta*, 1973, **30**, 217.

state vibrational energies of the inner sphere, expressed in terms of the degrees of excitation in the isolated species needed to conform to the aggregate state. In other words, the formation of an aggregate state of the electron-transfer system requires some (perhaps substantial) modification of the inner solvation or co-ordination spheres of the individual (initially isolated) reactants. This degree of modification required for each reactant can be 'seen' by examining the vibrational ground state of the aggregate system in terms of basis functions for the isolated species. The degree of distortion needed to form the aggregate state can be inferred in terms of amounts of vibrational excitation needed in the isolated species in order to conform to the actual eigenvalue problem. The analysis described is the familiar variational approach to the determination of state energies and functions for a complicated system in terms of basis functions associated with the component particles. A more detailed illustration of this process is the following.

We assume that an appropriate Born–Oppenheimer analysis has been applied to the solvated or co-ordinated electroactive reactants in their respective states of isolation (*i.e.* isolation with respect to the electron-transfer transition). Consequently, in the harmonic-oscillator limit, it is possible to write an individual vibrational Hamiltonian operator as equation (202) for the species A, and similar terms

$$H^\infty_{\text{vib}(A)} = T_A + \tfrac{1}{2} \sum_{a,a' \in A} (R_a - R_a^0) \cdot W^\infty_A \cdot (R_{a'} - R_{a'}^0) \quad (202)$$

for other species. If only two reactants A and B are considered, then a vector separation between the two centres of mass is specified by R_{AB}. The attitude, as well as the separation magnitude, will be expressed in the complete system frequency–mass tensor W through its dependence on the spherical harmonic functions.[90,91] The vibrational potential-energy operator for the aggregate state of the electron-transfer reaction system is, as before, given by equation (203). The total vibrational

$$V = \tfrac{1}{2} \sum_{n,n' \in N} (R_n - R_n^0) \cdot W \cdot (R_{n'} - R_{n'}^0) \quad (203)$$

Hamiltonian operator can be expressed as equation (204). It is possible, however, to express this operator as shown in equation (205), in which the operator V' is a

$$H_{\text{vib}} = T_A + T_B + V \quad (204)$$

$$H_{\text{vib}} = H_{\text{vib}(A)} + H_{\text{vib}(B)} + V' \quad (205)$$

perturbation [see equation (206) for its definition]. The vibrational eigenvalue problems of equations (207a) and (207b) are assumed to be completely solved.

$$V' = V - \tfrac{1}{2} \sum_{a,a' \in A} (R_a - R_a^0) \cdot W^\infty_A \cdot (R_{a'} - R_{a'}^0) - \tfrac{1}{2} \sum_{b,b' \in B} (R_b - R_b^0) \cdot W^\infty_B \cdot (R_{b'} - R_{b'}^0) \quad (206)$$

$$H^\infty_{\text{vib}(A)} X^\infty_A = E^\infty_A X^\infty_A \quad (207a)$$

$$H^\infty_{\text{vib}(B)} X^\infty_B = E^\infty_B X^\infty_B \quad (207b)$$

The vibrational eigenvalue problem of the complete, aggregate system is (208).

---
[90] P. P. Schmidt, *J. Phys.* (*B*), submitted for publication.
[91] P. P. Schmidt, *J.C.S. Faraday II*, 1977, **73**, 755.

Hence, if one assumes a trial wavefunction of the form (209), then the variational calculation (210) yields values for the expansion coefficients α_v and $\beta_{v'}$. These coefficients can be taken as measures of the degree of vibrational excitation needed in each defined reactant species in order to achieve the aggregate state.

$$H_{vib}X = EX \tag{208}$$

$$X = \left(\sum_v \alpha_v X_{A(v)}^\infty\right)\left(\sum_{v'} \beta_{v'} X_{B(v')}^\infty\right) \tag{209}$$

$$\delta\{\langle X|H_{vib}|X\rangle - E\langle X|X\rangle\} = 0 \tag{210}$$

It is clear from the general form of the quantum-mechanical variational calculation that the total energy E will be a function of the expansion coefficients in equation (209). Thus, the total energy can be factored into the energies of the isolated species together with vibrational interaction energies. For the ground vibrational state of the aggregate, the energy E_0 is simply given by equation (211). The excited vibrational states of the aggregate system are given similarly.

$$E_0 = \langle X_0|H_{vib}|X_0\rangle / \langle X_0|X_0\rangle \tag{211}$$

The second route to the resolution of the individual reactant vibrations and their mutual interactions, *viz.* diagonalization of the complete aggregate system vibrational Hamiltonian operator, can be carried out in terms of the normal modes of the isolated species. The result of the operation is a new set of system normal modes which express the normal vibrations of the reactants as modified by their mutual interaction. The conclusions to be drawn are essentially the same as with the variational calculation. This second method of approach to the analysis of the vibrations of the aggregate state is particularly useful for the evaluation of the statistical sums which occur in the transition probability expression for the electron-transfer transition.

The consequence of this enlarged view of the electron-transfer system in the non-adiabatic limit is important in the consideration of the energetics of these reactions. Hitherto, in almost all non-adiabatic treatments of the electron-transfer reaction, the work terms associated with the formation of the aggregate state, and ultimately the transition state, have been ignored.[83] In a recent paper by Efrima and Bixon[92] these work terms can be extracted, but do not appear in a direct form. Essentially, in previous treatments, the use of the simplest form of the Born–Oppenheimer separation together with only a two-level model of the adiabatic electronic states has accounted for the omission.[83] Kestner, Logan, and Jortner[31] provided for an account of the work terms, but omitted them in their final rate-constant expressions.

It is not difficult to see that the Taylor series expansion needed to define equation (197) in the harmonic-oscillator limit yields a vibronic state energy of the form shown in (212), in which $J_\gamma(\{R_\gamma^0\})$ and $w_\gamma(\{R_\gamma^0\})$ are defined as shown in equations (213) and (214).

$$E_{\gamma(s,u)} = J_\gamma(\{R_\gamma^0\}) + w_\gamma(\{R_\gamma^0\}) \tag{212}$$

$$J_\gamma(\{R_\gamma^0\}) = E_{el(\gamma)}(\{R_\gamma^0\}) - E_{p(\gamma)} \tag{213}$$

$$w_\gamma(\{R_\gamma^0\}) = \langle\gamma|V_{int(s)}(\{R_\gamma^0\})|\gamma\rangle + S_{\gamma\alpha}^{-1}\langle\alpha|V_{int(s)}(\{R_\gamma^0\})|\gamma\rangle \tag{214}$$

[92] S. Efrima and M. Bixon, *J. Chem. Phys.*, 1976, **64**, 3639.

It is shown elsewhere that the term w_y contains all the classical work contributions as well as quantum-mechanical modifications associated with the formation of the aggregate state.[83] It is clear that in none of these terms is there only a dependence upon the energy of a solvated reactant in isolation from its electron-exchange partner. This observation has an important bearing on the evaluation of the electron-transfer rate constant.

Let E_i be the total energy of the electron-transfer initial state and E_f that of the final state for the system in its aggregate configuration. These energies are therefore the energies identified by means of the BOH analysis discussed in the preceding paragraphs. Let E_i^∞ be defined as the total energy of the reactant system in a state of complete inter-reactant separation. It is characteristic of the theoretical treatments of reactions in general and of the electron-transfer reaction in particular that the energetics of the system be referred to the infinite-separation limit for the reactants (and products). The reason for this, of course, is to be able ultimately to relate the energy quantities in the activation-energy expression to thermodynamically defined and measurable states. This clearly is the objective of the linear and non-linear free-energy analyses (*viz.* the Bronsted plots). If theoretical rate-constant expressions are to be related to thermodynamic quantities defined for thermodynamic equilibrium initial and final states, then it is essential that the appropriate measures be taken in the evaluation of the transition probability which is required for the determination of the rate constants. We can therefore define the initial state for an electron-transfer transition within a given transfer-system aggregate state in terms of the *vibronic* binding energy for that state: *viz.* $E_i - E_i^\infty$.

The transition-probability expression for the electron transfer can be expressed as a convolution (215),[83] where $[S \otimes F]_{if}$ is defined in equation (216), and $f_{if}(t)$ and

$$W_{if} = (2\pi/\hbar)|\bar{V}_{if}|^2 [S \otimes F]_{if} \qquad (215)$$

$$[S \otimes F]_{if} = (1/2\pi\hbar Z)\exp[-\beta(J_i + w_i - J_i^\infty - e_i^\infty)]$$
$$\times \int_{-\infty}^{\infty} dt \exp[it(\Delta J + w_f - w_i)] f_{if}(t) s_{if}(t) \qquad (216)$$

$s_{if}(t)$ are the Fourier transforms of the structure factors associated respectively with the inner-sphere vibrations and the solvent [equation (217)], with a similar term for $s_{if}(t)$. The energy e_i^∞ is the vibrational energy of the species in the initial state of infinite inter-reactant separation.

$$f_{if}(t) = \sum_{u,v} \langle X_u | X_v \rangle|^2 \exp[-(\beta + it/\hbar)E_u + itE_v/\hbar] \qquad (217)$$

From the first exponential term in equation (216) it is possible to see that the transition-probability expression depends upon the work necessary to form the aggregate state from the initially infinitely separated reactant state. The second exponential term in equation (216) indicates that the final form of the evaluated transition-probability expression will depend upon $\Delta J + w_f - w_i$. This dependence now brings the general non-adiabatic, quantum-mechanically based treatment into a form which agrees with Marcus's statistical-mechanical treatments.[5,16] The energy difference ΔJ is not, however, simply the difference between initial and final infinitely-separated transfer electronic states. The individual components of ΔJ are defined in terms of the set of equilibrium atomic co-ordinates associated with a given transfer aggregate state. It is possible to extract from ΔJ a quantity ΔJ^∞ which can be identified with the difference in solvated transfer electron energies for the reactants and products in states of infinite separation. This quantity ΔJ^∞ can be thermodynamically defined. The difference $\Delta J - \Delta J^\infty$ is therefore an additional work quantity associated with the work necessary to deform the reactants and products into aggregate-state configurations.[83]

Finally, it is necessary to point out that the expression for the rate constant itself follows from the direct integration of the transition probability over all possible aggregate state configurations. No additional potential energy term(s) need be inserted,[83] as in fact is done in the usual treatments.[1—11] Thus we can write equation (218),[83] which is more simple, and where R is the optimum inter-reactant separation and $\{\boldsymbol{R}^*_{0i}\}$ is the optimum set of atomic co-ordinates for the system.

$$k_{el} = (4\pi R^3/3) W_{if}(\{\boldsymbol{R}^*_{0i}\}) \tag{218}$$

4 The Collective-continuum Treatment of the Inner-sphere Complex

In the last section, with the use of the Born–Oppenheimer–Holstein analysis,[31,32] in principle we defined an accurate inner-sphere vibrational potential-energy function. In particular, it was possible there to specify the harmonic oscillator limit. The assumption of this limit, however, is a restriction on the theory which limits its applicability to the outer-sphere reaction. The fact that, with the use of the work terms which arise, a considerable amount of distortion of the inner sphere can be considered does not alter the limit of applicability of the harmonic-oscillator form. In order to be able to consider electron-transfer reactions in which there is in fact a considerable degree of inner-sphere reorganization with respect to configuration and composition, it is necessary to consider the full potential-energy function. To do so, at this time, is still a mathematically intractable problem. Consequently, alternative, semi-phenomenological methods have been sought recently in order to be able to cast the treatment of the inner-sphere reaction into some initially suitable and acceptable form.

The general characteristics of the inner-sphere reaction have been outlined in Part I. There it was indicated that not only does one have to contend with the electron-transfer event, but also one has to consider alterations in the chemical bonding and atomic compositions of the reactants and products. The difficulties which attend any attempt to formulate the simultaneous description of the two connected processes, *viz.* the electron-transfer and bond-alteration processes, are certainly greater than those associated with the description of the electron-transfer step alone. It is difficult, for example, to find a suitable co-ordinate system that is capable simultaneously of defining the initial, the intermediate activated-complex, and the final product states. For diatomic molecules, it is possible to effect a complete analysis of the low-energy, harmonic vibrations as well as the high-energy, dissociated states. For polyatomic molecules, on the other hand, the same type of analysis is not as clear-cut. Low-energy harmonic molecular normal modes are well characterized. The molecular fragments in the dissociated state can also be well characterized for the species in their vibrational ground and low excited states. However, a good characterization of the dissociated state depends upon the use of separate co-ordinate systems for the individual fragments. It is assumed that the molecular fragments are effectively infinitely separated. The connection between the good representations of the molecule and its fragments is difficult to find. Nevertheless, for gas-phase reactions, in particular, this problem is an active area of interest (cf. ref. 93).

For condensed-phase reactions, there is another approach that can be used. It is an approach which is not as well-defined and rigorous mathematically and physically as is much of the current gas-phase work.[93] It is an approach which is phenomenological. Consequently, there is a possibility of using this alternative approach as a means of understanding the general behaviour of the inner-sphere reaction. It is hoped that at least comparative behaviour can be understood. Specifically, it can be assumed that the reactive, and often vibrationally highly excited, states of the electron-transfer donor and acceptor species in a polar solvent can be represented in terms of their respective mass and charge densities.[12—14,94] The density representations can be viewed as fluids of continuous (but not necessarily uniform) distribution and extent. The model for the electron-transfer system reactants is therefore analogous to the two-fluid model of the atomic nucleus.[15] Thus, the charge density associated with the electroactive species and the density of the solvent or co-ordination species in the first shell can be represented as two interpenetrating fluids. In the treatment,[12—14] account is taken of the fact that the distributions of the two fluids are not uniform, but must conform to the general spatial representations of the entire solvated or co-ordinated species. The interaction between the two fluid systems is given a general form, and as a result of this interaction, harmonic displacements of the two fluids relative to each other are possible.[12—14] By analogy to the similar vibrational modes of the atomic nucleus,[15] these oscillations of the fluid representation components of the electroactive species are given the name 'giant acoustic resonances'.[12] The first giant acoustic dipole resonance, or 'acoustic gion', has been associated with the far-i.r. absorption line of the simple solvated alkali-metal cations.[12]

The advantage of this approach over an attempt to define a universal co-ordinate system for the reactants and products is simply contained in the fact that the co-ordinate system for the treatment has its origin in the electroactive species. This co-ordinate system does not depend upon the charge state or upon the state of solvation of the electroactive species.[12—14] Because this is the case, the inner-sphere vibrations (as modelled in terms of the acoustic gions) of an electroactive species in its differing charge and solvation or co-ordination states differ only in terms of frequency and equilibrium displacements of fluid-system co-ordinates.[13] Consequently, a single co-ordinate system applies to the initial and final electron-transfer states of an electroactive species.

The Collective Model of Highly Excited Molecular Vibrational States.—A detailed, but at this time necessarily qualitative, account of the collective representation of the molecular excited vibrational states has been given.[93] In the following paragraphs a summary of the treatment is given, together with intuitive physical and chemical arguments to justify the introduction of such a treatment.

[93] M. S. Child, cf. ref. 46.
[94] P. P. Schmidt, J.C.S. Faraday II, 1976, 72, 1074.

With respect to the vibrations of a diatomic molecule, it has been known[95—97] for some considerable time that the molecular potential-energy function (operator) is not simply parabolic. Strong, short-range repulsive interactions between the atoms force a steep rise in energy as the atoms approach each other closer than a certain distance. On the other hand, for increasing inter-atomic separation in the diatomic molecule, the energy first rises past the equilibrium separation, then asymptotically approaches the zero energy value. The true potential-energy function is definitely non-uniform and definitely non-parabolic. Nevertheless, as all treatments show, low-energy vibrational states are reasonably well represented by the harmonic-oscillator analysis. The more exact treatments[95,96] possess the harmonic-oscillator energies as limiting forms.

Although many treatments of the vibrational states of complicated molecular systems rely upon the use of the Morse,[95] or other, equivalent potential-energy expressions, there is no guarantee that a superposition of diatomic anharmonic potential-energy functions is wholly accurate. More to the point is the fact that the eigenfunctions of the Morse problem, as basis functions for a complicated molecular vibrational analysis, are extremely difficult to work with. As a consequence, most anharmonic contributions are considered in terms of the additional cubic, *etc.*, terms found in a Taylor-series expansion of a given molecular vibrational potential-energy function. The analysis of anharmonic contributions is carried out in terms of the simple harmonic oscillator functions as basis functions for the calculation.

The limitation of any particular analysis of a vibrating system to harmonic-oscillator-limit states severely restricts the number of phenomena which can be considered. This is particularly true for the consideration of chemical reactions in general, and the inner-sphere electron-transfer reaction in particular. Thus, as was indicated in the last section, the assumption of the harmonic oscillator potential-energy function for the inner spheres of the electron-transfer aggregate state restricts the analysis to reactions which are essentially outer-sphere in nature; this is the case even though it is possible to consider reasonably large distortions in the frequencies and equilibrium positions of co-ordinates in the harmonic limit.

An attempt to overcome this difficulty in considering vibrational anharmonicity, in particular with respect to chemical reactions, has recently been made.[94] The basic idea of the approach is not new, but it seems to be new with respect to chemical applications. Specifically, if one considers a complicated, vibrating polyatomic molecular system, and if one recognizes that anharmonic interactions (to all orders) play a dominant role in the mixing of the well-defined, low-energy harmonic normal modes, then it is possible in some instances to give the excited states a continuum, fluid representation. The situation envisaged is illustrated in Figure 2. If we consider two well-defined molecular normal modes, then as we consider vibrational excitation within these two modes it is possible to see that the effect of anharmonic interactions is to distort the parabolic character of these modes. At some point, the degree of distortion is sufficiently severe as to obliterate the character of a particular normal mode. In that state of excitation, one normal mode merges into another; essentially, the state of excitation verges on the dissociation limit for the molecule.

[95] P. M. Morse, *Phys. Rev.*, 1929, **34**, 57.
[96] E. Fues, *Ann. Physik*, 1926, **80**, 367.
[97] A. Kratzer, *Z. Physik*, 1920, **3**, 289.

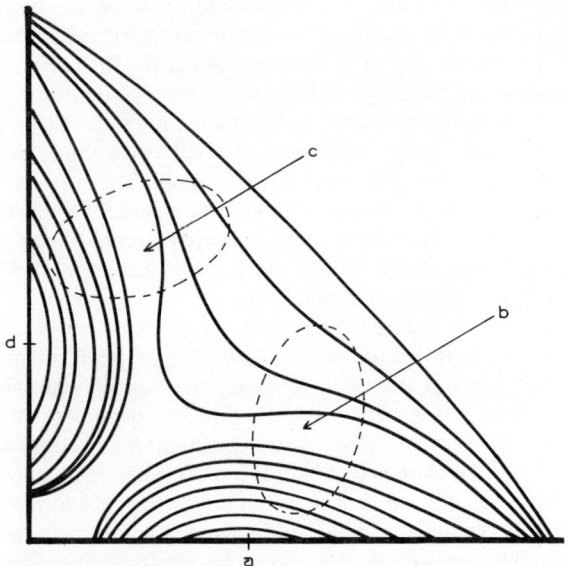

Figure 2 *A schematic representation of the interaction of two molecular vibrational normal modes. The points a and d label the vibrational ground-state equilibrium positions for the two normal modes. The regions labelled b and c indicate the portions of the normal-co-ordinate space associated with states of high vibrational excitation. In these regions the discrete point-group character of the normal modes is presumed to change to a continuous representation. The regions b and c are dominated by large amounts of anharmonic interactions.*

However, before any dissociation state has been reached, anharmonic interactions, which operate to couple and mix the defined normal modes, have essentially spread the sharp point-group representation of the original normal mode into a diffuse, continuous representation. The diffuse representation, as is known from nuclear physics,[15] is typical of a system of strongly interacting particles present in large numbers. The continuous groups can be used to handle the many-body strongly interacting systems. The low-energy vibrational states of discrete polyatomic molecules, as is well known, can be handled with a point-group representation. However, it is also known that the point groups are contained as subgroups within the continuous rotation group. As a consequence, for a vibrating polyatomic molecule, it is possible to see that as the molecular vibrations increase in level of excitation within a well-defined normal mode, in the anharmonic limit the states of vibration, which now are considered in a collective, continuous sense, must preserve the essential character of the point group from which the molecule was excited. Near any molecular fragmentation or dissociation limit, however, high-energy vibrational anharmonic interactions can mix these fluid-state modes, thus obliterating traces of the original normal mode from which the molecule originally was excited.

The discussion to this point can be summarized simply by stating that, for a complicated polyatomic molecular system, it appears to be possible to consider two representations of the states of vibration. For states of low vibrational excitation, the familiar normal-mode analysis applies. For states of high, anharmonic excitation a new, continuous, or fluid, representation seems suitable. Thus, it remains necessary to consider the form of the synthesis of these two representations in order to be able to provide an accurate mathematical analysis of all the states of molecular vibration.

As the analysis of the last part of the last section indicates, it is possible to carry out the BOH analysis of the electron-transfer aggregate state with the objective of obtaining an inner-sphere vibrational potential-energy operator. This quantity can be expressed exactly in terms of the solvent-modified electronic operators and basis functions (to the level of accuracy of the treatment of the electronic energy states of the aggregate, *e.g.* HF LCAO–MO, *etc.*). With the use of the complete molecular vibrational potential-energy operator obtained, it is also possible to define the harmonic-oscillator-limit operator by means of the Taylor expansion. Consequently, the harmonic-oscillator normal modes of the aggregate state can be defined.

If we now assume that it is possible to obtain a *model* representation of the states of high vibrational excitation in terms of some adequate continuous representation, then it is possible to effect a coupling of the two representations (one for the states of low excitation, the other for states of high excitation) with the use of the complete molecular vibrational potential-energy operator as defined by the BOH analysis indicated above. Let the vibrational Hamiltonian operator for the aggregate state be defined as shown in equation (219). The potential-energy operator can be expressed as equation (220), where the purturbation $\mathbf{U}'$ is simply defined [equation (221)]. This may seem to be a trivial partitioning, but the result is that if $\mathbf{U}(\{R\})$ is

$$\mathbf{H}_{vib} = \mathbf{T}_N + \mathbf{U}(\{R\}) \qquad (219)$$

$$\mathbf{U}(\{R\}) = \tfrac{1}{2} \sum_{n \in N} (R_n - R_n^0) \cdot \mathbf{W} \cdot (R_{n'} - R_{n'}^0) + \mathbf{U}'(\{R\}) \qquad (220)$$

$$\mathbf{U}'(\{R\}) = \mathbf{U}(\{R\}) - \tfrac{1}{2} \sum_{n \in N} (R_n - R_n^0) \cdot \mathbf{W} \cdot (R_{n'} - R_{n'}^0) \qquad (221)$$

known, together with the harmonic potential, $\mathbf{U}'$ is known. The complete vibrational Hamiltonian operator now can be defined as shown in equation (222), where H^0 has the meaning shown in equation (223), and the eigenfunctions and eigenvalues of this operator are known [equation (224)].

$$\mathbf{H}_{vib} = H^0 + \mathbf{U}'(\{R\}) \qquad (222)$$

$$H^0 = \mathbf{T}_N + \tfrac{1}{2} \sum_{n \in N} (\;_n - R_n^0) \cdot \mathbf{W} \cdot (R_{n'} - R_{n'}^0) \qquad (223)$$

$$H^0 \chi_n^0 = E_n^0 \chi_n^0 \qquad (224)$$

It is clear from the definition of $\mathbf{U}'$ that the matrix elements with respect to the basis functions χ_n^0 should be nearly vanishing for low values of n: $\langle \chi_n^0 | \mathbf{U}' | \chi_n^0 \rangle = 0$. Nevertheless, anharmonic distortion of the vibrational ground and low excited states can be considered, when needed, with the use of the usual quantum-mechanical variational techniques. More important, however, is the fact that the anharmonic distortion is important in states of high vibrational excitation, and this limit cannot be treated accurately with the use of equation (222), together with the basis functions of (224). There is a variational treatment which can be used which was developed by Skyrme for application to problems in nuclear physics.[98] The general application of Skyrme's method to chemical problems is discussed in ref. 94.

In particular, we assume that a model solution to the problem of the states of high vibrational excitation is known.[94] It has the form of equation (225), and can

$$h\phi_n(\zeta) = e_n\phi_n(\zeta) \qquad (225)$$

be solved. A specific model for the Hamiltonian operator h will be considered shortly for the electron-transfer system. Before examining the model, however, we need to establish the link between the complete Hamiltonian operator, the operator for the low-energy normal modes, and the model problem.

Skyrme assumed that the energy difference ε [equation (226)] is essentially constant.[98] A variational calculation is then constructed to determine ε together with

$$E_n - e_n = \varepsilon \qquad (226)$$

functions. In particular, it is possible to express the solutions to equation (224) as equation (227)[94,98] in which the kernel $F(\{R\},\zeta)$ is given by equation (228). If the

$$\chi_n^0(\{R\}) = \int d\zeta F(\{R\},\zeta)\phi_n(\zeta) \qquad (227)$$

$$F(\{R\},\zeta) = \sum_n \chi_n(\{R\})\phi_n^*(\zeta) \qquad (228)$$

partitioning of the total Hamiltonian operator is exact, then equations (229) and (230) may be written, where X is a set of co-ordinates for the excited state. In general,

$$\chi_n(\{R\}) = \chi_n^0(\{q\})\phi_n(X) \qquad (229)$$

$$F(\{R\},\zeta) = \chi_n^0(q)\delta(\zeta - X) \qquad (230)$$

the partitioning is not exact; the model representation may be accidentally close to the true representation. However, with the choice of some reasonably sharply peaked function in place of the delta function in (230), it is possible to optimize the model representation to the point at which agreement with the true energy quantities and functions is close.

The analysis of the non-linear least-squares problem (231),[93,98] subject to the appropriate boundary and normalization conditions, yields the optimum values of the variational coefficients and the energy difference ε.

$$\delta \iint d\{R\}d\zeta F^*(\{R\},\zeta)(H_{vib} - h - \varepsilon)^2 F(\{R\},\zeta) = 0 \qquad (231)$$

The form of the above analysis indicates that, if it is possible to introduce some form of model representation of the states of high molecular vibrational excitation, then it is possible to determine the form of the partitioning of the energy and basis functions. In order to be useful, the model Hamiltonian operator for the vibrationally excited states must be chosen on the basis of physical compatibility with the real system and mathematical compatibility with the total molecular vibrational

---

[98] T. H. R. Skyrme, *Proc. Roy. Soc.*, 1957, **A239**, 399.

potential-energy operator. It seems to be possible to satisfy both of these requirements.

We now turn our attention to the problem of determining a suitable physical and mathematical representation of the highly excited vibrational states of the electron-transfer aggregate state complex.

Given a molecular system, it is possible to define the low-energy molecular harmonic vibrations executed in the energy region of the ground state. As was indicated above, as the total energy increases, vibrational anharmonicity must be considered, and the effect of the anharmonic terms in the vibrational potential function increases in importance as the total vibrational energy increases. Vibrational anharmonic interactions mix the well-defined normal modes. As a consequence of this mixing, the normal modes lose their discrete character (in terms of point-group representations) with respect both to energy and position. In other words, as the anharmonic interactions become important, the original, low-energy, discrete normal-mode representation transforms into a more diffuse, continuous representation. For systems in contact with an energy reservoir at finite temperature, the assumption of a continuous, fluid-like representation of the system is reasonable. The assumption that the highly excited states of the molecules and ions which make up the electron-transfer reaction system can be represented by fluid models is central to the following discussion of the inner-sphere reaction.

The fluid representation of the highly excited vibrational states of solvated or co-ordinated ions has been worked out in some detail.[12] A fluid representation for neutral solvated species also can be formulated; however, only the rough details have been given at this time.[94] As a result, the following discussion concentrates on the consideration of the excited vibrational states of ionic complexes.

The theory which has been developed for the solvated ionic system is based on an analogy between these species and the atomic nucleus. In particular, collective-continuum treatments of the giant resonances of the atomic nucleus[15,99] are based on a model consisting of two interpenetrating fluids contained within a well-defined spherical (or distorted spherical) boundary. One of the fluids (the proton fluid) is charged, the other being neutral (the neutron fluid). The interaction between these two fluids is expressed in terms of the von Weisäcker potential, which can be expressed as equation (232), in which K is a constant, n_Z and n_N are respectively the proton

$$U = K \int_V d^3r(n_Z - n_N)^2/n_0 \qquad (232)$$

and neutron density functions, n_0 is the total density $(n_Z + n_N)$, and V is the volume of the atomic nucleus. The above potential-energy function expresses the fact that, especially for even–even nuclei, strong attractive and repulsive restoring forces operate to preserve a uniform, stable distribution of nuclear matter in the atom.

This concept is easily extended through slight modification to apply to the solvated (or co-ordinated) ion. The ion can be represented as the charged fluid component. Any solvate or co-ordinated ligand molecular species in the well-defined first (and perhaps second) shell about the ion can be represented as a neutral fluid. [It should be noted that if the ligand is charged, as well as the central species (the

---

[99] M. Danos, *Ann. Physik*, 1952, **6**, 265.

ion), then the two-fluid representation is essentially that of a spherically contained plasma.] A unique feature of the atomic system, in contrast to the typical solvated molecular one, is the fact that the total number of charged species is the same as the total number of neutral species. For a solvated ion, the situation is analogous but not identical. Specifically, for a given charge state of an ion, there is generally a fixed co-ordination number for the solvent in the inner sphere. This applies, of course, to co-ordinated species in the same sense. As a result, a potential-energy quantity U_s can be defined for the solvated or co-ordinated ionic system which takes into consideration the fixed co-ordination number [equation (233)], where again K

$$U_s = K \int_{R(r)} d^3r (n_c - \alpha n_s)^2 / n_0 \qquad (233)$$

is a constant that is characteristic of the system, n_c is the number density of charges, n_s the number density of the solvent or ligand species, α is the reciprocal of the co-ordination number, n_0 is the total number density, and finally $R(r)$ is the equation of the surface of the solvated system. It is easy to see that, for a given stationary state of the system, n_0 is constant.

The potential-energy function U_s [equation (233)] expresses the effect of attractive and repulsive forces in the molecular system which tend to preserve an equilibrium number and configuration of species in the complex. The constant K in this parametric representation must therefore depend significantly upon the nature and strength of bonding in the complex. Consequently, in principle, it should be possible to relate the parametric constant K to specific molecular bond properties as determined on the basis of quantum-chemical calculations.[12]

A more important aspect of the fluid representation, especially as parameterized by the potential-energy function (233), is the following. If one attempts to formulate a specific quantum-mechanical calculation for an inner-sphere electron-transfer reaction, then it is necessary to attempt to find some form of universal co-ordinate system applicable to all states of the reacting system. This task, for almost all processes, is virtually impossible. The reason for the great difficulty connected with specific calculations lies with the requirement to consider each individual species together with its own co-ordinate system. It is certainly possible in many cases to find co-ordinate transformations such that the co-ordinates of a particular species can be expanded about a single co-ordinate origin. However, the manipulations are not simple; this is especially clear when one examines the methods used to evaluate electrostatic and other matrix elements for quantum-chemical calculations.[100] The clear advantage of the fluid representation lies with the fact that it is ionocentric. That is, the origin of the co-ordinate system for a particular ionic electroactive species is at the centre of mass for the system. The single co-ordinate system applies both to the charged and to the neutral fluids. Moreover, the same co-ordinate system can be used even if the electroactive species changes charge state, mass, or co-ordination number. As we indicate shortly, these changes in charge state, mass, and co-ordination number are accounted for by means of changes in effective frequency quantities and density distributions for a given species. As a consequence, it is possible to derive a representation of the inner-sphere reaction for which all these

---

[100] H. Silverstone, *cf.* ref. 66.

changes can be handled. In overall form, the contributions due to these changes enter the expression for the activation energy in essentially the same manner as the solvent-repolarization contributions.

In the region of the electroactive species the structure of the system differs substantially from the average random structure of the bulk of the solution. Therefore, with reference to the electroactive species, an arbitrary boundary exists beyond which the remainder of the solvent system is uniform and continuous. Generally, this boundary is taken to be defined by the first solvation or co-ordination shell, but there is no reason why it cannot extend to enclose a larger volume. Indeed, when the second solvation or co-ordination shell about an ion is nearly as well-defined and ordered as is the first, then it is mandatory to include this layer. The boundary surface which defines the electroactive species is generally not simply spherically symmetric and uniform. The surface, in fact, is permanently distorted, to conform to the distribution of solvent or ligand species about the central ion. In addition, in the interior of the complexed species the distribution of matter is non-uniform, but continuous. In reality, the distribution of matter is discontinuous. However, the effect of thermal agitation through the contact between the electroactive species and its environment is to distribute matter. The result is an apparently continuous distribution.

The analysis of the vibrational states of the two-fluid system proceeds first with a treatment of the completely uniform, continuous model. From this treatment it is possible to define fundamental oscillatory modes. An expression for the total energy of this uniform representation can be derived. Next, an interaction operator is defined, the effect of which is to distort the initially uniform system into a configuration which matches the actual distribution of matter in the real electroactive species. Once these two initial aspects of the analysis have been completed, it is possible to proceed with the evaluation of the electron-transfer transition-probability expression and to obtain expressions for the inner-sphere reorganization energies.

Wave propagation in hydrodynamic and acoustic problems is well known.[101,102] The hydrodynamic system is specified by its mass-density functions, the velocities of fluid components, the equations of continuity, and boundary conditions. The appropriate variational calculation leads to the Euler equation, and this equation, through subsequent manipulation, yields a wave equation. The solutions to the wave equation (subject to the boundary conditions) indicate the existence of stationary-wave states in the system. Under certain conditions, the waves can propagate. Finally, the total hydrodynamic energy can be given the form of a Hamiltonian function (classical) or operator (quantal).

The total charge associated with an ionic electroactive species is given by Z [equation (234)], in which n_c is the number density associated with the charged

$$Z = \int_{R(\hat{r})} d^3r\, n_c(r) \tag{234}$$

fluid. The mass density is given simply by $\rho_c = m_c n_c$, where m_c is the mass per particle. The classical model of an ion, that due to Born,[45] satisfies equation (234),

---

[101] A. Sommerfeld, 'Mechanics of Deformable Bodies', Academic Press, New York, 1950.
[102] P. M. Morse and K. U. Ingard, 'Theoretical Acoustics', McGraw-Hill, New York, 1968.

but it does not satisfy the requirements of the hydrodynamic representation. On the other hand, the continuous, non-uniform quantal and classical charge distributions considered recently[49–54] do satisfy all the conditions.

The two-fluid interpenetrating system is specified by the equations of continuity [equations (235) and (236)] for the individual fluids, together with the boundary conditions (237) and (238), where n is the unit vector normal to the surface of the

$$\frac{\partial \rho_c}{\partial t} + \nabla \cdot (\rho_c v_c) = 0 \tag{235}$$

$$\frac{\partial \rho_s}{\partial t} + \nabla \cdot (\rho_s v_s) = 0 \tag{236}$$

$$\left. n \cdot v_c \right|_{R(\hat{r})} = 0 \tag{237}$$

$$\left. n \cdot v_s \right|_{R(\hat{r})} = 0 \tag{238}$$

two-fluid system. As we are interested in the internal system degrees of freedom, it is possible to isolate the centre-of-mass velocity and eliminate it from the remaining considerations. The transformations are (239a) and (239b). The total energy E is

$$V_{cm} = \frac{1}{\rho_0}(\rho_s v_s + \rho_c v_c) \tag{239a}$$

$$v = v_c - v_s \tag{239b}$$

expressed in terms of the kinetic energy T [equation (240)], with ρ_{red} defined by equation (241), and the potential energy, which is the sum of two terms. The first,

$$T = \tfrac{1}{2} \int d^3 r (\rho_0 |V_{cm}|^2 + \rho_{red}|v|^2) \tag{240}$$

$$\rho_{red} = \rho_c \rho_s / \rho_0 \tag{241}$$

U_s, is given by equation (242), with β defined by equation (243). The second contribution to the potential energy is an electrostatic self energy, which can be expressed in terms of the charge number density n_c as shown in (244).

$$U_s = (K/m_c^2 n_0) \int d^3 r (\rho_c - \beta \rho_s)^2 \tag{242}$$

$$\beta = \alpha m_c / m_s \tag{243}$$

$$U_c = e^2 \int d^3 r\, d^3 r' \frac{n_c(r) n_c(r')}{|r - r'|} \tag{244}$$

The variational principle (Hamilton's principle), applied to the Lagrange function $L = T - U$, yields, after a certain amount of manipulation,[12] a wave equation (245) for the oscillations of the two fluids with respect to each other, where $\Box^2$ is the acoustic d'Alembertian operator [as defined in equation (246)] and u is the effective velocity of sound in the medium [equation (247)]. Because of the relation (248), once the solution to equation (245) has been found, equation (248) can be used to find ρ_s.

$$\Box^2 \rho_c = 0 \tag{245}$$

$$\Box^2 = \nabla^2 + (iu)^{-2} \partial^2/\partial t^2 \tag{246}$$

$$u^2 = 2K(1+\beta)^2 \rho_c^{(0)} \rho_s^{(0)}/m_c^2 n_0 \rho_0 \tag{247}$$

$$\rho_0 = \rho_c + \rho_s = \text{constant} \tag{248}$$

The separation of space and time variables yields the Helmholtz equation for the determination of the stationary states of oscillation of the system (249), where $\rho_c(r,t)$ is as defined in equation (250) and $\rho_c(0)$ is the equilibrium density of charge.

$$\nabla^2 \xi + k^2 \xi = 0 \tag{249}$$

$$\rho_c(r,t) = \rho_c(0)[1 + \xi(r,t)] \tag{250}$$

The time-dependence of $\xi(r,t)$ is given by equation (251), and (252) may therefore be written. The Helmholtz equation, equation (249), is satisfied by solutions of the

$$\xi(r,t) = \xi(r)\exp(-i\omega t) \tag{251}$$

$$k^2 = \omega^2/u^2 \tag{225}$$

form $j_l(k_i^{(n)}r) Y_{lm}(\hat{r})$, where $j_l(x)$ is the spherical Bessel function of the first kind[65] and $Y_{lm}(\hat{r})$ is the spherical harmonic function.[65] The eigenvalues $z_i^{(n)} = k_i^{(n)} R_0$, where R_0 is the radius of the (spherical) volume enclosing the two fluids, are found as the roots to the equation (253).[12] The lowest eigenvalue corresponds to a giant acoustic

$$\left.\frac{\partial j_l(k_i^{(n)}r)}{\partial r}\right|_{R_0} = 0 \tag{253}$$

monopole resonance, the second to a giant dipole resonance, and so on.[12] The giant dipole resonances are of importance with respect to the far-i.r. spectrum of simple solvated ions.[12]

The explicit form of the solution to equation (249), subject to the boundary conditions, is equation (254),[12] where $F_l^{(n)}$ is a normalization coefficient [see equation (255)] and $[q^{(n)[l]}(t) \times Y^{[l]}]^{(0)}$, in the Fano–Racah[103] notation, is defined by equation (256). As a result, it is possible to transform the total energy function into a

$$\xi(r,t) = \sum_{n,l=0}^{\infty} F_l^{(n)}(-1)^l j_l(k_i^{(n)} r)(2l+1)^{\frac{1}{2}}[q^{(n)[l]}(t) \times Y^{[l]}]^{[0]} \tag{254}$$

$$F_l^{(n)} = \frac{\sqrt{2}(z_l^{(n)})^{\frac{1}{2}}}{\{z_l^{(n)}[j_l(z_l^{(n)}) - j_{l-1}(z_l^{(n)})j_{l+1}(z_l^{(n)})]\}^{\frac{1}{2}}} \tag{255}$$

$$[q^{(n)[l]}(t) \times Y^{[l]}]^{[0]} = \frac{(-1)^l}{(2l+1)^{\frac{1}{2}}} \sum_m (-1)^m q_{lm}^{(n)}(t) Y_{lm}(\hat{r}) \tag{256}$$

[103] U. Fano and G. Racah, 'Irreducible Tensorial Sets', Academic Press, New York, 1959.

Hamiltonian function of the form (257), where $B_l^{(n)}$ and $C_l^{(n)}$ are given by equations (258) and (259). The frequency $\omega_l^{(n)}$ is simply given by (260).

$$H_{gr} = \tfrac{1}{2}\sum_{n,l}(-1)^l(2l+1)^{\frac{1}{2}}\{B_l^{(n)}[\dot{q}^{(n)[l]} \times \dot{q}^{(n)[l]}]^{[0]} + C_l^{(n)}[q^{(n)[l]} \times q^{(n)[l]}]^{[0]}\}$$

$$= \sum_{n,l,m} \hbar\omega_l^{(n)}\{b_{lm}^{(n)\dagger}b_{lm}^{(n)} + 1/2\} \tag{257}$$

$$B_l^{(n)} = \rho_c^2(0)/\rho_{red}^{(0)}(k_l^{(n)})^2 \tag{258}$$

$$C_l^{(n)} = 2K(1+\beta)^2 \rho_c^2(0)/m_c^2 \rho_0 \tag{259}$$

$$\omega_l^{(n)} = \{C_l^{(n)}/B_l^{(n)}\}^{\frac{1}{2}} \tag{260}$$

The analysis developed to this point assumes a completely simple spherical volume enclosing the inner-sphere system. Real systems, in contrast, are characterized by non-uniform distributions of matter, both within the inner-sphere system and with respect to the surface. The distortions of the surface modes, which have not yet been considered, interact with the internal giant acoustic resonance modes to bring about modifications of the fundamental frequencies just defined. This effect is recognized and documented in nuclear physics.[15,104] A similar effect will apply to the solvated or co-ordinated electroactive species modelled as a two-fluid system. A detailed analysis has not yet been carried out. The second distortion effect, which has been considered in detail,[12,13] concerns the replacement of the uniform distribution developed above by a distribution which mirrors the real distribution of matter within the complex. This effect is fundamental to the consideration of the inner-sphere reaction, and is considered in the following paragraphs.

An examination of the fundamental modes of excitation in the originally defined uniform system indicates that it is possible to consider individual excitations or linear combinations of excitations which possess the same group-theoretical transformation properties as the real system. By drawing an analogy to the familiar distortion of vibrational potential-energy surfaces in molecular vibronic excited states, it is possible to see that, if one considers some form of interaction which can permanently distort the initially uniform distribution of matter in the two-fluid system, then for a given stationary state a reasonable representation of the system can be found. For the solvated or complexed ionic systems which have been considered so far, this distortion interaction is relatively simple to specify.[12,13]

Associated with the uniform fluid distribution of charge considered above is an induction field $D_c(r)$. If this charge distribution is allowed to interact with another charge distribution $D_d(r)$, a distortion field, located at the same co-ordinate origin, the response of the fluid system can be found in terms of the elementary excitations defined for that system, viz. giant monopole, dipole, etc. excitations. The magnitude of each contribution can be determined. In form, the determination of the degree of distortion of the model system needed to conform to the actual distribution of matter in the inner-sphere system is similar to the determination of the splitting of the d-electron energy levels in a crystal field, as is usually done for the transition-metal compounds. The interaction can be specified as equation (261). If the Fourier

[104] H. Arenhövel, M. Danos, and W. Greiner, *Phys. Rev.*, 1967, **157**, 1109.

$$H_{int} = \frac{1}{8\pi} \int d^3 r \, \boldsymbol{D}_d(\boldsymbol{r}) \cdot \boldsymbol{D}_c(\boldsymbol{r}) \tag{261}$$

transforms of each charge distribution are known, then, by virtue of Parseval's theorem,[65] (261) can be written as (262), since both charge distributions share the

$$H_{int} = \frac{1}{2\pi^2} \int d^3 k \, k^{-2} n_d^*(\boldsymbol{k}) n_c(\boldsymbol{k}) \tag{262}$$

same co-ordinate origin. The Fourier transform of $n_c(\boldsymbol{r})$ is easily found to be $n_c(\boldsymbol{k})$ [equation (263)].

$$n_c(\boldsymbol{k}) = 4\pi n_c(0) \left[2\sqrt{\pi R_0^3} j_1(kR_0)/(kR_0) \right. \\ \left. + \sum_{n,l,m} (F_l^{(n)})^{-1}(-i)^l (2l+1)^{\frac{1}{2}} [q^{(n)[l]} \times Y^{[l]}]^{[0]} \delta_{k,k_l^{(n)}} \right] \tag{263}$$

The actual distribution of charge within the electroactive species is governed by the appropriate solutions to the electronic wave equation for the system. That is, given an electronic state of the system, the charge density at any point is the complex product of the wavefunctions. As was indicated earlier in this chapter, a considerable amount of attention has recently been focused on the formulation and application of continuous charge-distribution representations of solvated-ion systems in order to obtain solvation free energies, *etc.*[49—54] The classical Slater-type distribution mirrors the accurate quantum-mechanical quantity, but does not admit some of the subtle quantum-mechanical interactions such as exchange interactions.[49] If we assume that a Slater-type distribution can be used as an accurate description of the charge distribution in the solvated or co-ordinated electroactive species, then it is possible easily to obtain an expression for the interaction energy which accounts for the distortion necessary to alter the uniform two-fluid representation.[12,13]

For the moment, we consider the simplest case: a 1s Slater-type classical charge-distribution model of the ion.[12] The normalized charge distribution is given by equation (264), where p is related to an effective Bohr radius[49] by equation (265). The Fourier transform of this distribution is easily found to be (266), where Z is the

$$n(\boldsymbol{r}) = \frac{Zp^3}{4\sqrt{\pi}} Y_{00}(\hat{r}) \exp(-pr) \tag{264}$$

$$p = 2/a \tag{265}$$

$$n(\boldsymbol{k}) = \sqrt{4\pi} \, Y_{00}(\hat{k}) Zp^4/(p^2 + k^2)^2 \tag{266}$$

effective charge on the ion. With the use of equations (258), (263), and (266), the distortion interaction energy is found to be given by equation (267). In this equation,

$$H_{int} = \frac{ZV_0^{\frac{1}{2}}}{\pi} \sum_{n=0}^{\infty} (6\hbar\mu/\omega_0^{(n)})^{\frac{1}{2}} \frac{z_{0n}^2}{(1+y_{0n}^2)^2} [j_0^2(z_{0n}) - j_{-1}(z_{0n})j_1(z_{0n})]^{\frac{1}{2}} q_{00}^{(n)} \tag{267}$$

μ is the reduced mass of the species, V_0 is the ionic volume, and y_{0n} is given by equation (268). The form of this interaction energy can be simplified with the use of the observation that $j_1(z_{0n}) \simeq 0$. Hence, H_{int} is given by the approximation (269).

$$y_{0n} = k_0^{(n)}/p = az_{0n}/2R_0 \tag{268}$$

$$H_{int} \simeq \frac{ZV_0^{\frac{1}{2}}}{\pi} \sum_{n=0}^{\infty} (6\hbar\mu/\omega_0^{(n)})^{\frac{1}{2}} \frac{z_{0n}^2}{(1+y_{0n}^2)^2} |j_0(z_{0n})| q_{00}^{(n)} \tag{269}$$

When the above form of interaction term is added to the general Hamiltonian operator for the acoustic giant resonance modes, the resulting expression can be written as (270). The co-ordinate shift quantity $\bar{q}^{(n)}$ is readily identified as (271). The

$$H_{gr} = H_{gr(l \neq 0)} + \tfrac{1}{2} \sum_{n=0}^{\infty} \hbar\omega_0^{(n)} [(p_{00}^{(n)})^2 + (q_{00}^{(n)} - \bar{q}_{00}^{(n)})^2] - \tfrac{1}{2} \sum_{n=0}^{\infty} \hbar\omega_0^{(n)} (\bar{q}_{00}^{(n)})^2 \tag{270}$$

$$\bar{q}_{00}^{(n)} = -\frac{ZV_0^{\frac{1}{2}}}{\pi} (6\mu/\hbar\omega_0^{(n)3})^{\frac{1}{2}} \frac{z_{0n}^2}{(1+y_{0n}^2)^2} |j_0(z_{0n})| \tag{271}$$

third term in equation (270) is essentially the self energy of the electroactive species with respect to the initially uniform distribution of fluids.

It has been shown[13] that in the normal region for the heat of reaction, the electron-transfer transition probability can be written as (272). The quantity ΔJ, even with

$$W_{if} = \frac{|H'_{if}|^2}{\hbar} \left(\frac{\pi\beta}{E_r^s + E_r^{(2)}} \right)^{\frac{1}{2}} \exp\left(-\frac{[\Delta J + E_r^s + E_r^{(1)}]^2 \beta}{4[E_r^s + E_r^{(2)}]} \right) \tag{272}$$

the use of the simple Born–Oppenheimer separation, contains the difference of the self energies (271) for the initial and final state as well as the difference in the solvation energies, *etc.* In the simple treatment used in ref. 13, consideration of the work terms for the formation of the transfer electroactive species from the reactants in an initial state of infinite separation was not considered. It is clear that these terms belong in the expression, and their inclusion can be effected; such a treatment, however, has not yet been carried out for the hydrodynamic representation. The energy quantity E_r^s is the solvent-subsystem repolarization energy, and it has the form of the repolarization energy discussed in Part I. If the disperse character of the polar solvent is taken into account,[40-44] the form of the repolarization energy will be modified slightly to the form given by Dogonadze and Kornyshev.[43] The other reorganization energy $E_r^{(i)}$ ($i = 1$ or 2) is connected with the hydrodynamic inner co-ordination or solvation shells.[13] The general term can be written as equation (273). The terms in this expression have the following definitions. The quantity $\gamma_l^{(n)}$ is the ratio of the final to the initial frequency of an individual hydrodynamic state [equation (274)]. The quantity $^i\phi_{lm}^{(n)}$ is defined as (275),[13] and the quantity

$$E_r^{(i)} = \tfrac{1}{2} \sum_{n,l,m} (\gamma_l^{(n)})^{i-1} \{^f\phi_{lm}^{(n)} - (\gamma_l^{(n)})^{\frac{1}{2}} \, ^i\phi_{lm}^{(n)}\}^2 \tag{273}$$

$$\gamma_l^{(n)} = \omega_l^{(n)f}/\omega_l^{(n)i} \tag{274}$$

$$^i\phi_{lm}^{(n)} = \frac{Z}{2\pi(1+R)} \left(\frac{3(Z+S)}{K} \right)^{\frac{1}{2}} i^{l+2m} |j_l(z_{ln})| g_{lm}^{(n)}(k_l^{(n)}) \tag{275}$$

$g_{lm}^{(n)}$ depends essentially upon the radial part of the Fourier transform of the distortion charge density evaluated at $k_i^{(n)}$.[13] The form of these terms is illustrated shortly, with the use of specific models.

The expression for the reorganization energy alone yields some interesting limit conditions, even without considering the specific form of the g-quantities introduced in equation (275). In particular, we see from equation (273) that, if $\gamma_i^{(n)} \to 0$, which implies $\omega_i^{(n)f} \ll \omega_i^{(n)i}$ for a particular state n,l,m, then $E_r^{(2)} \to 0$. This limit implies that the reorganization energy reduces essentially to the self energy of the inner sphere of the final state with respect to its formation from the initially spherical two-fluid distribution. This energy quantity subtracts from the similar energy quantity in ΔJ, and the entire expression for the activation reduces to the outer-sphere limit. Yet, interestingly enough, it is possible that the process described by this representation could represent a case for which there is substantial reorganization of the chemical character of the solvation or co-ordination shell in a particular reactant or pair of reactants. Reactions of this type have been called vibrationally barrierless.[13] This, of course, does not mean that the reaction is barrierless overall.

There is a second condition for which the reaction can be defined either as vibrationally barrierless or activationless. In particular, if $E_r^{(i)} = 0$ for $i = 1$ and 2, then the ratio of the final and initial state frequencies is nearly equal to the ratio of the squares of the displacements [see (276)]. Essentially, this means simply that the

$$\omega_l^{(n)f}/\omega_l^{(n)i} \simeq (^f\phi_{lm}^{(n)}/^i\phi_{lm}^{(n)})^2 \qquad (276)$$

distortion of the inner-sphere potential in one state is so great in comparison to the other that it requires only a low-order vibration of the species in the inner shell to reach the transition-state configuration.

The assessment of the magnitude of the contribution of the inner co-ordination shell to the activation energy requires the consideration of all the pertinent terms in the general expression (273). Clearly, this term can be written as (277), in which each term $E_r^{(i)}(n,l,m)$ is defined by the summand of equation (273). It is obvious

$$E_r^{(i)} = \sum_{n,l,m} E_r^{(i)}(n,l,m) \qquad (277)$$

that, although the above-mentioned limits may be satisfied by some of the terms $E_r^{(i)}(n,l,m)$, it is by no means certain that all the terms will obey the same limit. The careful consideration of the inner-shell contributions to the activation process, therefore, requires the consideration of the form of the quantities $g_{lm}^{(n)}(k_i^{(n)})$, as defined by equation (275), for each state. This we now consider by examining a specific model representation.

Assume that the co-ordination of ligand or solvent about a given ionic electroactive species is tetrahedral. In view of this distribution of neutral solvent or ligand, it is reasonable to assume that the central electronic charge distribution will be distorted to a degree, and that this degree of distortion has the same space-group representation as the ligand distribution. An argument of this type has been used, with success, in the consideration of the solvation energy of simple ionic systems.[52] The charge density for the distortion field is given by equation (278). The p-quantity in this case is related to the effective Bohr radius for the distribution by equation (279). The quantity b is adjustable, and represents the degree of tetragonal distortion

$$(nr) = \frac{Zp}{4!\sqrt{4\pi}} r^2 \exp(-pr)\{Y_{00}(\hat{r}) - (br)(i/\sqrt{2})[Y_{32}(\hat{r}) - Y_{3-2}(\hat{r})]\} \quad (278)$$

$$p = 4/a \quad (279)$$

needed in the classical phenomenological soft charge density in order for that charge density to be an accurate representation of the actual charge density. The Fourier transform of the charge density (278) is given by equation (280). When this quantity

$$n(k) = \sqrt{4\pi}\left\{\frac{p^6}{(p^2+k^2)^3}\left(1 - \frac{2p^2}{(p^2+k^2)}\right)Y_{00}(\hat{k}) + \frac{8bp^6k^3}{(p^2+k^2)^5}[Y_{32}(\hat{k}) - Y_{3-2}(\hat{k})]\right\} \quad (280)$$

is inserted into the expression for H_{int} [equation (269)] and the indicated integrations are carried out, one may then identify the g-quantities as shown in equations (281) and (282). It is not difficult to see that the above g-quantities are readily evaluated in

$$g_{00}^{(n)}(k_1^{(n)}) = -\sqrt{4\pi}\frac{1}{(1+y_{0n}^2)^3}\left(1 - \frac{2}{1+y_{0n}^2}\right) \quad (281)$$

$$g_{32}^{(n)}(k_3^{(n)}) = g_{3-2}^{(n)}(k_3^{(n)}) = -2\sqrt{4\pi}(ab)\frac{y_{3n}^3}{(1+y_{3n}^2)^5} \quad (282)$$

terms of experimentally available crystallographic and solvation radii. Moreover, it is possible to establish a set of conditions which, at the most, indicate whether a reaction is outer-sphere or not, and typically indicate whether the inner-sphere reorganization energy contributes substantially to the activation energy.

A general condition for an outer-sphere reaction is expressed in the statement that the inner co-ordination shell does not undergo any substantial chemical or physical modification. If this is to be the case, then it is reasonable to expect that the distortion charge density, which provides the modification of the two-fluid model needed to make it realistic, is the same in the initial and final states. Specifically, the functional form of the charge density is the same in these two states; although there may be a general overall expansion or contraction due to the change in charge state. This assumption simply means that equation (283) is valid, or, equivalently, equation (284), where the crystallographic radii are used for the effective Bohr radii, a,

$$^f g_l^{(n)} \cong \, ^i g_l^{(n)} \quad (283)$$

$$a_f/a_i \cong R_0^f/R_0^i \quad (284)$$

and R_0 is the solvation radius. With the use of the condition expressed by equation (283) it is possible to write the individual reorganization terms $E_r^{(i)}(n,l,m)$ as shown in equation (285), where Z_f, Z_i are the charges on the species in the final and initial

$$E^{(i)}(n,l,m) = (3/4\pi)^{\frac{1}{2}} i^{l+2m} |j_l(z_l^{(n)}/pR^i)|(\gamma_l^{(n)})^{i-1}$$

$$\times \left\{\frac{(Z_f+S_f)^{\frac{1}{2}}}{R_0^f K_f^{\frac{1}{2}}(1+\beta_f)} - (\gamma_l^{(n)})^{\frac{1}{2}}\frac{(Z_i+S_i)^{\frac{1}{2}}}{R_0^i K_i^{\frac{1}{2}}(1+\beta_i)}\right\}^2 \quad (285)$$

states, and $S_{i(f)}$ is the solvent or ligand co-ordination number. In general, the phenomenological constant K, which is identified in equation (233), is essentially an energy per ion–ligand bond, and is constant and independent of the charge state of the central ion. As a result, this quantity also can be removed from the expression in braces in equation (285). The terms which remain essentially cancel one another for an outer-sphere reaction. Consequently, the entire term vanishes. Moreover, for an outer-sphere electron-transfer reaction, all terms of the form (285) vanish. This is partly due to the fact that the terms within the braces in equation (285) are relatively insensitive to the changes in charge state, solvation radius, and changes in frequency which can be allowed for the outer-sphere reaction. It is possible to summarize the outer-sphere limit in the relationship (286),[13] where V_i^0 is the ionic partial molar

$$a_i/a_f = \left(\frac{V_i^0 + \tfrac{1}{2}S_i V_s^0}{V_f^0 + \tfrac{1}{2}S_f V_s^0}\right)^{1/3} \tag{286}$$

volume, V_s^0 the partial molar volume of the solvent, and $S_{i(f)}$ is the co-ordination number of the solvent (or ligand) in the first layer. For the Fe ^{II}/Fe ^{III} system the ratio of crystallographic radii is 1.2, and the term on the right of equation (286) is 1.11. As a consequence, one would expect the inner-sphere contribution in that system to be small.[13]

The analysis outlined above can be carried out with other types of distortion charge-density distributions. Gaussian distributions, which find extensive use today in quantum-chemical calculations, have been applied to the electron-transfer system in the sense implied by this hydrodynamic theory.[13] The results have a simpler form, generally, than is the case with the use of the Slater functions. Although, thus far, the hydrodynamic representation of the inner-sphere degrees of freedom has been investigated with the use only of the classical, phenomenological charge densities,[49—52] it is entirely possible to carry out the analysis with the use of the actual charge densities found by any of the usual quantum-chemical means. The quantum-chemical charge density is given by equation (168) in the LCAO–MO approximation. As the hydrodynamic analysis is most conveniently carried out in a Fourier-transform representation, it is possible to make use of the fact that a general function $F(\boldsymbol{R})$ [see equation (287)] can be developed in terms of an expansion about a

$$F(\boldsymbol{R}) = \frac{1}{(2\pi)^3} \int d^3k\, f(\boldsymbol{k}) \exp[-i\boldsymbol{k}\cdot(\boldsymbol{R}_1 - \boldsymbol{R}_2)] \tag{287}$$

different co-ordinate origin; $\boldsymbol{R} = \boldsymbol{R}_1 - \boldsymbol{R}_2$, and $f(\boldsymbol{k})$ is the Fourier transform of $F(\boldsymbol{R})$. Thus, if $\boldsymbol{R}_1$ is a co-ordinate directed from the centre of mass of the hydrodynamic representation of the electroactive species, then the distortion charge density can be expanded in terms of the position vectors of all the species with respect to the centre-of-mass vector.

As was indicated earlier, an advantage of this hydrodynamic representation of the electron-transfer system is tied to the fact that all reference states are described in terms of a single co-ordinate system. The form of the inner-sphere reorganization energy, as indicated by equation (272), is virtually the same as the repolarization energy due to the solvent-system modes. That is, the reorganization energy is defined in terms of boson operators, and their expectation value, and it enters the transition-probability expression in the same manner as the solvent-subsystem repolarization energies. Consequently, the assessment of the contributions of the inner co-ordination shell to the activation energy, even for the case in which there is gross chemical change in the electroactive species, is straightforward. The evaluation of these terms is greatly simplified by the use of this analysis.

It must be borne in mind, however, that this analysis applies only to systems in which the binding of the ligands to the central species is weak (*viz.* solvated ions and labile complexes) or to systems in which the gross chemical change which accompanies the electron transfer involves highly excited states of the inner sphere. Systems for which there is gross chemical change in species which are nominally stable, *viz.* tightly bonded complexes, will take place probably by means of nuclear tunnelling of the ligands. The form of the transition probability, and rate constant, will be considerably different compared to the expression (272). This problem will be discussed briefly in the next sections, which deal with the formal aspects of the formulation of the transition-probability and rate-constant expressions.

To conclude this discussion of the collective treatment of the inner solvation shell, we examine the dynamics of the solvation boundary. This aspect of the representation is not as important as the internal giant resonance modes with respect to the activation process for an electron transfer. That is, surface-mode reorganization energies generally contribute only small energy terms to the activation energy. The surface modes, and their permanent distortions brought about by the distortion charge densities considered above, are important, however, in the determination of the frequencies of the internal giant resonances, and they are important with respect to the form of the interaction potential-energy function which operates between reactants.

A solvated electroactive species is reasonably expected to consist of a central active species surrounded by a shell of ligand, which may be solvent. The space defined by the electroactive species, together with its first co-ordination shell, is expected reasonably to differ substantially from the rest of the system. The electroactive species generally consists of an ordered arrangement of molecules in a small volume defined by the solvation or co-ordination radius. Outside this volume, the remainder of the system may be expected to be random. The volume external to the electroactive species therefore appears as a continuum.

Several forces operate within the electroactive species to preserve its structured independence. It is clear that electrostatic and quantum-mechanical bonding interactions are responsible for the internal order, yet the objective of the continuum hydrodynamic representation of the co-ordinated, electroactive species is to provide the means for carrying out calculations of transition probabilities, *etc.*, for complicated systems in terms of simple, phenomenological (but accurate) model representations. Thus it is necessary to define a surface for the solvated electroactive species, and it is necessary to be able to estimate the orders of magnitude of the parameters associated with the model. For a solvated ionic species, the surface can be defined in terms of the total charge of the system and a surface-tension parameter. It is essential to note that this surface-tension quantity is not rigorously defined in the same sense as a typical macroscopic surface tension, but it can be defined for the solvated system in essentially the same sense as it is used in nuclear physics.[15] In particular, a surface tension defined for a complex solvated ionic system represents the balance of forces operating within the ionic system and in the bulk of the solution to preserve the local internal molecular order.[14] When defined in this manner, it is possible for the electrochemical system, as it is for the atomic nucleus,[105] to relate the surface tension to molecular properties of the system.[12,14] Although it is clear that this is the case, detailed calculations have not yet been carried out. Nevertheless, estimates of the magnitudes of the surface-tension quantity have been made.[12,14]

The collective surface modes are defined for an electroactive species in essentially the same manner as they are defined for the atomic nucleus.[12] The distortion of an initially uniform, spherical to an accurate geometric model of the actual distribution of molecular species is similar to that used to treat the internal giant resonance modes.[14] A sketch of the analysis of the surface modes is now given.

The surface modes are defined in part in terms of the hydrodynamic velocity potential ϕ, which for an irrotational and incompressible fluid gives[106] the relationship shown in equation (288). The potential, therefore, is a solution to the Laplace

$$v = \nabla \phi \tag{288}$$

$$\phi = \sum_{L,M} (R_0^2/L)\dot{\alpha}_{LM}^*(r/R_0) Y_{LM}(\hat{r}) \tag{289}$$

equation.[106] Thus, the general solution can be written as equation (289).[12,15] The kinetic energy is given by equation (290),[15] with B_L defined as shown in equation (291). The coefficient B_L is referred to as an 'inertia' parameter.[15]

$$T = \tfrac{1}{2}\rho_0 \int d^3r |\nabla \phi|^2 = \tfrac{1}{2} \sum_{L,M} B_L |\dot{\alpha}_{LM}|^2 \tag{290}$$

$$B_L = \rho_0 R_0/L \tag{291}$$

The potential-energy function consists of two parts. One is associated with the surface tension and the other with the charge. The origin of this term is with Rayleigh,[107] who first obtained it in an investigation of capillary waves on charged droplets. The potential-energy function can be expressed as (292). The first term depends on the variation of the surface area with change in the function defining the surface, $R(\hat{r})$ [equation (293)]; γ is the surface tension. ΔS is the change in surface area, and it is a function of the quantity ζ, which is defined by equation (294). The

$$V = V_{\text{sur}} + \delta V_c \tag{292}$$

$$V_{\text{sur}} = \gamma \Delta S \tag{293}$$

$$\zeta = R(\hat{r}) - R_0 = R_0 \sum_{L,M} \alpha_{LM} Y_{LM}^*(\hat{r}) \tag{294}$$

variation of the electrostatic charge with ζ is found by means of a variational calculation.[15] The results for the system can be expressed as the potential-energy function (295),[15,108] with C_L, called the 'stiffness' parameter, given by equation (296); Z is the total charge on the electroactive species.

[105] F. Flugge, *Ann. Physik*, 1941, **39**, 373.
[106] L. D. Landau and E. M. Lifshitz, 'Fluid Mechanics', Pergamon Press, London, 1959.
[107] Lord Rayleigh, *Proc. Roy. Soc.*, 1879, **29**, 71; *Phil. Mag.*, 1882, **14**, 184.
[108] M. Fierz, *Helv. Phys. Acta*, 1943, **16**, 365.

$$V = \tfrac{1}{2} \sum_{L,M} C_L |\alpha_{LM}|^2 \qquad (295)$$

$$C_L = (L-1)\left((L+2)R_0^2 \gamma - \frac{3e^2 Z^2}{4\pi(2L+1)R_0}\right) \qquad (296)$$

The total energy, $T+V$, defines the Hamilton function. Thus, for the surface modes we have the Hamiltonian operator of equations (297). The fundamental surface mode frequency, ω_L, is related to the quantities B_L and C_L by equation (298).

$$H_{\text{sur}} = \tfrac{1}{2} \sum_{L,M} \{B_L |\dot{\alpha}_{LM}|^2 + C_L |\alpha_{LM}|^2\} \qquad (297a)$$

$$= \sum_{L,M} \hbar \omega_L (a^\dagger_{LM} a_{LM} + 1/2) \qquad (297b)$$

$$\omega_L = (C_L/B_L)^{\tfrac{1}{2}} \qquad (298)$$

The Hamiltonian operator, equation (297), is appropriate to use in the analysis of the contribution of the surface modes to the activation process.[14]

In order to carry out an analysis of the surface-mode contribution to the activation process, it is necessary first to determine how the above initially uniform spherical representation can be deformed into a configuration which models the actual distribution of matter and the surface which can be defined for it. The surface distortions and oscillations are given by the function ζ defined by equation (294). It is clear that a suitable linear combination of individual permanent distortions can, in most instances, duplicate the actual molecular surface. As a result, we apply to the surface a simple Galilean transformation which has the form (299), for which

$$\alpha_{LM} \to \alpha_{LM} + \beta_{LM} \qquad (299)$$

α_{LM} remains an operator, but β_{LM} is a c-number (*i.e.* it is not an operator). Moreover, the quantity β_{LM} is independent of time; thus, $\dot{\beta}_{LM} = 0$. As a result, the Hamiltonian operator for the surface in a state of permanent distortion can be written as equation (300)[14] where q_{LM} is related to α_{LM} by equation (301).[14] Similarly, $\bar{q}_{LM}$ can be related

$$H_{\text{sur}} = \tfrac{1}{2} \sum_{L,M} \hbar\omega_L \{|p_{LM}|^2 + |q_{LM} - \bar{q}_{LM}|^2\} \qquad (300)$$

$$q_{LM} = (2B_L \omega_L/\hbar)^{\tfrac{1}{2}} \alpha_{LM} \qquad (301)$$

$$\bar{q}_{LM} = (2C_L/\hbar\omega_L)^{\tfrac{1}{2}} \beta_{LM} \qquad (302)$$

to β_{LM} by equation (302).[14] We see, therefore, that if it is possible to determine the β-quantities, it is possible ultimately to determine the surface contribution to the activation energy.

The surface activation-energy contributions are given by equation (303),[14] with $E_r^{\text{sur}}(L,M)$ defined as in equation (304), where $\Delta \bar{q}_{LM}$ is the change in the distortion of

$$E_r^{\text{sur}(i)} = \sum_{L,M} \Gamma_L^{1-i} E_r^{\text{sur}}(L,M) \qquad (303)$$

$$E_r^{\text{sur}}(L,M) = \tfrac{1}{2} \hbar \omega_L |\Delta \bar{q}_{LM}|^2 \qquad (304)$$

the surface for the initial and final electron-transfer states. The factor Γ_L in this case is the ratio of surface mode frequencies [equation (305)]. In the same manner as was done for the internal giant resonance modes,[13] for the surface modes we can write equation (306),[14] where $\Phi_{LM}^{(i)}$ is defined as shown in (307).[14]

$$\Gamma_L = \omega_L^{(f)}/\omega_L^{(i)} \tag{305}$$

$$E_r^{sur}(L,M) = |\Phi_{LM}^{(f)} - \Gamma_L^{\frac{1}{2}}\Phi_{LM}^{(i)}|^2 \tag{306}$$

$$\Phi_{LM}^{(i,f)} = [C_L^{(i,f)}]^{\frac{1}{2}}\beta_{LM} \tag{307}$$

If the surface of an initially uniform sphere is permanently displaced, that displacement is signified by $\Delta R(\hat{r})$. The equation for the surface itself under a displacement condition is expressible as (308).[14] The surface displacement is therefore given by equation (309). Generally, the surface displacement ΔR is a known function.[14]

$$R(\hat{r}) \to R_0\{1 + \sum_{L,M}(\alpha_{LM} + \beta_{LM})Y_{LM}^*(\hat{r})\} \tag{308}$$

$$\Delta R(\hat{r}) = R_0 \sum_{L,M} \beta_{LM} Y_{LM}^*(\hat{r}) \tag{309}$$

Hence, it is possible simply to invert the relationship (309) in order to express the displacement coefficients β_{LM} in terms of the actual equation for the surface, giving (310), and $d\Omega_r$ is the element of surface area.

$$\beta_{LM} = \frac{1}{R_0} \int_{4\pi} d\Omega_r \Delta R(\hat{r}) Y_{LM}(\hat{r}) \tag{310}$$

As an example, we consider a permanent tetragonal surface displacement.[14] The expression for the distortion can be formed from the angular part of the charge distribution used to distort the internal giant resonance modes. Here, however, monopole terms are not involved, as the system is volume-conserving. Specifically, the tetragonal distortion is given as shown in equation (311).[14] Consequently, the β terms are identified immediately as (312). The contribution to the reorganization energy from the surface modes is given as (313).[14] An estimate of the magnitude of this quantity has been made, based upon the use of the relationship (314), which

$$\Delta R(\hat{r})/R_0 = \frac{i}{\sqrt{2}}(bR_0)^3\{Y_{3-2}(\hat{r}) - Y_{32}(\hat{r})\} \tag{311}$$

$$\beta_{32} = \beta_{3-2} = -\frac{i}{\sqrt{2}}(bR_0)^3 \tag{312}$$

$$E_r^{sur(i)} = \Gamma_3^{i-1}C_3^f(b_fR_f)^6\{1 - \Gamma_3^{-\frac{1}{2}}(R_i/R_f)^4(b_i/b_f)^3(m_i/m_f)^{\frac{1}{2}}\}^2 \tag{313}$$

$$b = 1/(a-c) \tag{314}$$

was derived from an examination of the solvation free energies of simple cations and anions.[52] The quantity a is taken to be a crystallographic radius, and c is a factor characteristic of a given group. As long as the type of distortion remains the same, the surface contribution to the activation energy is small: of the order of 40 J mol^{-1}. That is, as long as the permanent displacement of the initially spherical surface remains, for example, tetragonal, even though the radial quantities, charge, and masses change, this energy generally will be small. On the other hand, if the surface undergoes a substantial change in the geometry of the permanent displacement, *e.g.* from tetrahedral to octahedral, it is reasonable to expect a substantial surface activation contribution to the electron transfer.[14]

The energies associated with the fundamental modes in the collective representation both of the internal and surface modes of the inner-sphere system are not large. For the giant resonance modes, the frequencies can be expressed, suggestively, as (315), in which N_0 is the total number of charges and solvent molecules, M_c is the

$$1/\lambda = 737.5 \left(\frac{K(1+\beta)^2 M_s}{N_0 M M_c} \right)^{\frac{1}{2}} (z_l^{(n)}/R_0) \qquad (315)$$

mass per unit charge for the charged species, M_s is the total solvent mass, and M is the total mass of the system. The far-i.r. spectra of a number of solvated ions lie in the range 300—500 cm^{-1}. An analysis of a few of these systems suggests that the value of K in (315) should be of the order of 30 eV.[13] On this basis, together with considerations about ionic solvation radii, it has been argued that the fundamental far-i.r. and Raman frequencies should change by about 100 cm^{-1} for a change in charge state of one e.s.u.

The fundamental frequencies of the solvation surface modes, on the other hand, are given by the formula (316),[12] where c is the velocity of light. It is seen from this

$$1/\lambda_s = \frac{4\pi}{3} \left(\frac{L(L-1)(L+2)}{\rho_0} R_0^2 \Gamma - \frac{e^2 Z^2}{(2L+1)R_0} \right)^{\frac{1}{2}} (1/2\pi c R_0) \qquad (316)$$

expression that the first allowed excitation is a quadrupole mode, $L = 2$. Monopole and dipole transitions are not allowed. On the basis of the examination of some of the simple solvated ionic systems, *viz.* ^{7}Li$^+$ to Al^{3+}, it was concluded that the minimum values of the effective surface tension can range from about 100 to 3000 erg cm^{-1}. The frequencies expected for the quadrupole modes in these cases range from 10 to 100 cm^{-1}. Thus, it is seen that the surface-mode frequencies are considerably smaller than the internal modes, and this is as it should be. Whether in all cases the difference in energy between these two degrees of excitation will be this large remains to be tested. The predicted difference in energy is sufficient, for the most part, to enable one to use a Born–Oppenheimer approximation to separate the two modes of excitation, and to treat them separately.

This hydrodynamic representation of the internal solvation or co-ordination states of the electroactive species is by no means complete. The analysis of the representation, sketched in this section, is admittedly complicated, but the complications are very much less than would be the case if one attempted a more exact treatment based on the usual particulate quantum mechanics.

A number of phenomenological, adjustable parameters have been introduced with this treatment: in particular, K, effectively an energy per ion–solvent bond, and the effective solvation surface tension. The quantity K can be estimated by comparison with experiment. As indicated, it is of the order of 30 eV. This may seem an excessively large number. However, one must remember that it enters as a harmonic-density-restoring 'force constant'. For small density displacements, the energy of restoration is generally small. The effective surface-tension quantity, on the other hand, does not yet seem to be an experimentally accessible quantity. It may be possible to extract it from conductivity data, if the theory of ionic conduction can be modified in such a manner as to include this collective representation of a deformable ion.[14]

It is possible, nevertheless, to relate these phenomenological quantities to molecular parameters such as vibrational force constants, bond orders, charge densities, *etc.*, in much the same manner as has been done in nuclear physics.[82,105] It is clear, for example, that the K term should be related to the exact molecular vibrational potential-energy function for the solvated or co-ordinated complex. The analysis anticipated depends substantially upon the careful BOH-type treatment discussed in the last section. Ultimately, it should be possible to consider the dynamics of the gross reorganization of the ligands in the co-ordination shells of the species that are undergoing electron-transfer reactions with considerable confidence.

The reorganization of ligands which accompanies many electron-transfer reactions can take place according to one of two principal limits: the first is the low-energy, tunnelling limit and the second is the high-energy collective limit. We have considered this second limit in this section. The low-energy tunnel transitions have been considered by several authors.[109,110] Whether a reaction follows a high- or low-temperature-limit transition depends upon the temperature at which the reaction is carried out (kT) and the strength of the bonding within the species. Thus, at ambient temperatures, we can consider the collective states of a simple, loosely solvated ionic system in the high-energy limit. In contrast, a tightly bound complexed ion at ambient temperatures would undergo change by tunnelling – a low-energy limit with respect to the bond energy in comparison to kT. The treatment of the tunnel transitions makes use of the actual molecular vibrational potential-energy functions. In most of the treatments, simple rectilinear reaction co-ordinates have been assumed. Most of the remaining elements of the potential-energy functions have been taken as adjustable parameters or ignored. In particular, in many cases only breathing modes have been considered. This is done on the basis of the fairly general observation that breathing modes constitute the major activation contribution in a wide variety of radiationless transition processes. The tunnel models remain substantially simple, and it is questionable just how representative they are of the true reaction path. The treatments, and numerical comparisons with experiment, which have been carried out are reasonably successful.[109,111]

Finally, we note that the continuous hydrodynamic representation of a solvated ion offers a possible basis for constructing interionic potential-energy functions for deformable species. Potentials of average force[112,113] are used extensively in studies of ionic equilibrium and conductivity. These potentials, however, generally are not sufficiently flexible to account for the deformation of the interior of the solvated or co-ordinated ionic system which is required in the evaluation of the electron-transfer transition-probability expression. It is certainly true that environmental deformation is considered in transport theories.[113,114] However, for the electron-

[109] M. N. Vargaftig, E. German, R. R. Dogonadze, and Y. K. Syrkin, *Doklady Akad. Nauk S.S.S.R.*, 1972, **206**, 370.
[110] P. P. Schmidt, *cf.* ref. 10.
[111] R. R. Dogonadze, J. Ulstrup, and Yu. I. Kharkats, *cf.* ref. 11.
[112] H. L. Friedman, 'Ionic Solution Theory', Interscience, New York, 1962.
[113] H. L. Friedman, *Physics*, 1964, **30**, 509, 537.
[114] H. S. Harned and B. B. Owen, 'The Physical Chemistry of Electrolyte Solutions', Reinhold, New York, 1958, 3rd edn.

transfer system, these deformations are outer-sphere contributions. The BOH analysis presented in the last section points the way to the derivation of accurate potential-energy functions for deformable ionic species. There, we showed that the vibrational states of the donor and acceptor species in an aggregate state can be modified considerably from the states defined for the species at infinite separation. It is not difficult to see that the same arguments apply to solvated ions described in the hydrodynamic representation.

The aim of the hydrodynamic, collective representation of an electroactive species outlined in this section is simply to find a means of treating the inner-sphere reaction which is both simple and accurate. The point of view contained in the analysis is that it is possible to replace the extremely complicated, detailed quantum-mechanical analysis by a simpler treatment which accounts for all the important and chemically meaningful properties of the system. We anticipate that the time when this approach can be applied extensively and in detail to electron transfer and electrochemical systems is not far off. The real test of the validity and usefulness of the representation certainly will be with its ability to distinguish the truly outer-sphere reactions from others, and with its ability to provide agreement with experiment. Ultimately, a theory and the models used to express it survives only to the degree that it enhances our understanding of the general system processes. The hydrodynamic representation certainly contains sufficient built-in flexibility to do this. It has been successful in its application in nuclear physics. It is hoped that within a few years it will prove equally successful in its application to the electron-transfer reaction.

5 The Evaluation of the Expression for the Transition Probability

Up to this point, we have considered the formal derivation of the electron-transfer transition-probability and rate-constant expressions, and we have considered models for the electron-transfer system in detail. We have not yet considered the specific evaluation of the general transition-probability expression in terms of the variables and operators associated with the models. This we now do.

There are several approaches which one can take for the evaluation of the transition-probability expression. The result, generally, is an explicit form which expresses the rate constant in the non-adiabatic limit. In this section we consider two principal approaches. The first is based on the Kubo–Toyozawa[21] generating-function method, as modified by Dogonadze, Kuznetsov, and Vorotyntsev.[19,20,115,116] The important extension of the Kubo–Toyozawa method introduced by these authors is the use of the multi-dimensional saddle-point integration.[32,117] The second approach to the evaluation of the transition-probability expression is the method developed primarily by Jortner and his colleagues for the treatment of general molecular radiationless transitions.[118—120]

---

[115] M. A. Vorotyntsev and A. M. Kuznetsov, *Vestnik. Moskov. Univ., Fiz.*, 1970, **2**, 146.
[116] R. R. Dogonadze, A. M. Kuznetsov, and M. A. Vorotyntsev, *Z. phys. Chem. (Frankfurt)*, 1976, **100**, 1.
[117] G. Rickayzen, *Proc. Roy. Soc.*, 1957, **A241**, 480.
[118] R. Englman and J. Jortner, *Mol. Phys.*, 1970, **18**, 145.
[119] A. Nitzan and J. Jortner, *J. Chem. Phys.*, 1972, **56**, 3360, 5200.
[120] A. Nitzan and J. Jortner, *J. Chem. Phys.*, 1972, **57**, 2870.

This approach has been applied to the electron-transfer reaction system by Ulstrup and Jortner[7] and by Efrima and Bixon.[92]

The methods used by these various authors share some common features. However, they differ in some substantial respects. For the most part, the results of either approach are the same. The choice of one method for the evaluation of the transition probability over another is dictated as much by the nature of the problem (system) under consideration as it is by personal preference. Generally, the Dogonadze–Kuznetsov–Vorotyntsev method leads to various limiting expressions which are more recognizably 'chemical' than is the case with Jortner's method. However, Dogonadze's method is intrinsically more complex in its general expression in that he allows for the entanglement of normal modes during the transition. This allowance for normal-mode entanglement certainly provides the greatest generality. In most applications, nevertheless, the system is partitioned into inner- and outer-sphere parts, and these are treated individually. The Ulstrup–Jortner[7] approach makes the partitioning assumption in the beginning of the analysis.

The Dogonadze–Kuznetsov–Vorotyntsev Analysis.—Vorotyntsev and Kuznetsov introduced the modified Kubo–Toyozawa[21] treatment of transitions in condensed media in 1970.[115] Dogonadze, Kuznetsov, and Vorotyntsev[19,20] expanded this treatment in 1972. These three papers treat problems of general radiationless transitions in condensed media. However, their easy application to the electron-transfer problem is evident. Dogonadze, Kuznetsov, and Vorotyntsev recently summarized a few of these applications.[116]

The analysis is applicable to multi-dimensional parabolic surfaces defined by the totality of system normal modes. This totality includes the solvent as well as internal molecular normal modes of the reactants. It is, however, restricted to the harmonic-approximation limit for all system vibrations. For the solvent subsystem, the harmonic approximation is embodied in the boson representation for the accessible states.

The system is characterized, in general, by an initial- and a final-state potential-energy surface. In terms of the totality of normal modes, this pair of potential-energy surfaces is represented as (317a) and (317b), in which ω_i and q_i are the

$$U_i = \tfrac{1}{2}\sum_i \hbar\omega_i q_i^2 \tag{317a}$$

$$U_f = \Delta J + \tfrac{1}{2}\sum_i \hbar\omega'_i(q'_i - q_i^{0\prime})^2 \tag{317b}$$

initial-state frequencies and normal modes, and ω'_i and $q_i^{0\prime}$ are those of the final state. ΔJ is the displacement (in energy units) of the minima of the two surfaces, and $q_i^{0\prime}$ is the position of the final-state minimum. The quantity ΔJ contains (in principle) all the solvation and work terms which need to be considered. It can be assumed to have been derived satisfactorily by means of the BOH analysis discussed earlier.

The individual potential-energy functions [equations (317)], together with the kinetic-energy operators for each normal mode, define harmonic oscillator wave equations (318) and a similar equation for the final state. The index n implies the

$$H_i\chi_{in} = E_{in}\chi_{in} \tag{318}$$

set of normal-mode occupation indices. As a consequence, the state energies are expressible as (319) and (320). [The subscript i on n and n' serves the purpose of

$$E_{in} = \sum_{n_i} \hbar\omega_i(n_i+1/2) \tag{319}$$

$$E_{fn} = \sum_{n_i} \hbar\omega_i'(n_i'+1/2)+\Delta J \tag{320}$$

identifying normal modes. In the energy quantity E_{in}, i identifies the initial state.] The solutions to (318) are the usual harmonic oscillator functions (321),[65] and $H_n(q)$ is the Hermite polynomial.

$$\chi_n(q) = \frac{1}{[\sqrt{\pi}2^n n!]^{\frac{1}{2}}} H_n(q)\exp(-q^2/2) \tag{321}$$

In the evaluation of the transition-probability expression, it is convenient to employ the harmonic oscillator density matrix, which is defined as shown in equation (322). The evaluation of the summation in equation (322) in terms of the wave-

$$\rho(x_1,x_2|\beta) = \sum_{n=0}^{\infty} \exp[-\hbar\omega(n+1/2)\beta]\chi_n^*(x_1)\chi_n(x_2) \tag{322}$$

functions (321) can be carried out (cf. Part I), and the result is shown in equation (323).

$$\rho(x_1,x_2|\beta) = [2\pi\sinh(\hbar\omega\beta)]^{-\frac{1}{2}}\exp\{-\tfrac{1}{4}[(x_1+x_2)^2\tanh(\hbar\omega\beta/2)$$
$$+(x_1-x_2)^2\coth(\hbar\omega\beta/2)]\} \tag{323}$$

We proceed with the evaluation of the transition-probability expression. For the electron-transfer reaction, this expression is written in the Condon approximation limit as equation (324). The matrix element, in terms of the electronic wavefunctions,

$$W = (2\pi/\hbar)|V|^2 \prod_i 2\sinh(\hbar\omega_i\beta/2)\{\sum_{n_i,n_i'} \exp[-(n_i+1/2)\hbar\omega_i\beta]$$
$$\times |\langle n_i'|n_i\rangle|^2 \delta(E_n - E_{n'})\} \tag{324}$$

of the driving perturbation is symbolized as V. The product involving the index i covers all normal modes of the total system of solvent and solute. The double summation over occupation numbers n_i and n_i' can be carried out with the use of the Fourier-transform representation of the delta function [equation (325)]. Thus, the

$$\delta(E) = \frac{1}{2\pi\hbar}\int_{-\infty}^{\infty} dt\exp(-itE/\hbar) \tag{325}$$

$$W = \frac{|V|^2}{2\pi\hbar}\int_{-\infty}^{\infty} dt\exp(it\Delta J/\hbar)\prod_i G_i(t) \tag{326}$$

transition probability is conveniently written in the form (326), where $G_i(t)$ is defined by equation (327). It is convenient in the subsequent analysis to make the

$$G_i(t) = \frac{4\pi}{\hbar}\sinh(\hbar\omega_i\beta/2) \sum_{n_i,n_i'} \exp[-(n_i+1/2)\hbar\omega_i(\beta+it/\hbar)$$

$$+it\omega_i'(n_i'+1/2)]\int_{-\infty}^{\infty}\int_{-\infty}^{\infty} dx_1 dx_2 \chi_{n_i}(x_1)\chi_{n_i}(x_2)\chi_{n_i'}(\bar{x}_1)\chi_{n_i'}(\bar{x}_2) \quad (327)$$

variable transformation $\theta = -it/\hbar\beta$, which has the effect of casting the time integration into the form of an inverse Laplace transformation. Consequently, the transition probability is expressed as equation (328), where in the contour integration, the path involving c is chosen to avoid the singularities in the integrand and to lie within the convergence limit of the integrand.[21] $G_i(\theta)$ is related to $G_i(t)$ by equation (329).

$$W = \frac{1}{2\pi i}|V|^2 \int_{c-i\infty}^{c+i\infty} d\theta \exp(-\beta\theta\Delta J)\prod_i G_i(\theta) \quad (328)$$

$$G_i(\theta) = \beta G_i(it/\hbar\beta) \quad (329)$$

On making the transformation indicated by equation (325), one finds with equation (327) that one obtains (330a) and (330b). The co-ordinates $\bar{x}_1$ and $\bar{x}_2$ refer to

$$G_i(\theta) = \frac{4\pi\beta}{\hbar}\sinh(\hbar\omega_i\beta/2)\iint dx_1 dx_2 \rho(x_1,x_2|(1-\theta)\hbar\omega_i\beta)\rho(\bar{x}_1,\bar{x}_2|\theta\hbar\omega_i'\beta) \quad (330a)$$

$$= \frac{2\beta\sinh(\hbar\omega_i\beta/2)(\omega_i'\omega_i)^{\frac{1}{2}}}{\hbar^2\{\sinh[(1-\theta)\hbar\omega_i\beta]\sinh(\theta\hbar\omega_i'\beta)\}^{\frac{1}{2}}}\iint dx_1 dx_2 \exp(-\tfrac{1}{4}\{(x_1+x_2)^2$$

$$\times \tanh[(1-\theta)\hbar\omega_i\beta/2]+(\bar{x}_1+\bar{x}_2)^2\tanh[\theta\hbar\omega_i'\beta/2]$$

$$+(x_1-x_2)^2\coth[(1-\theta)\hbar\omega_i\beta/2]+(\bar{x}_1-\bar{x}_2)^2\coth[\theta\hbar\omega_i'\beta/2]\}) \quad (330b)$$

the final state. The final-state co-ordinates are displaced with respect to the minimum of the potential-energy surface. In addition, they can be rotated with respect to the initial-state co-ordinates. Thus, a transformation (331) can be written, linking the

$$\bar{q}_k = \sum_s a_{ks} q_s \quad (331)$$

final- and initial-state co-ordinates. The final-state co-ordinate displacements are written as q_k^0. In general, these quantities are operators in a quantum-mechanical sense. The variables x in equation (330) are not operators. They are essentially the expectation values of the co-ordinate operators. The transformation (331) applies to these co-ordinate variables as well. Therefore, the integral in equation (330) can be written as (332) for one normal mode. When the product of integrals of the form

$$\iint dx_1 dx_2 \exp(-\tfrac{1}{4}\{(x_1+x_2)^2\tanh[(1-\theta)\hbar\omega_i\beta/2]$$

$$+\sum_{s,t} a_{is}a_{it}(x_{1s}+x_{2s}-q_s)(x_{1t}+x_{2t}-q_t)\tanh[\theta\hbar\omega_i'\beta/2]$$

$$+(x_{1i}-x_{2i})^2\coth[(1-\theta)\hbar\omega_i\beta/2]$$

$$+\sum_{s,t} a_{is}a_{it}(x_{1s}-x_{2s})(x_{1t}-x_{2t})\coth[\theta\hbar\omega_i'\beta/2]\}) \quad (332)$$

(332) is evaluated, the result is that for a standard matrix integral of Gaussian form.[121] In order to simplify the resulting expression, however, it is useful to introduce the co-ordinate transformations (333a) and (333b); the transformation

$$x_1 = \xi + \eta \tag{333a}$$

$$x_2 = \xi - \eta \tag{333b}$$

Jacobian is 2. Thus, for each normal mode, the surface element $dx_1 dx_2$ transforms to $2d\xi d\eta$. The resulting 2^N-dependence (where N is the total number of normal modes) is cancelled in the entire resulting expression by virtue of the normalization of the Boltzmann distribution.

The expression for the transition probability is written as (334a), which can

$$W = \frac{2\beta}{\pi i \hbar^2}|V|^2 \int_{c-i\infty}^{c+i\infty} d\theta \exp(-\beta\theta\Delta J) \prod_{i,j} \frac{(\omega_i \omega_i')^{\frac{1}{2}} \sinh(\hbar\omega_i \beta/2)}{\{\sinh[(1-\theta)\hbar\omega_i \beta/2]\sinh(\theta\hbar\omega_i'\beta/2)\}^{\frac{1}{2}}}$$
$$\times \iint d\xi_i d\eta_j \exp(-\{\xi_i^2 \tanh[(1-\theta)\hbar\omega_i\beta/2] + (\bar{\xi}_i - 2q_i^0)^2 \tanh[\theta\hbar\omega_i'\beta/2]$$
$$+ \eta_j^2 \coth[(1-\theta)\hbar\omega_j\beta/2] + \sum_{s,t} a_{js}a_{jt}\eta_s \eta_t \coth[\theta\hbar\omega_j'\beta/2]\}) \tag{334a}$$

$$W = \frac{1}{2\pi_i} \int_{c-i\infty}^{c+i\infty} d\theta g(\theta) \int_{-\infty}^{\infty} \prod_i d\xi_i \exp[-H(\Delta J, \theta, \xi_i)] \tag{334b}$$

alternatively be written as (334b), where $g(\theta)$ and $H(\Delta J, \theta, \xi_i)$ are as defined in equations (335) and (336). At this point the analysis is very much like that carried out

$$g(\theta) = |V|^2 \prod_i \frac{4\beta(\omega_i'/\omega_i)^{\frac{1}{2}} \sinh(\hbar\omega_i\beta/2)}{\hbar[\sinh[(1-\theta)\hbar\omega_i\beta/2]\sinh(\theta\hbar\omega_i'\beta/2)]^{\frac{1}{2}}}$$
$$\times (\det||\{\coth[(1-\theta)\hbar\omega_j\beta/2] + \sum_j a_{js}a_{jt}\coth[\theta\hbar\omega_j'\beta/2]\}||)^{-\frac{1}{2}} \tag{335}$$

$$H(\Delta J, \theta, \xi_i) = \beta\theta\Delta J + \sum_{i=1}^{N} \{\xi_i^2 \tanh[(1-\theta)\hbar\omega_i\beta/2] + (\bar{\xi}_i - q_i^0)^2 \tanh[\theta\hbar\omega_i'\beta/2]\} \tag{336}$$

for various types of radiative and radiationless processes by O'Rourke,[122] and particularly by Vasileff[123] and Lin.[124] The difference between these works, including that of Kubo and Toyozawa,[21] and the Dogonadze–Kuznetsov–Vorotyntsev theory[19,20,115,116] is that, in the last-mentioned theory, account has been taken of changes in frequency and entanglement of the normal modes. For this reason, the co-ordinate integrations in equation (334) have not been completed, whereas they are carried out in the other works mentioned.

The co-ordinate integrations, together with the θ-integration, are carried out by means of a multi-dimensional saddle-point method.[19,20,115,116] There are a number of distinct advantages in carrying out the integration in this manner, as

---

[121] A. C. Aitkin, 'Determinants and Matrices', Interscience, New York, 1954, example 25, p. 138.
[122] R. C. O'Rourke, *Phys. Rev.*, 1953, **91**, 265.
[123] H. D. Vasileff, *Phys. Rev.*, 1954, **96**, 603.
[124] S. H. Lin, *J. Chem. Phys.*, 1966, **44**, 3759.

opposed to attempting to complete the formal space and time integrations. For one, it is clear that the restrictive approximations concerning co-ordinate entanglement need not be made. For another, restrictive approximations concerning allowed frequencies and densities of states need not be made in order to complete the time integrations. However, apart from the liberation from these restrictions, the method provides a clear connection with the adiabatic-limit statistical-mechanical treatments through the saddle-point value of θ, namely θ^*. Dogonadze et al.[19,20,115,116] have been able to show that the saddle-point value θ^* can be identified with the transfer coefficient. This quantity has a relationship to the m parameter (a Lagrange multiplier) used by Marcus in his theory.[5,16] The treatment at this point has reached a stage of compatibility with the semi-classical treatment investigated by Dogonadze and Urushadze.[125]

The co-ordinate integrations in equation (334) depend upon real variables. Consequently, the Laplace integration method (steepest descent in real space) can be used if the width of the distribution, $\Delta_\xi(\theta)$, is a monotonically changing function of $|\xi_i - q_i^0|$ for all normal modes.[116] In one of their most recent publications, Dogonadze, Kuznetsov, and Vorotyntsev[116] have sought to make the analysis at this point as transparent as possible. In view of the quantities defined in association with equation (334), we now write the transition-probability expression as (337), where we have completed the co-ordinate integrations (indicated by the N-dimensional vector ξ^*) using the Laplace method. The function $B(\theta)$ is identified as (338).

$$W = \int_{c-i\infty}^{c+i\infty} d\theta \, (B\theta) \exp[-H(\Delta J, \theta, \xi^*)] \tag{337}$$

$$(B\theta) = \frac{1}{2\pi i} g(\theta) \Delta_\xi(\theta) \tag{338}$$

Our primary interest at this point is in H, as it serves to determine the transition state through the specification of the activation energy and a tunnel factor. The B function is a generally slowly varying quantity with respect to the variable θ.

The general form of H is further made transparent with the definition (339), where $F(\theta)$ is defined by equation (340) The specific forms of the functions w_i and w_f can be determined by comparison with equation (336). In terms of these quantities, the transition configuration is given by the set of equations (341).

$$H(\Delta J, \theta, \xi^*) = \beta\theta\Delta J + \beta F(\theta) \tag{339}$$

$$F(\theta) = (1-\theta)w_i(\xi^*, \theta) + \theta w_f(\xi^*, \theta) \tag{340}$$

$$\partial H / \partial \xi_i \big|_{\xi_i^*} = 0 \tag{341a}$$

$$(1-\theta)\frac{\partial w_i}{\partial \xi_k} + \theta \frac{\partial w_f}{\partial \xi_k} = 0 \tag{341b}$$

Because θ is complex, the remaining integration over θ requires the use of the saddle-point method. The saddle-point itself is determined by the equations (342).

[125] R. R. Dogonadze and Z. D. Urushadze, *J. Electroanalyt. Chem.*, 1971, **32**, 235.

$$\left.\partial H/\partial \theta\right|_{\theta^*} = 0 \tag{342a}$$

$$w_i - w_f - \Delta J = (1-\theta)\frac{\partial w_i}{\partial \theta} + \theta\frac{\partial w_f}{\partial \theta} \tag{342b}$$

Thus, the final form of the transition probability is equation (343). The differentiation of $\ln W$ by ΔJ determines the transfer coefficient [equation (344)].

$$W = [2\pi/F''(\theta^*)]^{\frac{1}{2}}|(B\theta^*)|\exp[-\beta\theta^*\Delta J - \beta F(\theta^*)] \tag{343}$$

$$\alpha = -\frac{1}{\beta}\frac{d\ln W}{d\Delta J} = \theta^* \tag{344}$$

At this point, in order to make the preceding results as physically and chemically transparent as possible, it is useful to consider the semi-classical analysis carried out by Dogonadze and Urushadze.[125] We use the initial- and final-state potential-energy surfaces given by equations (345). The individual potential-energy functions are general, and not restricted to parabolic surfaces.

$$U_i(q) = \sum_k U_i(q_k) \tag{345a}$$

$$U_f(q) = \sum_k U_f(q_k') + \Delta J \tag{345b}$$

The free energy of activation for an adiabatic reaction is given by equation (346), where Z is the configurational integral shown in (347). The integration over S^* in

$$\Delta G^* = -kT\ln[(2\pi\hbar\beta)^{\frac{1}{2}}Z^{-1}\int dS^* \exp(-\beta U_i)] \tag{346}$$

$$Z = \int \prod_k dq_k \exp(-\beta U_i) \tag{347}$$

equation (346) involves the reaction hypersurface in the same sense as in Marcus's theory.[5] In particular, the integration is carried out on the surface defined by $U_i(q) = U_f(q)$. As was discussed earlier, there is an infinity of points on S^*, but only a relatively small number of points define the region on S^* for which U_i assumes a minimum value. It is evident that, in the region of the minimum value of U_i the integrand in (346) makes its largest, and principal, contribution to ΔG^*. Hence, if we seek to determine the activation free energy with equation (346) by means of the Laplace method (assuming a real potential-energy surface), then the maximum value assumed by the integrand is specified for the minimum value of $U_i(q_k)$. The problem is equivalent to the determination of the activation energy ΔE^*. The condition $U_i(q) = U_f(q)$, expressed as equation (348), together with the minimization condition (349), where λ is a Lagrange multiplier, serves to determine the activation energy ΔE^* once $q_k^*(\lambda)$ and λ have been determined. It is clear that the activation energy can be identified as shown in (350).

$$U_i(q_k) = U_f(q_k) + \Delta J \quad \text{for all } q_k \tag{348}$$

$$(1-\lambda)\frac{\partial U_i(q_k^*)}{\partial q_k^*} + \lambda\frac{\partial U_f(q_k^*)}{\partial q_k^*} = 0 \tag{349}$$

$$\Delta E^* = U_i[q_k^*(\lambda)] \tag{350}$$

The variation of the activation energy with reaction heat (here, effectively ΔJ) can be written as equation (351a) or (351b). Using the equality between the last two

$$\frac{\partial \Delta E^*}{\partial \Delta J} = \frac{\partial U_i(q_k^*)}{\partial \Delta J} = \sum_k \frac{\partial U_i(q_k^*)}{\partial q_k^*} \frac{\mathrm{d} q_k^*}{\mathrm{d}\Delta J} \tag{351a}$$

$$= 1 + \sum_k \frac{\partial U_f(q_k^*)}{\partial q_k^*} \frac{\mathrm{d} q_k^*}{\mathrm{d}\Delta J} \tag{351b}$$

terms in this expression, together with equation (349), it is possible to show that equation (352) is valid. That is, the Lagrange multiplier is equivalent to the transfer coefficient.

$$\frac{\partial \Delta E^*}{\partial \Delta J} = \alpha = \lambda \tag{352}$$

For the simple outer-sphere reaction, assuming a non-disperse solvent, the initial- and final-state potential-energy functions are particularly simple; they are shown in equations (353). The application of equation (349) and the condition $U_i = U_f$ yields equations (354) and (355). Both of these forms are well known from the

$$U_i = \tfrac{1}{2} \sum_k \hbar\omega q_k^2 \tag{353a}$$

$$U_f = \tfrac{1}{2} \sum_k \hbar\omega (q_k - q_k^0)^2 + \Delta J \tag{353b}$$

$$\lambda = \tfrac{1}{2} + \Delta J / [\hbar\omega \sum_k (q_k^0)^2] = \alpha, \; q_k^* = \lambda q_k^0 \tag{354}$$

$$\Delta E^* = [\Delta J + \tfrac{1}{2}\hbar\omega \sum_k (q_k^0)^2]^2 / 4[\tfrac{1}{2}\hbar\omega \sum_k (q_k^0)^2] \tag{355}$$

simple outer-sphere transfer theory of Marcus, Hush, and Levich and Dogonadze.[16-18] The quantity $\tfrac{1}{2}\hbar\omega \sum_k (q_k^0)^2$ is identified as the repolarization energy for the solvent-subsystem modes; *cf.* Part I.

We return to the analysis of the general quantum-mechanical transition probability; in particular, that associated with equation (343). Immediately, two interesting limits in addition to the normal limit [which is expressed semi-classically by equations (354) and (355)] can be identified. The first limit is associated with activationless transitions. This limit corresponds to a value of $\theta^* = 0$. From equation (341b) it is possible to see that $\partial w_i/\partial \xi_k^* = 0$. Since, for $\theta^* = 0$, the effective temperature tends to infinity, the w_i function reduces to the potential-energy surface of the initial state. In the formal analysis, the potential-energy surface of the initial state is taken as the reference system. The final-state potential is displaced from the initial-state one both with respect to co-ordinates and energy (expressed in terms of ΔJ). The condition $\theta^* = 0$ therefore means that w_i is minimal, and as it is taken as the reference surface, it must be zero at $\xi_k^* = \xi_{ki}^0$. If, at the co-ordinate saddle-point $\xi_k^*, w_i = 0$, then with $\theta^* = 0$ we see that $H^* = 0$. The activation energy in this limit effectively vanishes. This condition applies to highly exothermic reactions.

The second limit follows in the opposite extreme, when $\theta^* = 1$. In this case, from equation (341b), we see that $\partial w_f/\partial \xi_k^* = 0$. The saddle-point intersects the minimum in the final-state potential-energy surface. In this case, effectively, H* = ΔJ. Reactions which satisfy this limit have been called barrierless.[19,20,115,116] The two situations are summarized in Figure 3.

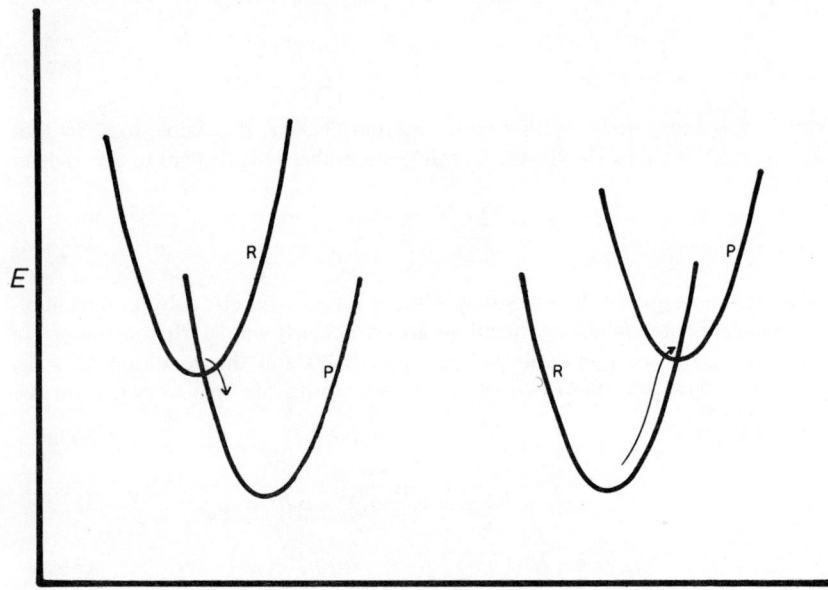

Reaction co-ordinate

Figure 3 (a) *The configuration for the intersection of the reactant and product potential-energy surfaces for an activationless electron transfer.* (b) *The configuration for the intersection of the reactant and product potential-energy surfaces for a barrierless electron transfer.*

This discussion of these two limits is, of course, overly simplistic. As Dogonadze, Kuznetsov, and Vorotyntsev pointed out recently,[116] the specific form of the transition-probability expression in the activationless limit, in particular, depends sensitively upon the nature of the analysis used to determine θ^*. Thus, for example, different forms for the transition probability follow different methods of handling the contributions from the solvent subsystem. Their treatment of the solvent contribution is based on their theory of the disperse polar medium. If a Debye-type cut-off spectrum is used, the result for the transition probability differs from a treatment based upon a continuous spectrum.[116] The Dogonadze–Kuznetsov–Vorotyntsev treatment is summarized by the two equations (356) and (357). In the first, when $|\Delta J| < E_r$ (E_r is the reorganization energy), and θ^* is not 'too' small, the transition probability is given by the familiar Gaussian form: viz. (356). As ΔJ approaches $-E_r(\theta^* \to 0)$, high-frequency modes of the system play an important part, and the transition-probability expression is given by equation (357), where

$$W = |V|^2 \sqrt{\pi} (\hbar \sqrt{kTE_r})^{-1} \exp[-\beta(\Delta J + E_r)^2/4E_r] \qquad (356)$$

$$W \cong \exp\{-(\Delta J + E_r)^2 (4\hbar\omega_D E_r)^{-1} [\ln(\beta\hbar\omega_D E_r/|\Delta J + E_r|)]^{-1}\} \qquad (357)$$

ω_D is a characteristic Debye frequency.[116] A similar situation has been examined by Fischer[29] with reference to the problem of the general molecular radiationless transition.

It can be seen from the last two equations that, as the reaction heat tends to the abnormal exothermic limit, the transition probability passes from a Gaussian to a simpler exponential form. Consequently, the activation energy, which is predicted to have a symmetric parabolic dependence on ΔJ in the statistical-mechanical treatment,[126] in fact tends from parabolic to linear. This has been seen in homogeneous reaction systems (cf. ref. 22, and additional references contained in that paper), but as yet it has not been proved to be applicable to the electrochemical system. The basic structure of the theory seems sound, and it is reasonable to expect that this effect should be found eventually for the electrochemical system. The difficulties in proving its existence may lie with physical and chemical factors peculiar to the electrode and the double layer, or they may lie with difficulties in experimental detection (F. C. Anson, personal communication). In any case, the experimental demonstration of the non-linearity of the transfer coefficient with overpotential remains as the first important step in the verification of the theory.

Generating-function Techniques from the Theory of Molecular Radiationless Transitions.—Within the framework of the non-adiabatic limit–quantum-mechanical treatment of the electron-transfer reaction, as more consideration was devoted to the contribution of inner-sphere vibrational degrees of freedom to the activation process, an increasing use of techniques from the general area of molecular radiationless transitions entered the theory. Basic to most of the treatments which exist to date is the use of generating-function techniques. These techniques are similar in many respects to the Kubo–Toyozawa analysis that was considered in the last section. Although, in many instances, saddle-point methods have been used to evaluate the time integrations involved in the determination of the transition probability, often analytical methods have been used. That is, closed forms for various parts of the transition probability have been found. Usually, these closed forms follow an analysis which requires the use of a number of approximations. In some cases, the approximations used can severely restrict the range of validity of the transition-probability expressions derived.

In this section, consideration is given to approximate treatments based on generating-function techniques. These treatments have been considered by a number of authors.[7,22,31,48,127–130] We focus on the work of Ulstrup and Jortner,[7] who have carried out a number of numerical analyses in order to observe the effect of various system parameters on the activation energy and transfer coefficient.

An assumption central to most of these treatments is that it is legitimate to regard the vibrational degrees of freedom of the inner solvation or co-ordination shells as essentially independent of the degrees of freedom of the solvent subsystem. As a

[126] R. A. Marcus, *J. Chem. Phys.*, 1965, **43**, 2654.
[127] A. A. Ovchinnikov and M. Ya. Ovchinnikova, *Doklady Akad. Nauk S.S.S.R.*, 1969, **186**, 76.
[128] P. P. Schmidt, *J.C.S. Faraday II*, 1973, **69**, 1104.
[129] P. P. Schmidt and H. B. Mark, jun., *J. Chem. Phys.*, 1973, **58**, 4290.
[130] W. Schmickler, *Electrochim. Acta*, 1976, **21**, 161.

result, it is possible simply to treat the two parts of the entire system separately, and to combine the results of the analyses to determine the final form of the transition-probability expression. As was pointed out earlier in this chapter, this assumption is not generally defensible, but it is nevertheless almost universally made. It remains to be seen how adequate the assumption is. The proof obviously depends upon extensive correlation with experimental data; a situation which is hopefully soon to be realized.

An advantage of this partitioning (or deconvolution) assumption is found in the fact that it is definitely easier to consider anharmonic effects in the co-ordination shell. As Ulstrup and Jortner have shown,[7] these effects can be important, at least as far as the formal theoretical analysis is concerned. An additional advantage in the deconvolution assumption is tied to the use of simpler analyses on smaller segments of the entire system.

The electron-transfer system consists, as usual, of an electron-donor and -acceptor species embedded in a continuum polar solvent. This system is referred to by Ulstrup and Jortner as a super-molecular complex,[7] which is equivalent to our definition of the aggregate state. The electron-transfer transition is then considered as a molecular radiationless transition. The perturbation driving the transition, however, is assumed to depend upon matrix elements of the electrostatic operators, as found with the BOH analysis, in contrast to the typical molecular radiationless process. [Generally, radiationless transitions within individual molecular species, even if embedded in a dense continuum, are driven by the non-adiabaticity operators.[131]] The initial and final states are associated with multi-dimensional potential-energy surfaces of the form (358). The solvent terms u^s are expressed in the usual

$$U_i(q) = u_i^D(q_i^D) + u_i^A(q_i^A) + u_i^s(q_i^s) \tag{358a}$$

$$U_f(q) = u_f^D(q_f^D) + u_f^A(q_f^A) + u_f^s(q_f^s) + \Delta J \tag{358b}$$

boson representation. The internal molecular vibrational terms u^D and u^A can be expressed in terms of harmonic (*i.e.* parabolic) surfaces or they may be allowed to assume a more general representation, *viz.*, a Morse potential form in the Ulstrup–Jortner work.[7] It is more important to note that the terms u_f^D and u_f^A are, respectively, acceptor and donor potential-energy functions for the reverse electron transfer defined for the system. In other words, the labelling of the potential-energy functions in equations (358) takes the initial-state donor and acceptor species designations as labels for the final state.

As we have already seen on a number of occasions in this chapter, the specification of the potential-energy surfaces for the electron-transfer system allows one to specify total energy functions for the initial and final states. Thus, one may write equations (359), and the labelling here preserves the sense of the labelling of the potential-energy surfaces.

$$E_i = \varepsilon_i^D + \varepsilon_i^A + \varepsilon_i^s \tag{359a}$$

$$E_f = \varepsilon_f^D + \varepsilon_f^A + \varepsilon_f^s + \Delta J \tag{359b}$$

Before continuing with a development of the details of the evaluation of the rate-

[131] K. F. Freed and J. Jortner, *J. Chem. Phys.*, 1970, **52**, 6272.

constant expression, it is important to note that the specification of the potential-energy functions, and the energies, given by equations (358) [and equations (359)] strictly implies that there is no significant interaction between the individual molecular degrees of freedom and that there is no interaction between the molecular degrees of freedom and the continuum modes of the solvent. The effect of the solvent on the system is included (approximately) in the determination of the quantity ΔJ, which expresses the difference in the system electronic levels of the solvated species. In physical terms, this approximation implies that one considers the species in the aggregate state effectively as fixed, but allowed to execute *small* oscillatory motions in the region of the fixed configuration. It is assumed further that this small oscillatory motion does not substantially alter the value of ΔJ. Strictly speaking, therefore, the only analysis allowed for this system is a harmonic analysis of the degrees of freedom of all parts of the system. However, if the aggregate state consists of donor and acceptor species which are reasonably far apart, such that anharmonic distortion in one species is effectively insulated by the distance from the other, then an account of anharmonic contributions in the individual species can be entered into the theory. On the other hand, if the separation between donor and acceptor reactants is not great, then anharmonic contributions still can be considered. In this case, however, vibrational states of the aggregate state considered as a single entity should be considered.

The rate-constant expression for the system is written simply as equation (360), in which, for clarity, we have isolated the matrix element of the driving perturbation,

$$k = \int_R^\infty d^3R \, |V(R)|^2 \exp[-w(R)] A(R) \qquad (360)$$

$V(R)$, using the Condon approximation, and we have isolated the work terms leading to the formation of the initial state, $w(R)$. The $A(R)$ terms depend upon a product of three Franck–Condon vibrational overlap factors, one each for donor, acceptor, and solvent system. It is expressed as equation (361), where Z is the system

$$A(R) = (2\pi/\hbar Z) \sum_{\{i,f\}} S_D(\varepsilon_i^D; \varepsilon_f^D) S_A(\varepsilon_i^A; \varepsilon_f^A) S_s(\varepsilon_i^s; \varepsilon_f^s)$$
$$\times \exp[-\beta(\varepsilon_i^D + \varepsilon_i^A + \varepsilon_i^s)] \delta(\varepsilon_i^D + \varepsilon_i^A + \varepsilon_i^s - \varepsilon_f^D - \varepsilon_f^A - \varepsilon_f^s - \Delta J) \qquad (361)$$

partition function. In view of the complete partitioning, it is simply given as the product of the individual subsystem partition functions: $Z = Z_D Z_A Z_s$. The FC overlap quantities are expressed as equation (362a) or (362b), where the product is constructed of all the normal modes for the particular subsystem.

$$S(\varepsilon_i^I; \varepsilon_f^I) = |\langle \chi_f^I | \chi_i^I \rangle|^2 \qquad [I = A, D, s] \qquad (362a)$$
$$= \prod_v |\langle \chi_{fv}^I | \chi_{iv}^I \rangle|^2 \qquad (362b)$$

In view of the fact that the delta function, $\delta(A-B)$, can be written as shown in equation (363), it is possible to express the system FC factor as a convolution of the solvent and internal vibrational (generally quantum) modes [equation (364)].

$$\delta(A-B) = \int dx \, \delta(A+x) \delta(x-B) \qquad (363)$$
$$A(R) = \int dx \, F_Q(x) F_s(\Delta J - x) \qquad (364)$$

This method of separating the subsystem contributions is used by Ulstrup and Jortner,[7] and also by Kestner, Logan, and Jortner.[31] It has been used in other treatments as well.[10,128] The solvent sybsystem FC factor is readily evaluated in terms of the simple polar solvent model or in terms of the improved model of Dogonadze, et al.[40–44] (the general form of the expression in either case is the same). The result is equations (365).

$$F_s(\Delta J - x) = (2\pi/\hbar Z_s) \sum_{s_i, s_f} \exp[-\beta \varepsilon_i^s] S_s(\varepsilon_i^s; \varepsilon_f^s) \delta(\varepsilon_i^s - \varepsilon_f^s - \Delta J - x) \quad (365a)$$

$$= (\pi\beta/\hbar^2 E_s)^{\frac{1}{2}} \exp\left(-\frac{[\Delta J + E_s + x]^2 \beta}{4 E_s}\right) \quad (365b)$$

The FC factor for the quantum vibrations of the co-ordination shell is evaluated as follows. The factor itself is defined by equation (366), where Z_Q is the complete

$$F_Q(x) = (1/Z_Q) \sum_{\varepsilon_i^Q, \varepsilon_f^Q} \exp(-\beta \varepsilon_i^Q) S_Q(\varepsilon_i^Q; \varepsilon_f^Q) \delta(x + \varepsilon_i^Q - \varepsilon_f^Q) \quad (366)$$

partition function for the vibrational modes in the co-ordination shells of the reactants. With the use of the integral representation of the delta function, equation (325), together with the definition of the density matrix for the harmonic oscillator, we can write $F_Q(x)$ as shown in equation (367). The product is taken over all normal modes of the complete system of vibrations of the co-ordination shell.

$$F_Q(x) = (1/2\pi\hbar Z_Q) \int_{-\infty}^{\infty} dt \exp(-ixt/\hbar) \prod_l \int_{-\infty}^{\infty}\int_{-\infty}^{\infty} dx_{l1} dx_{l2} \rho(\bar{x}_{l1}, \bar{x}_{l2}; -i\omega_l' t)$$
$$\times \rho[x_{l1}, x_{l2}; (\beta + it/\hbar)\hbar\omega_l)] \quad (367)$$

The evaluation of equation (367) is simplified to a considerable extent if we assume the following. First, assume that the initial- and final-state frequencies for each normal mode are the same. Second, assume that there is no entanglement of normal modes. Third, assume that the normal mode frequencies in one of the reactants are essentially the same as in the other. The second of these assumptions allows us to complete all the co-ordinate integrations in a formal sense. Consequently, a closed expression for the generating function for the FC factors can be obtained. The first and third assumptions simplify the subsequent evaluation of the time integral in that all individual generating functions are treated in essentially the same manner.

As a result of the application of these assumptions to the evaluation of (367), we find that equation (368) may be written, where $g(t)$ is given by equation (369).[132]

$$F_Q(x) = (1/2\pi\hbar) \int_{-\infty}^{\infty} dt \exp(-ixt/\hbar) g(t) \quad (368)$$

$$g(t) = \exp(\sum_l (\zeta_l^2/2)\{\coth(\hbar\omega_l \beta/2)[\cos(\omega_l t) - 1] + i \sin(\omega_l t)\}) \quad (369)$$

The quantity ζ_l^2 is the sum of the squares of the (dimensionless form) displacements of the normal co-ordinate q_l with respect to the initial and final states (see equation

---

[132] P. P. Schmidt, *Theor. Chim. Acta*, 1975, **40**, 263.

(370)],[31] where a and b label the reactants (and products); Δ_a is given by equation (371), and there is a similar term for Δ_b. The R_l^0 are co-ordinates associated with the normal mode l.

$$\zeta_l^2 = \Delta_{al}^2 + \Delta_{bl}^2 \tag{370}$$

$$\Delta_a = (M_a \omega_{al}/\hbar)^{\frac{1}{2}}[R_l^0(A^{(N+1)+}) - R_l^0(A^{N+})] \tag{371}$$

The generating function can be manipulated in the following transparent fashion.[132] Define $g(t)$ by equation (372), with $\gamma(t)$ defined by equation (373). As a consequence of this, it is possible to consider the expansion (374). The first term

$$g(t) = \exp[\gamma(t) - \gamma(0)] \tag{372}$$

$$\gamma(t) = \sum_l (\zeta_l^2/2)[\coth(\hbar\omega_l\beta/2)\cos\omega_l t + \mathrm{i}\sin\omega_l t] \tag{373}$$

$$g(t) = \exp[-\gamma(0)] \sum_n^\infty [\gamma(t)]^n/n! \tag{374}$$

clearly gives the delta function $\delta(x)$ upon carrying out the time integration in equation (368). In terms of the definition of the transition shape function, $G(x)$ [equation (375)], it is possible to define individual multi-phonon contributions as $\mathscr{G}_n(x)$ [equation (376)] such that $G(x)$ is now defined as shown in equation (377). Further

$$G(x) = (1/2\pi\hbar) \int_{-\infty}^\infty dt \exp(-\mathrm{i}xt/\hbar) g(t) = F_Q(x) \tag{375}$$

$$\mathscr{G}_n(x) = (1/2\pi\hbar) \int_{-\infty}^\infty dt \exp(-\mathrm{i}xt/\hbar)[\gamma(t)/\gamma(0)]^n \tag{376}$$

$$G(x) = \exp[-\gamma(0)] \sum_n [\gamma^n(0)/n!] \mathscr{G}_n(x) \tag{377}$$

manipulation reveals that the function $\mathscr{G}_n(x)$ can be expressed in terms of an infinite series of Hermite–Gauss functions (harmonic oscillator functions).[132] The result of the manipulations is expressed as equation (378), where the $c_v^{(n)}$ are expansion coefficients, and Δ_0 is defined by equation (379) and y_n by equation (380). In view

$$\mathscr{G}_n(x) = \frac{2^{n/2}}{\hbar\Delta_0\sqrt{2\pi n}} \sum_{v=0}^\infty c_v^{(n)} \exp\{-y_n^2/2\Delta_0^2 n\} H_v(y/\Delta_0\sqrt{n}) \tag{378}$$

$$\Delta_0 = \frac{1}{4\gamma(0)} \sum_l \omega_l^2 \zeta_l^2 \coth(\hbar\omega_l\beta/2) - \frac{1}{\gamma(0)} \sum_{l,l'} \omega_l \omega_{l'} \tag{379}$$

$$y_n = x/\hbar + n\Sigma[\omega_l/\gamma(0)] \tag{380}$$

of the definition of the delta function, it can be seen that the first term $\mathscr{G}_0$ in equation (378) is in fact the delta function. As a result, it is possible to write the shape function as (381). Bearing in mind the form of the $\mathscr{G}_n$ function given by equation (378),

$$G(x) = \exp[-\gamma(0)]\left(\delta(x) + \sum_{n=1}^\infty [\gamma^n(0)/n!]\mathscr{G}_n(x)\right) \tag{381}$$

it can be seen that the multi-phonon contributions to the general shape function are skewed with respect to a symmetric Gaussian distribution. The skewing of the transition envelope for radiative F-centre transitions was recognized long ago by Lax[133] and O'Rourke.[122] If one ignores skewing effects, which generally are not pronounced for transitions near the vibrational ground state, then it is possible to write equation (382), where $n\Omega$, defined by equation (383), is the centre of gravity of the transition, and E_0 is effectively the width of the envelope [equation (384)]. It

$$\mathcal{G}_n(x) = \frac{1}{\hbar\Delta_0\sqrt{2\pi n}} \exp\left(-\frac{[x+n\Omega]^2}{2nE_0^2}\right) \quad (382)$$

$$n\Omega = n\sum_l \hbar\omega_l/\gamma(0) \quad (383)$$

$$E_0 = \hbar\Delta_0 \quad (384)$$

is possible to compare equations (378) and (382) in order to generate an expression for correction terms which account primarily for skewness effects.[132] This correction is not needed here.

When these results of the preceding analysis are convoluted into the expression for the solvent-subsystem contributions, we obtain an expression for the complete transition probability. A general expression for $A(R)$ is given by equation (385). This

$$A(R) = \exp[-\gamma(0)]\left\{(\pi\beta/\hbar^2 E_s)^{\frac{1}{2}}\exp\left(-\frac{[\Delta J + E_s]^2\beta}{4E_s}\right)\right.$$
$$\left. + \sum_{n=1}^{\infty}\left(\frac{\pi\beta}{\hbar^2(2E_s+n\beta E_0^2)}\right)^{\frac{1}{2}}\exp\left(-\frac{[\Delta J + E_s + n\Omega]^2\beta}{4E_s}\right)\right\} \quad (385)$$

expression is somewhat simpler than that given by Ulstrup and Jortner,[7] but it is useful nevertheless in the same sense. In particular, it illustrates the situation proposed by Efrima and Bixon[134] that vibrational excitation of quantum modes may lead to activationless transitions for strongly exothermic reactions ($-\Delta J$). It is seen in equation (385) that if the quantum modes satisfy the condition $\hbar\omega_l \gg kT$ (all l), then the first term in the expression dominates. In this case, the transition-probability expression, and the expression for the rate constant, assume the form for an outer-sphere reaction. The condition usually claimed to define the quantum limit is $\hbar\omega_l > kT$, although in several papers Dogonadze et al. have assumed $\hbar\omega_l > 4kt$.[111,135] In most of their recent papers, Dogonadze et al. have assumed simply that $\hbar\omega_l > kT$. The choice of limit is not a trivial matter. This is especially true for the consideration of the vibrational modes in loosely bound complexes and simple solvated systems for which, typically, the breathing modes range from 200 to 600 cm^{-1}.[136] The limit assumed by Kestner, Logan, and Jortner[31] and by Ulstrup and Jortner[7] requires that one consider these degrees of freedom as quantal. Obviously, the results, in the form of activation-energy expressions, will be substantially different if the classical limit is used.

[133] M. Lax, *J. Chem. Phys.*, 1952, **20**, 1752.
[134] S. Efrima and M. Bixon, *Chem. Phys Letters*, 1974, **25**, 34.
[135] R. R. Dogonadze and A. M. Kuznetsov, *Itogi Nauki, Elektrokhimiya*, 1967 (published, 1969); cited in ref. 111.
[136] D. E. Irish, in 'Ionic Interactions', ed. S. Petrucci, vol. II, Academic Press, New York, 1971.

In an extensive number of computer calculations, using an expression for $A(R)$ which accounts only for displacement of the normal co-ordinates of the co-ordination shell, Ulstrup and Jortner[7] have shown that the form of the activation energy is essentially parabolic, up to a maximum (at $\Delta J = 0$). However, in the abnormal region (and in particular, the strongly exothermic region) the dependence changes. They have found, as have others,[22,116] that the activation energy falls off much more slowly than it would if it were quadratic in ΔJ. Dogonadze, Kuznetsov, and Vorotyntsev[116] indicate that the dependence shifts to a logarithmic one. In addition to the different behaviour of the activation energy with respect to the heat of reaction in the abnormal region, Ulstrup and Jortner[7] find that, under certain circumstances, there can be oscillatory behaviour in the dependence of the activation energy on reaction heat. They claim that this is consistent with similar behaviour in molecular radiationless transitions. Whether this behaviour is real for the electron-transfer system, and further, whether it is indeed important, remains to be seen. The experimental evidence on homogeneous systems found to date[137] seems to show a smooth behaviour. This is consistent with the high-temperature limit.[7]

The assumption that the normal-mode frequencies do not change during the electron-transfer step is obviously restrictive, and limits the validity of the expressions derived. As was argued earlier in this chapter, the hydrodynamic representation[12—14] indicates that for a change in charge state of 1 e.s.u., the change in frequency of the inner-shell modes is of the order of 100 cm^{-1}. It is possible to examine the effect of frequency changes in an elaborate and formal sense;[119,120] however, the resulting expressions are not particularly physically transparent. In their investigation, Ulstrup and Jortner[7] made use of a simpler approach based on an investigation of transition which take place from the lowest initial vibrational state to all possible final states. As a result, they were able to make use of the analysis advanced by Siebrand[138,139] for estimating Franck–Condon contributions to molecular radiationless transitions.

Specifically, for the normal modes of the co-ordination shell, Ulstrup and Jortner[7] use the FC factor in the form of equation (386), in which each individual contribution is given by equations (387)—(391)[138,139,7] for each normal mode of the co-

$$S(0,v) = \prod_l S_Q(0,v_l) \tag{386}$$

$$S_Q(0,v_l) = F(0)(\xi/2)^{v_l}|H_{v_l}(iz)|^2/v_l! \tag{387}$$

$$F(0) = (1-\xi)^{\frac{1}{2}}\exp(-\Delta_c^2/2) \tag{388}$$

$$\xi = (\omega_{ci}^l - \omega_{cf}^l)/(\omega_{ci}^l + \omega_{cf}^l) \tag{389}$$

$$\Delta_c^2/2 = (k_i k_f)^{\frac{1}{2}}(q_b^0 - q_a^0)^2/\hbar(\omega_{ci}^l + \omega_{cf}^l) \tag{390}$$

$$z = [\Delta_c^2(1-\xi)/4\xi]^{\frac{1}{2}} \tag{391}$$

ordination shells. In the limit where the frequency change vanishes, $\xi \to 0$ and $S_Q(0,v_l)$ tends to the limit shown in equation (392). It is not difficult to show that this

---

[137] D. Rehm and A. Weller, *Israel J. Chem.*, 1970, **8**, 259.
[138] W. Siebrand, *J. Chem. Phys.*, 1967, **46**, 440.
[139] W. Siebrand, *J. Chem. Phys.*, 1967, **47**, 2411.

$$\lim_{\xi \to 0} S_Q(0, v_l) = \exp(-\Delta_c^2/2)(\Delta_c^2/2)^{v_l}/v_l! \qquad (392)$$

limit also is found from equation (381) in the low-temperature limit. In equation (387), $H_n(iz)$ is the Hermite polynomial of imaginary argument. Consequently, the modulus squared of this quantity yields non-vanishing contributions to S only for v_l even.

A particularly simple expression for the transition probability results if one assumes that there is distortion (*i.e.* frequency change) but no displacement of the normal modes. In this case, $\Delta_c^2 = 0$. The expression for $A(R)$ is found to be equation (393). In this expression there is a strong oscillatory behaviour of the activation energy as a function of ΔJ. $\overline{\Delta J}$ contains the difference in the zero-point energies: $\overline{\Delta J} = \Delta J + \hbar(\omega_{cf} - \omega_{ci})/2$.

$$A(R) = (\pi \beta/\hbar^2 E_s)^{\frac{1}{2}} (1 - \xi^2)^{\frac{1}{2}} \sum_{v_l \text{ (even)}} \xi \frac{v_l(v_l-1)!!}{v_l!!}$$
$$\times \exp[-\beta(\overline{\Delta J} - E_s - n\hbar\omega_{cf})^2/4E_s] \qquad (393)$$

The range of allowed values of ξ is defined by $0 \leq |\xi| \leq 1$. As was indicated, the zero value corresponds to an absence of frequency change during the electron transfer. The value 1, a limiting case, corresponds to a complete bond break or to the establishment of a new bond. The Ulstrup–Jortner[7] results indicate (i) that the oscillatory behaviour of the activation energy as a function of reaction heat is characteristic of very low temperature ranges (generally inaccessible to normal experimentation) and (ii) that for increasing values of $|\xi|$ the decrease in the activation energy with respect to reaction heat in the abnormal region is lessened.

Ulstrup and Jortner[7] have also investigated the effect of distortion and displacement on the transition probability. They have also investigated the effect of anharmonicity on the transition probability. In all cases, they found that all of these effects can be important in determining the magnitude of the behaviour of the activation energy with respect to ΔJ. The amounts by which these various contributions influence the activation energy, in particular, differ from case to case. Nevertheless, the overall pattern is evident. The consideration of displacement, distortion, and anharmonicity individually and together indicates that all of these effects can influence the magnitude of the rate constant, particularly in the abnormal region. In the normal region these effects are not so pronounced. There, the behaviour of the rate constant and activation energy is essentially that predicted by the simpler outer-sphere-type expression.

There is, in addition to the physical effects of distortion, displacement, and anharmonicity, a strong temperature-dependence of the activation energy, as defined by equation (394). This is evident in the expression (385), which depends upon the

$$E_a = -\frac{1}{k}\frac{dk}{d\beta} \qquad (394)$$

quantity $\gamma(0)$ defined by equation (373). Generally speaking, the contribution of the solvent modes to the activation process can almost always be considered in the high-temperature limit. Over the temperature range accessible to normal experimentation, the temperature-dependence of the contributions of the solvent mode in the high-temperature limit cannot be seen. For the quantum modes, on the other hand, the temperature-dependence may be seen even at ambient temperatures. This should be especially true for complexed or solvated transition-metal species.

The Ulstrup–Jortner analysis is interesting, and valuable in that it raises a number of points which are accessible to further substantial investigation. In particular, the limit $\xi \to 1$, which amounts to the bond break–formation limit, is very important for a large class of reaction systems. This limit, as they have used it, applies to individual normal modes. In the complete expression for the transition probability, equation (393), it is clear that the summation is taken over all normal modes in the system. Consequently, although one mode may undergo alteration, there remain a large number of other modes which are substantially unchanged. It is true, of course, that the frequencies of these remaining modes can shift considerably if one mode corresponds to a bond break. The overall effect of this bond break–formation step, however, may not be great, as far as it can be manifested in the magnitude of the activation energy. The hydrodynamic representation discussed earlier is in accord with this observation. In particular, the relationship (386) applies if the analysis embodied in equations (387)—(391) is couched in the terms of the hydrodynamic analysis.

6 Inner-sphere Reactions and Bridge-assisted Outer-sphere Transfers

The classification of inner- and outer-sphere reactions originally was proposed to distinguish reactions in which the electron-transfer step was assisted by the sharing of a ligand between the reactants, and all others. The archetypal inner-sphere example is generally considered to be:[140—142]

$$[Cr(H_2O)_6]^{2+} + [ClCo(NH_3)_5]^{2+} \to [(H_2O)_5CrCl]^{2+} + [Co(H_2O)_6]^{2+} + 5NH_3$$

To some degree, this distinction has been used in a more relaxed manner to include reactions in which there is substantial alteration of the inner co-ordination shell during the electron-transfer step (*cf.* Part I; the Reporter is perhaps as guilty as anyone in adopting this relaxed usage). It is clear that a revision of definitions is needed. However, such a task is not attempted here. In this section, *inner-sphere* reaction is taken to be its usual meaning. Some of the discussion applies to these reactions. In addition, we give consideration to reactions which, by the traditional definition, are outer-sphere, but assisted in some sense by a bridging group or body (*viz.* a metal or polymer).

Two treatments of the true inner-sphere reaction have been published.[110,111] Both of these treatments assume simple rectilinear reaction co-ordinates for the normal mode which suffers bond break or formation. Although the assumptions concerning the nature of the electron-transfer step in the transition state differ, the results are essentially the same. The first treatment, by Schmidt,[110] uses the nonadiabatic quantum-mechanical formulation, and the second, due to Dogonadze, Kharkats, and Ulstrup,[111] uses the formalism developed by Dogonadze and

---

[140] H. Taube, H. Myers, and R. L. Rich, *J. Amer. Chem. Soc.*, 1953, **75**, 4118.
[141] H. Taube, and H. Myers, *J. Amer. Chem. Soc.*, 1954, **76**, 2103.
[142] W. L. Reynolds and R. W. Lumry, 'Mechanisms of Electron Transfer', Ronald Press, New York, 1966.

Urushadze,[125] which is particularly useful in the (approximate) treatment of adiabatic reactions.

Chronologically, the first paper, by Schmidt,[110] is the simpler of the two treatments, and it contains some restrictive assumptions. Interestingly enough, these restrictive assumptions are easily relaxed, as was done in the second paper, by Dogonadze et al.,[111] and the relaxation does not materially alter the initial analysis. The reason this is so lies entirely with the (similar) methods of handling the bond-alteration co-ordinate in the two treatments.

The basic type of reaction considered is summarized in the simple notation of equation (395). The assumption of ref. 110 is that this reaction proceeds by means

$$AX^+ + B \underset{k_2}{\overset{k_1}{\rightleftarrows}} A + BX^+ \qquad (395)$$

of the formation of a state AXB^+. This state is metastable in the sense that it has a lifetime somewhat greater than the uncertainty lifetime. Nevertheless, it is sufficiently short-lived to satisfy the usual stationary-state approximation requirement of phenomenological kinetic analyses. Hence, the rate law is specified by equation (396), where the constants k apply to the sequence of reactions shown in (397). Attention focuses, therefore, on an effective rate constant which describes the simultaneous electron transfer and bond alteration in the species AXB^+.

$$\frac{d[A]}{dt} = -\frac{k_a k_b}{k_{-a} + k_b}[AX^+][B] \qquad (396)$$

$$AX^+ + B \underset{k_{-a}}{\overset{k_a}{\rightleftarrows}} AXB^+ \underset{k_{-b}}{\overset{k_b}{\rightleftarrows}} A + BX^+ \qquad (397)$$

As in other treatments of the electron-transfer process, we let V be the matrix element of the perturbation responsible for the transition. In the case of the inner-sphere reaction which takes place in a quasi-stationary metastable state of the form AXB, the driving perturbation may in fact be the non-adiabaticity operator(s) of the Born–Oppenheimer separation.[48] The transition-probability expression is written as a convolution (398), where $B(-\varepsilon)$ defines a Franck–Condon factor for the

$$w = (2\pi/\hbar)|V|^2 \int d\varepsilon\, B(-\varepsilon)F(\varepsilon) \qquad (398)$$

bond-break mode (including, in principle, changes in the co-ordination shell modes) and $F(\varepsilon)$ is the usual environmental contribution. $F(\varepsilon)$ is found to be given by equation (365), and need not be considered further. The evaluation of the bond-break mode requires special consideration.

In this model system representation, the initial state consists of the species AXB^+ and the final state is $A + BX^+$. By means of the application of a simple Born–Oppenheimer analysis, it was shown[110] that a vibrational Hamiltonian operator for the system AXB^+ in its initial state can be formed. This operator H_v^i applies to the system before electron transfer, and it is specified as shown in equation (399), where ω_b is the frequency associated with the normal mode which undergoes fission. The quantities p and q are conjugate momentum and co-ordinate variables

$$H_v^i = p^2/2m + \tfrac{1}{2}m\omega_b^2 q^2 \qquad (399)$$

for the bond co-ordinate. In this treatment,[110] specific consideration was not given to the changes in the remaining normal modes of the co-ordination shell. However, it is not difficult to include these changes. An account of changes in the co-ordination shell has been given by Dogonadze, Kharkats, and Ulstrup.[111] We therefore defer consideration of the contributions from the co-ordination shell to the discussion of the Dogonadze theory which follows this.

The final state, $A + BX^+$, requires special consideration in order to find a set of co-ordinates compatible with the initial-state set. For gas-phase reaction systems this is a truly difficult task, as complicated collision effects must be included in the analysis. For the solution phase, on the other hand, it is possible (at least loosely) to define a 'diffusion model'.[110] This model is in fact central to the treatments of both ref. 110 and 111. In the diffusion model, one assumes that in the final state the species A and BX^+ occupy positions in the solvent medium for times of the order of the characteristic diffusion times for these species. Specifically, we associate frequencies ω_1 and ω_2 with the species A and BX^+ in the final state, and assume that these species oscillate within cages with these frequencies. Consequently, the final-state, vibrational Hamiltonian operator can be written as shown in equation (400). By

$$H_v^f = p_1^2/2m_2 + p_2^2/2m_2 + \tfrac{1}{2}m_1\omega_1^2 x_1^2 + \tfrac{1}{2}m_2\omega_2^2 x_2^2 \qquad (400)$$

means of a transformation to relative and centre-of-mass co-ordinates [equations (401)], equation (400) is transformed into equation (402). The co-ordinate q' is

$$x_1 = R - (m_2/M)q' \qquad (401a)$$

$$x_2 = R + (m_1/M)q' \qquad (401b)$$

$$M = m_1 + m_2 \qquad (401c)$$

$$\mu = m_1 m_2/(m_1 + m_2) \qquad (401d)$$

now comparable with q in the initial state. It is, in fact, related to the initial-state co-ordinate by means of a simple displacement d.

$$H_v^f = P^2/2M + p^2/2\mu + \tfrac{1}{2}(m_1\omega_1^2 + m_2\omega_2^2)R^2 + (\mu^2/2m_1m_2)(m_2\omega_1^2 + m_1\omega_2^2)q'^2$$
$$+ \mu(\omega_1^2 - \omega_2^2)Rq' \qquad (402)$$

At this point, we shall ignore the centre-of-mass degrees of freedom by assuming that the centre of mass of the reacting system remains stationary. This is indeed not the case, as can be seen easily on consideration of the different diffusional characteristics of the species involved. However, the centre-of-mass contribution is expected to be much smaller than the bond-break contribution to the activation energy.[111]

The generating function for the bond-break mode Franck–Condon factor is written simply as equation (403), where the bracket notation implies the statistical

$$b(t) = \langle \exp[iH_v^f t/\hbar] \exp[-iH_v^i t/\hbar] \rangle \qquad (403)$$

average (trace) over the operators. In the classical limit for the vibrations, we can ignore quantum effects, and directly combine the Hamiltonian functions in the exponential. Since the final-state co-ordinate q' is related to the initial state q by a displacement d, we can carry out all the integrations simply, and write equation

(404), where $\tilde{d}$ is the dimensionless form of the co-ordinate d. The quantity λ accounts for the change in frequency $\Delta\omega$ which occurs [equation (405)].

$$w = |V|^2[\pi\beta/\hbar^2(E_s + \lambda\hbar\omega_b\tilde{d}^2/2)]^{\frac{1}{2}} \exp\left(-\frac{[\Delta J + E_s + \frac{1}{2}\lambda\hbar\omega_b\tilde{d}^2]^2\beta}{4(E_s + \frac{1}{2}\lambda\hbar\omega_b\tilde{d}^2)}\right) \quad (404)$$

$$\lambda = 1 + \Delta\omega/\omega_b \quad (405)$$

It is seen that the activation energy contains information about the bond-break step which is of the same form as the information about solvent repolarization. The critical distance d has been estimated by Dogonadze et al.,[111] and is considered next, with reference to their treatment.

This treatment is obviously deficient in that no specific consideration is given to the ligand-shell rearrangement energies. However, as a first step, and as a proposed model, it does illustrate the nature of the bond-break contribution to the activation energy. It is worth noting that this form of treatment compares well with the hydrodynamic treatment [12-14] which was developed to handle these types of systems as well as others. In the hydrodynamic approach, the outcome (the form of the activation energy) is very much the same as in equation (404). The difference is simply that the displacement quantity depends upon the difference in structure of the initial- and final-state species in a collective sense. The requirement that special co-ordinate systems be developed which are compatible in the initial and final states is relaxed. This relaxation, as we have noted, is brought about by the consideration of co-ordinate systems centred on the electroactive species. The gain or loss of matter in the co-ordination shells can be taken into account in terms of the free energies of solvation of these ligand species.

In addition to the treatment of the AXB^+ decomposition considered in ref. 110, a similar treatment has been proposed[48] in which the electron transfer and bond break take place simultaneously, in a reasonably strongly bound complex. There it was assumed that the driving perturbation is indeed the non-adiabaticity operator of the Born–Oppenheimer analysis. As a result, the transition-probability (and ultimately the rate-constant) expression can be handled almost entirely in terms of the framework of the molecular radiationless transition theory. The results are very much similar to those found in ref. 110. The critical bond-break distance d is there assumed (tacitly) to be some distance within the complex, past which molecular decomposition occurs.

The assumption of the formation of a complex of the form AXB is in general restrictive, and perhaps in many cases objectionable. There is probably little evidence for the formation of such a complex. Consequently, Dogonadze, Kharkats, and Ulstrup[111] extended the treatment by omitting the objectionable assumption of the formation of a transition-complex AXB, and dealt with the system in its chemically definable initial and final states. They do assume, however, that some form of transition-state AXB can be defined at the top of the system potential-energy barrier. This state has a lifetime consistent with the uncertainty lifetime.

The Dogonadze, Kharkats, and Ulstrup (DKU) treatment[111] is semi-classical, and employs the method of analysis considered by Dogonadze and Urushadze.[125] In particular, they define initial- and final-state potential-energy functions, find suitable compatible co-ordinate systems, and determine the saddle point, using the condition $U_i(q^*) = U_f(q^*)$, together with minimization. The results, more complete than those in ref. 110, bear a close similarity in form to the earlier treatment of Schmidt.[110]

The semi-classical analysis of DKU[111] uses the initial- and final-state potential-energy functions of equations (406) and (407). The co-ordinates q_l apply to the

$$U_i = \tfrac{1}{2}\hbar \sum_l \Omega_l q_l^2 + \tfrac{1}{2}\mu_{AX}\omega_{AX}^2(r_1 - r_{10}^i)^2 + \tfrac{1}{2}\mu_{AX,B}\omega_{AX,B}^2(r_2 - r_{20}^i)^2$$

$$+ \tfrac{1}{2}\sum_{\text{lig}} m_{L1}\omega_{L1}^{i2}(\rho_1 - \rho_{10}^i)^2 + \tfrac{1}{2}\sum_{\text{lig}} m_{L2}\omega_{L2}^{i2}(\rho_2 - \rho_{20}^i)^2 \quad (406)$$

$$U_f = \tfrac{1}{2}\hbar \sum_l \Omega_l (q_l - q_{l0})^2 + \tfrac{1}{2}\mu_{BX}\omega_{BX}^2(r_1' - r_{10}^f)^2 + \tfrac{1}{2}\mu_{A,BX}\omega_{A,BX}^2(r_2' - r_{20}^f)^2$$

$$+ \tfrac{1}{2}\sum_{\text{lig}} m_{L1}\omega_{L1}^{f2}(\rho_1 - \rho_{10}^f)^2 + \tfrac{1}{2}\sum_{\text{lig}} m_{L2}\omega_{L2}^{f2}(\rho_2 - \rho_{20}^f)^2 + \Delta J \quad (407)$$

solvent subsystem. The co-ordinates applicable to the bond formation and break for the inner-sphere transfer are r_1, r_2, r_1', and r_2'. The ligand co-ordinates are

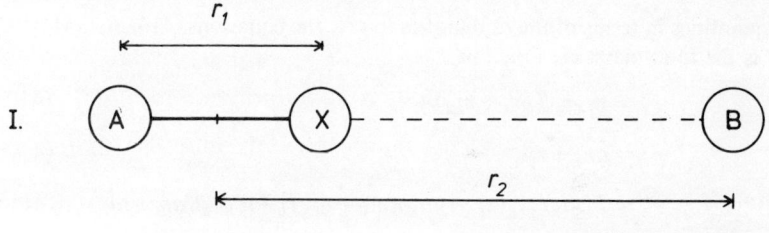

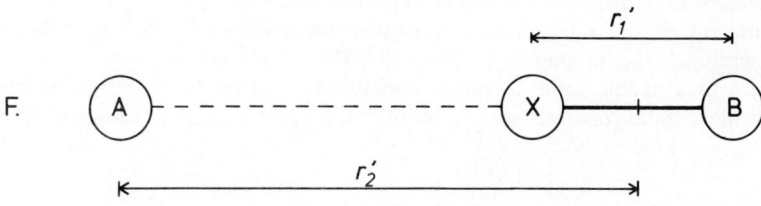

Figure 4 *I. The initial state: A is bonded to X. The co-ordinate r_1 is associated with the A—X bond. The co-ordinate r_2 defines the separation between the centre of mass of A—X and B. F. The final state: here the situation is reversed with respect to the initial state. B is bonded to X. The co-ordinate r_1' is associated with the B—X bond, and r_2' is defined as the separation between A and the centre of mass of B—X.*

signified by ρ. The situation treated is illustrated in Figure 4. The reduced masses μ_{AX}, μ_{BX}, $\mu_{A,BX}$, and $\mu_{AX,B}$ are associated with the various fragments, individually and in groups, also illustrated in the Figure. The various frequencies ω_{AX}, *etc.*, are associated with relative motions in the system. In particular, ω_{AX} and ω_{BX} are frequencies associated with stretching modes along the AX and BX bonds. The frequencies $\omega_{A,BX}$ and $\omega_{AX,B}$ are associated with the relative motions of the fragments A and BX, and AX and B. In magnitude, these frequencies are assumed to be of the order of the reciprocal of the diffusion lifetime for the relative motion.

Further development of the theory, leading to expressions for the activation energy and rate constant, depends upon the determination of a set of co-ordinates compatible with the initial and final states. Such a set can be found from the solution to the equations (408) and (409). Equation (409) specifies that the centre of mass remains constant during the reaction. This set of equations is readily solved for the

$$r_1 = r_X - r_A \tag{408a}$$

$$r_1' = r_X - r_B \tag{408b}$$

$$r_2 = (m_A r_A + m_X r_X)/(m_A + m_X) - r_B \tag{408c}$$

$$r_2' = (m_B r_B + m_X r_X)/(m_B + m_X) - r_A \tag{408d}$$

$$(m_A r_A + m_B r_B + m_X r_X)/(m_A + m_B + m_X) = 0 \tag{409}$$

r' quantities in terms of the r quantities to give the equations (410) and (411), where M is the total mass, $m_A + m_B + m_C$.

$$r_1' = m_A r_1/(m_A + m_X) + r_2 \tag{410a}$$

$$= ar_1 + r_2 \tag{410b}$$

$$r_2' = M m_X r_1/[(m_A + m_X)(m_B + m_X)] - m_B r_2/(m_B + m_X) \tag{411a}$$

$$= a'r_1 + br_2 \tag{411b}$$

At this point, the final-state potential-energy function [equation (407)] can be expressed in terms of the co-ordinates of the initial-state quantity [equation (406)]. Thus, with the use of the equality of the two potential-energy functions at the multi-dimensional saddle point, $U_i(Q^*) = U_f(Q^*)$, where Q^* represents the multi-dimensional saddle point, U_i can be minimized to determine the activation energy. The minimization proceeds with the use of the Lagrange multiplier [equation (412)].[125]

$$(1-\alpha)\frac{\partial U_i}{\partial \xi_k} + \alpha\frac{\partial U_f}{\partial \xi_k} = 0 \quad (\xi_k \in Q^*) \tag{412}$$

The analysis at this point is algebraic, and although moderately complicated, is nevertheless straightforward. One must find values for the co-ordinates q_1^*, p_1^*, p_2^*, r_1^*, and r_2^*, which correspond to the saddle-point co-ordinates. Dogonadze et al.[111] have introduced several approximations, based upon three definable models, which simplify the analyses. We shall specify the models, but not the details of the analyses. The results, in the form of the activation energies, for each model also will be given.

In their first model, DKU assume that all the masses of A, X, and B are of the same order of magnitude. This particular case requires the determination of the saddle point and the transfer coefficient α from a complicated set of algebraic equations. The analysis is simplified further with the assumption that the ligand frequencies are the same in the initial and final states. As a result, the expression for the activation energy is found to be given by (413), in which the factors P are defined by equations (414)—(417). The quantity E_s is the usual expression for the solvent repolarization energy.

$$E_a = P_i + (E_s + P_1^i + P_2^i + P_f - P_i + \Delta J)^2 / 4(E_s + P_1^i + P_2^i) \quad (413)$$

$$P_i = \tfrac{1}{2}\mu_{AX,B}\delta_{AX,B}^2(r_{10}^f - r_{10}^{'f})^2 \quad (414)$$

$$P_f = \tfrac{1}{2}\mu_{A,BX}\omega_{A,BX}^2(r_{20}^{'i} - r_{20}^f)^2 \quad (415)$$

$$P_1^i = \tfrac{1}{2}\sum_{\text{lig}} m_{L1}\omega_{L1}^2(\rho_{10}^f - \rho_{10}^i)^2 \quad (416)$$

$$P_1^i = \tfrac{1}{2}\sum_{\text{lig}} m_{L2}\omega_{L2}^2(\rho_{20}^f - \rho_{20}^i)^2 \quad (417)$$

The second case considered by DKU assumes the masses of A and B to be much greater than X. Consequently, A and B can be considered fixed. This situation corresponds to a model for the transfer of a light ion such as F$^-$. In this case, the initial- and final-state potential-energy functions can be simplified considerably. In particular, equations (418) and (419) can be written. Three sub-cases can be

$$U_i = \tfrac{1}{2}\hbar\sum_l \Omega_l q_l^2 + \tfrac{1}{2}m_X\omega_{AX}^2(r - r_{10})^2 \quad (418)$$

$$U_f = \tfrac{1}{2}\hbar\sum_l \Omega_l(q_l - q_{l0})^2 + \tfrac{1}{2}m_X\omega_{BX}^2(r - r_{20})^2 + \Delta J \quad (419)$$

investigated. Their results, in terms of expressions for the activation energies, are the following: (i) $\omega_{AX} = \omega_{BX}$, E_a is defined by equation (420). Here, apart from P_X, all the P quantities are as defined above. P_X is the reorganization energy of the ligand X, defined by equation (421).

$$E_a = (E_s + P_X + P_1^i + P_2^i + \Delta J)^2 / 4(E_s + P_1^i + P_2^i) \quad (420)$$

$$P_X = \tfrac{1}{2}m_X\omega_{AX}^2(r_{20} - r_{10})^2 \quad (421)$$

(ii) The second sub-case is expressed in terms of the inequality $\omega_{BX} \gg \omega_{AX}$, for which the activation energy is given by equation (422).

$$E_a = P_X + (E_s + P_1^i + P_2^i + \Delta J)^2 / 4(E_s + P_1^i + P_2^i) \quad (422)$$

(iii) Finally, when the ratio of the frequencies can be written $\omega_{AX}/\omega_{BX} = 1 + \delta$, equation (423) is valid, with K's defined by equations (424)—(426).

$$E_a = E_a(\omega_{AX} = \omega_{BX}) + (K_1 + K_2 + K_3)E_a(\omega_{AX} = \omega_{BX})\delta \quad (423)$$

$$K_1 = 4(E_s + P_1^i + P_2^i - P_X + \Delta J)/(E_s + P_1^i + P_2^i + P_X) \quad (424)$$

$$K_2 = 4P_X/(E_s + P_1^i + P_2^i + P_X + \Delta J) \quad (425)$$

$$K_3 = P_X(E_s + P_1^i + P_2^i + P_X - \Delta J)/2(E_s + P_1^i + P_2^i + P_X)^2 \quad (426)$$

If $P_X \gg E_s$, ΔJ, and the ligand frequencies, then one may write the equation (427), where $\gamma = (\omega_{BX}^f/\omega_{AX}^i)_a^0$, and E_a is the activation energy without X reorganization.

$$E_a = E_a^0 + 4E_s P_X E_a^0 / [E_s(1+\gamma) - \Delta J(1-\gamma)]^2 \qquad (427)$$

The last model considered by DKU assumed that m_X was much greater than the masses of A or B. As those authors did not carry out a numerical investigation of the activation energies using this model, we shall not reproduce the formal activation-energy expressions.

Dogonadze *et al.* applied their models A and B, *i.e.* the first two listed above, to an investigation of the following three reaction systems:

$$[(H_2O)_5FeX]^{2+} + [Fe(H_2O)_6]^{2+} = [Fe(H_2O)_6]^{2+} + [(H_2O)_5FeX]^{2+}$$
$$[(NH_3)_5CoX]^{2+} + [Fe(H_2O)_6]^{2+} \rightarrow [Co(H_2O)_6]^{2+} + [XFe(H_2O)_5]^{2+} + 5NH_3$$
$$[(NH_3)_5CoX]^{2+} + [Eu(aq)]^{2+} \rightarrow [Co(H_2O)_6]^{2+} + [Eu(aq)]^{3+} + 5NH_3 + X$$

where X = F, Cl, or Br. With the first reaction system, agreement between theoretical and experimental values for the activation energy was not found. The authors attribute this to the fact that the reaction may not proceed by the mechanism suggested.[111] However, for the others the agreement was reasonably good. The best agreement was found for the second reaction, using the second model.

These authors observe that the reorganization energy associated with the transferred ligand is small compared to the solvent repolarization energy E_s. A substantial contribution to the activation energy comes from the energy necessary to reorganize the co-ordination shells. Finally, the transfer coefficient for the $[Co(NH_3)_5X]^{2+}/Fe^{2+}$ system lies in the normal range of 1/2. This corresponds to $|\Delta J| \ll E_s$.[24] For the $[Co(NH_3)_5X]^{2+}/Eu^{2+}$ reaction, on the other hand, the transfer coefficient is much smaller.[111] This reaction is abnormal, and corresponds to the activationless type.[111]

It can be seen from the appealingly simple analysis of these authors that it is possible to make some progress in the analysis of complicated inner-sphere reactions using fairly simple models. The one-dimensional rectilinear approach may not always be useful in the analysis of these systems. However, it seems to offer a potentially useful method for calculating activation energies for a number of systems. A similar approach was been used in the past few years by Dogonadze and his colleagues in the investigation of ligand-substitution reactions.[143]

The second type of reaction system, which is similar in many respects to the true inner-sphere one, is in effect an outer-sphere transfer which is assisted by an intermediate or bridging species. Such a species might be a solvent molecule of the medium, or an ion of the supporting electrolyte, or an electrode surface. The reaction sequence is pictured as being essentially outer-sphere in the sense that the transfer electron in the initial donor state transfers to a state of the intermediate and thence to the final state of the true acceptor species. Consequently, the reaction potential-energy surface is constructed from three individual potential-energy functions for the initial, the intermediate, and the final state. Two formal intersections are involved, as is illustrated in Figure 5. The analysis of the system, therefore, requires the application of the Born–Oppenheimer separation to the separate parts of the system. From such an analysis, one finds that there are two driving perturbations for the electron transfer, one for the transition from the initial to the intermediate state, and one for the transition from the intermediate to the final state.

[143] M. N. Vargaftig *et al.*, *cf.* ref. 109.

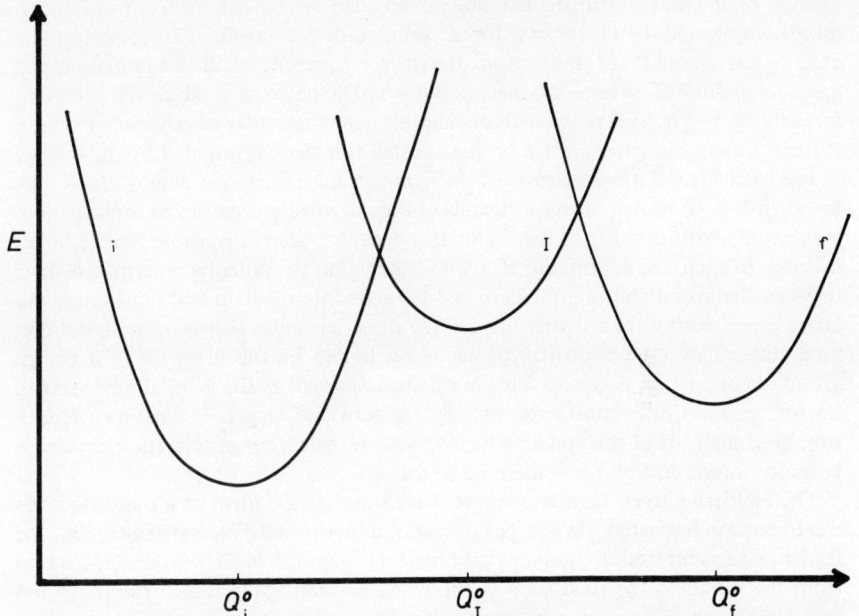

Figure 5 *A representation of the situation which applies to a bridge-assisted electron-transfer reaction. The initial and final states are indicated by i and f. The intermediate, bridge, state is labelled I. Transitions to and from this intermediate state I need not be energy-conserving. When not energy-conserving, such transitions are called 'virtual'.*

Two distinct reaction situations can be recognized. In the first, each electron-transfer transition is distinct, and conserves energy. As a result, the rate constant for the overall transfer will depend upon some function of the rate constants for the individual transfers. Generally, in terms of the usual steady-state approximation of phenomenological kinetic theory, the overall transition will be governed by the smallest rate constant (*i.e.* the slow, rate-determining step). Usually, the rate constant for the rate-determining step is associated with the step with the highest activation energy, but this need not be the case.

The second distinct situation is perhaps alien to classical kinetic theory; certainly it is alien to the classical phenomenology. In this case, the transfer to the intermediate state is *not* energy-conserving. Such transitions are generally termed 'virtual', in the sense of quantum-mechanical perturbation theory.[37] The transition probability in this case must be determined with the use of higher-order time-dependent perturbation theory.[144—152] It takes little effort to imagine that the

[144] P. P. Schmidt, *Austral. J .Chem.*, 1969, **22**, 673.
[145] M. V. Vol'kenshteyn, R. R. Dogonadze, M. V. Madumarov, and Yu. I. Kharkats, *Doklady Akad. Nauk S.S.S.R.*, 1971, **199**, 124.
[146] V. G. Levich, Yu. I. Kharkats, and R. R. Dogonadze, *Doklady Akad. Nauk S.S.S.R.*, 1971, **203**, 1315.
[147] Yu. I. Kharkats, *Elektrokhimiya*, 1972, **8**, 1300.

analysis of an electron-transfer reaction governed by this type of mechanisms is much more complicated than is the case for the simpler, direct transfers. This is indeed the case. In the remainder of this section, therefore, we present merely a summary of the essential points of several of the analyses which have been given.[145–152] Unfortunately, it is not yet possible to distinguish unambiguously whether any reaction systems satisfy the criteria for a bridge-assisted transfer as provided by the theory.

The basic idea of a bridge-assisted electron-transfer reaction system is that it may be possible in some instances for ligands of the intermediate species to overlap more successfully with orbitals of the donor and acceptor than is possible for the direct overlap. In such a situation, therefore, it is likely that the potential-energy crossings between the initial and intermediate and between intermediate and final states will be of lower energy than is the case for the direct crossing between the initial and final state. Thus, the electron transfer is facilitated by the presence of a bridge group, a conduction pathway. This mechanism, as well as the general inner-sphere transfer process, was considered first by Halpern and Orgel.[153] They provided a simplified analysis of the system which was nevertheless suggestive and provided a basis for interpretation for some time to follow.

The Halpern–Orgel[153] model immediately suggests the form of a more rigorous treatment. As indicated, these types of reaction involve intermediate states, but the transitions to and from these states are not energy-conserving. The overall transition from the initial to the final state is energy-conserving. As a result, the transition probability can be expressed in terms of the expression for a second-order process [equation (428)],[144–152] where V_1 is the perturbation responsible for the initial to the intermediate state transition, and V_2 is responsible for the transition to the final state.

$$w = (2\pi/\hbar)\sum_{i,f}\exp[-\beta E_i]/Z \left| \sum_I \frac{\langle f|V_2|I\rangle\langle I|V_1|i\rangle}{E_i - E_I + i\eta} \right|^2 \delta(E_i - E_f) \quad (428)$$

The first attempt to evaluate this expression[144] was not successful, and should not be considered as valid in any sense.

In 1971 Vol'kenshtein, Dogonadze, Madumarov, and Kharkats[145] published a quasi-classical treatment of the expression (428) for the case of linear initial-, intermediate-, and final-state potential-energy functions. The form of linear potential-energy function used is shown in equation (429) for the state i, with similar forms

$$U_i = I_i - \varepsilon_i |q - q_i^0| \quad (429)$$

for the other states I, and for state f. ε_i is a numerical quantity which gives the slope of the linear part of the potential. I_i is the minimum. This potential-energy function contains terms only for the outer-sphere contributions from the solvent. Consequently, the spectrum is quasi-continuous, and the transition probability can be rewritten as equation (430),[145] where $M(E)$ is given by equation (431).

[148] R. R. Dogonadze, J. Ulstrup, and Yu. I. Kharkats, *J. Electroanalyt. Chem.*, 1972, **30**, 47.
[149] R. R. Dogonadze, J. Ulstrup, and Yu. I. Kharkats, *J. Electroanalyt. Chem.*, 1972, **43**, 161.
[150] R. R. Dogonadze, J. Ulstrup, and Yu. I. Kharkats, *J. Theor. Biol.*, 1973, **40**, 259.
[151] R. R. Dogonadze, J. Ulstrup, and Yu. I. Kharkats, *J. Theor. Biol.*, 1973, **40**, 279.
[152] Yu. I. Kharkats, A. V. Madumarov, and M. A. Vorotyntsev, *J.C.S. Faraday II*, 1974, **70**, 1578.
[153] J. Halpern and L. E. Orgel, *Discuss. Faraday Soc.*, 1960, **29**, 23.

The Theory of Electron-transfer Reactions in Polar Media

$$w = (2\pi/\hbar)|V_1V_2|^2 \left[\int_{I_i}^{\infty} dE_i \rho_i(E_i) \exp(-E_i)\right]^{-1}$$

$$\times \int_{\max\{I_i, I_f\}} dE_i \exp[-\beta E_i] \rho_i(E_i)\rho_f(E_i)|M_{if}(E_i)|^2 \quad (430)$$

$$M_{if}(E_i) = \int_I^{\infty} dE_I \rho_I(E_I) \frac{\langle f(E_i)|I(E_I)\rangle\langle I(E_I)|i(E_i)\rangle}{E_i - E_I + i\eta} \quad (431)$$

The evaluation of this expression is carried out in the semi-classical limit with the use of the semi-classical (JWBK) wavefunctions for the solvent modes. The wavefunction for the state i can be written as shown in equation (432), where $P(q)$ is defined by equation (433), C_i is a normalizing coefficient, and z_i is a left-hand turn-

$$\chi_i(q) = \frac{C_i}{[\pi P_i(q)]^{\frac{1}{2}}} \cos\left(\int_{z_i}^q dsP_i(s) - \pi/4\right) \quad (432)$$

$$P(q) = \{2[E - V(q)]/\hbar\omega\}^{\frac{1}{2}} \quad (433)$$

ing point. The overlap integrals are evaluated with the use of the stationary-phase methods.[145–147,154] By means of arguments outlined by the Russian authors,[145–147] the energy integration in equation (431) is carried out, assuming that the intersection point between the second and third potential-energy surface is greater than the first intersection; the converse can be assumed as well. As a result, the matrix element M, [equation (431)] can be shown to assume the form shown in equation (434), where the Φ are the phases of the cosine functions in the overlap integrals

$$M_{if}(E_i) = -i\pi \sum_{l,k} D_{lk} \exp(-i\Phi_{Ii}^l) \cos\Phi_{fi}^k \quad (434)$$

evaluated at the stationary-phase points, D is a function of the normalization factors, *etc.*, for the wavefunctions and the overlap matrix elements, and the indices l and k denote various arrangements of stationary-phase points with respect to the equilibrium co-ordinates. The modulus squared of $M_{if}(E_i)$ enters the transition-probability expression. Thus, equation (435) may be written. As a result, Vol'kenshtein *et al.*[145]

$$|M_{if}(E_i)|^2 = \pi^2 |D_{lk}|^2 \cos^2\Phi_{fi} \quad (435)$$

find equation (436) for W, where E_{fi} is the energy at the intersection point between the intermediate- and final-state potential-energy surfaces.

$$W = \frac{3\pi^3}{4\hbar}|V_1V_2|^2 \frac{(\hbar\omega_I)^3(\varepsilon_i\varepsilon_I\varepsilon_f)^2[(\hbar\omega_I)^3/(E_{fi}-E_{Ii})]}{[(\varepsilon_i\hbar\omega_I)^3+(\varepsilon_I\hbar\omega_i)^3][(\varepsilon_f\hbar\omega_I)^3+(\varepsilon_I\hbar\omega_f)^3]} \beta\exp[-\beta(E_{fi}-I_i)] \quad (436)$$

It is not difficult to see that the intersection between the initial- and final-state potential-energy surfaces will be considerably greater in energy than the intersections between the initial- and intermediate-state and the intermediate- and final-state terms. Consequently, it is argued, reasonably, that the presence of a bridging

---

[154] H. Jeffreys and B. S. Jeffreys, 'Methods of Mathematical Physics', Cambridge University Press, London, 1972, 3rd edn.

group through which electron transfer can be facilitated can lead to increased electron-transfer rates by virtue of providing a path with a lower activation energy. It is interesting to note, however, that this conclusion, which depends completely on a complicated quantum-mechanical argument together with its associated analysis, is physically and chemically identical to the conclusion which can be drawn assuming two consecutive energy-conserving transitions. The rate-determining step, in the usual argument, probably is associated with the transfer for which the activation energy is the greatest.

Levich, Madumarov, and Kharkats,[146] and later Kharkats alone,[147] examined the second-order virtual-state transfer problem with the use of harmonic potential functions for the solvent modes. The results are similar to those obtained with the linear potentials used in the first treatment. For the simplest treatment, Levich, Madumarov, and Kharkats[146] obtained the expression (437) for the transition

$$W = 2\pi^{3/2}|V_1 V_2|^2 \frac{\beta^{\frac{1}{2}}}{\hbar^3 \omega_f \omega_I (q_I^0 - q_i^0)(q_f^0 - q_I^0)(E_{If} - E_{iI})^{\frac{1}{2}}} \exp[-\beta(E_{If} - I_i)] \tag{437}$$

probability. Kharkats,[147] in a complete and complicated treatment, assumed multi-dimensional harmonic potential-energy surfaces. The results, much more complicated in form than those quoted here, nevertheless show the same general behaviour.

The energies E_{iI} and E_{fI} of the two intersection points for the three potential-energy functions can be estimated in terms of the outer-sphere reaction model.[145—152] Thus, if the second intersection is greater than the first, E_{fI} is given by equation (438). Dogonadze, Kharkats, and Ulstrup[148—151] have investigated the application

$$E_{fI} = (I_f - I_I + E_s)^2 / 4E_s \tag{438}$$

of these second-order transition-probability expressions to various electron-transfer and electrochemical problems. With the use of the energy (438) together with various averaging methods required to obtain expressions for the rate constants, they have identified a number of limiting situations. In particular, barrierless and activationless reaction limits occur for bridging systems as well as for the simple case of direct transfer. In addition, certain aspects of metallic catalysis can be considered. Unfortunately, it is not yet at all clear that this approach provides an acceptable picture for any real process. Experimental substantiation is needed, but it has not yet appeared in an unambiguous form.

To date, the final chapter on the theory of these second-order processes has been provided by Kharkats, Madumarov, and Vorotyntsev.[152] Whereas the first three treatments of this problem made use of semi-classical wavefunctions, and the somewhat complicated contour integrations needed in the evaluation of the overlap matrix elements,[145—147] these authors have provided a quantum-limit treatment. This treatment is based upon the use of harmonic-oscillator wavefunctions and the density matrices associated with these functions for the various subsystems involved.[152] The analysis is lengthy, and will not be duplicated here. It is sufficient to say that the analysis is similar in most respects to the treatment of the simple direct electron transfer discussed earlier in this chapter. The Kharkats, Madumarov, and Vorotyntsev analysis[152] differs from the simpler treatment in two significant aspects. First, they use the integral transform (439) to eliminate the energy depend-

$$\frac{1}{E+i\eta} = \beta \int_0^{-i\infty} d\tau \exp[-\beta\tau(E+i\eta)] \qquad (\eta \to +0) \qquad (439)$$

ence in the denominator of the second-order transition-probability expression, equation (428). Consequently, an integration involving four vibrational density matrices can be constructed. In the limit of no change in frequency and a simple displacement of oscillator co-ordinate between the initial and final states, the four one-dimensional co-ordinate integrations can be carried out.

The second significant step in the Kharkats–Madumarov–Vorotyntsev analysis[152] is the formulation of a proper saddle-point analysis for the complex time integrations. The result of their analyses, the transition-probability expression, is of the same form as was obtained with the use of the semi-classical wavefunctions. In addition, they obtained an expression which applies when the two intersection points for the three overlapping potential-energy surfaces are of the same energy. This expression, worth quoting, is (440), where ΔJ_{di} is the difference between the

$$W = \frac{2\pi^2}{\hbar^2 \omega k T E_s} |V_1 V_2|^2 \ln\left(\frac{4E_s kT}{(E_s - \Delta J_{di})}\right)^2 \exp[-\beta(E_s + J_{di})^2/4E_s] \qquad (440)$$

electronic terms in the initial and intermediate states. This energy equals ΔJ_{fb}. The equality of the intersection points is reflected in the pre-exponential factor as the logarithm of a large number $(kT/\hbar\omega)$.

The theory of the bridge-assisted transfer certainly indicates that electron-transfer rates can be influenced substantially by the passage of the transfer electron through the intervening chemical species. It seems reasonable to expect that the bridge-assisted transfer should be of importance in a number of instances. The search for conclusive evidence of this type of mechanism may prove of value in the process of unravelling the complicated sequence of electron-transfer events associated with some types of electrochemical transfers at electrode surfaces.

7 Conclusion

In Part I, and in this Part as well, the focus of attention has been on the treatment of the electroactive species in electron-transfer and electrochemical reactions. Reference has been made primarily to the case of the homogeneous electron-transfer reaction. The electrochemical, heterogeneous transfer has not been considered in any detail.

The electrochemical transfer is definitely very important. However, up until very recently, the detailed treatment of this type of reaction continued to be given essentially in the same form as originally presented by Dogonadze and Chizmadzhev in 1962.[155] The electrochemical electron transfer is well summarized by Dogonadze in his review article; cf. ref. 156. There are some new developments with respect to

---

[155] R. R. Dogonadze and Y. A. Chizmadzhev, *Doklady Akad. Nauk S.S.S.R.*, 1962, **144**, 1077.
[156] R. R. Dogonadze, in 'Reactions of Molecules at Electrodes', ed. N. S. Hush, J. Wiley and Sons, New York, 1971, pp. 135—228.

the electrochemical transfer. In particular, Schmickler[157] has recently examined the effect of oxide layers on electrochemical rates, obtaining interesting results which agree with experiment. Nevertheless, the fundamental approach to the treatment of the electron transfer remains the same as originally given by Dogonadze and Chizmadzhev.

It is perhaps remarkable that the electrode system, for the most part, can be treated in a simple manner, and that differences between good conducting metals bear so little an influence on the rate of transfer. Surface states and metallic band structures appear to have little substantial influence on the transfer, particularly if it is an outer-sphere transfer. For semiconductor electrodes, on the other hand, the dependence of the transfer process on properties of the electrode is much more pronounced.

The electrochemical transfer is influenced substantially by the work terms which enter the activation-energy quantity. Discreteness of charge effects, and the structure of the double layer in general, seem to play an interesting and important part in the overall transfer process. One can anticipate a considerable amount of attention directed toward these aspects of the electrochemical process in the next few years.

Overall, the electron-transfer process, in both its homo- and hetero-geneous forms, requires the careful theoretical examination of all the contributions to the activation process from the degrees of freedom of the electroactive species and of the electrode metal and the interface. In this chapter we have concentrated our attention on the careful examination of the electroactive species. This aspect of the electron-transfer process in all its forms has been the most extensively studied, and continues to be examined. It is, of course, the aspect of the process which is most readily examined in considerable detail. Those parts of the electrochemical problem which are not closely related to the structure of the electroactive species need further examination. In particular, the interfacial, double-layer effects need to be examined in considerable detail. Apart from being able ultimately to determine electrochemical exchange currents and activation energies with confidence and accuracy, an examination of these interfacial state factors now is needed in order to be able to resolve the nagging problem of the untestability of the theory to date. The primary test so far has been to attempt to find the non-linearity of the Tafel plots (equivalently, the potential dependence of the transfer coefficient). It seems that, for the homogeneous case, the non-linearity of the free-energy plots has been seen.[22] The electrochemical case has, so far, defied demonstration. The search continues.[158] Certainly, with respect to the simplest form of the electrochemical exchange theory, for an outer-sphere transfer it should be possible reasonably easily to see the non-linear effect. That this has not yet been the case may be tied to events and properties of the interfacial state that are not connected with the primary electron-transfer event. If these factors can be untangled, and the primary transfer event isolated, then it may prove possible to see the potential dependence of the transfer coefficient and to substantiate the prediction of the theory. If not, then a substantial revision of the entire theory will apparently be necessary.

In this chapter, and in Part I, we have dealt extensively with the formulation of the

---

[157] W. Schmickler, American Chem. Soc. National Meeting, New Orleans, March 1977, and personal communication.
[158] F. Anson, personal communication.

electron-transfer theory, particularly as it depends upon the electroactive species. At this time it is possible to summarize the state of development of the theory of the electron-transfer process as follows. First, with respect to the formal aspects of the theory which lead to general forms for the rate constant, it is possible only to obtain expressions valid in the non-adiabatic limit. This, as noted earlier in this chapter, is clear from the nature of the approximations needed in order to be able to evaluate the Franck–Condon factors, or in a more general case the correlation functions, which describe the time evolution of the reaction system. It is formally possible to construct an infinite-order perturbation-series representation for the rate constant. If this series can be handled, and the series summed, then it is possible to obtain expressions for the rate constant that are valid to any order, adiabatic to non-adiabatic. It has been possible to do this only in one instance, for a model two-level system,[24] yet the results in the case of this model-system calculation substantiate the less formal, heuristic arguments usually used to argue the adiabatic limit, given expressions for the transition probability and rate constant for the non-adiabatic limit.

The second, and (for the immediate future development of the theory) the most important development, is the formulation of the treatment of the electroactive species in a manner which enables us to examine all the important dynamical contributions to the activation process. As we have shown in this chapter, it is possible to obtain a representation which provides not only the pertinent solvated electronic energy levels and states of the electroactive species, but also provides accurate vibrational potential-energy functions for these species in configurations leading to the electron-transfer step. In addition, terms associated with the work of forming a particular electron-transfer initial- and final-state configuration can be found. Within the dictates of the disperse polar solvent model, for example, it is clear how these terms are to be evaluated for any particular real system.

Finally, recent examinations of the time integrations involved in the evaluation of the final form of the transition-probability expression have shown that, in addition to the normal electron-transfer region (in terms of the heat of reaction), two abnormal regions can be defined. These abnormal regions are associated with activationless and barrierless transitions. Several authors have shown that the dependence of the activation energy with respect to the heat of reaction does not follow the simple parabolic dependence originally predicted by Marcus. The substantiation of this behaviour of the activation energy *versus* the heat of reaction seems reasonably well established for homogeneous electron-transfer reactions.[22]

There are a number of finer details of the theory of electron-transfer reactions which have not been covered in this chapter or in Part I. The contributions of the Russian school, particularly Dogonadze's recent contributions, are extensive. Some of the additionally important problems which they have considered, but which we have not reviewed in any detail, concern multiple-electron transfers,[159] and the effect of a cut-off Debye spectrum for the polar solvent.[160] The nature of the effect of a continuous spectrum of the polar solvent on the form of the interactions between the charges in the electroactive species can be an important factor to

[159] M. A. Vorotyntsev and A. M. Kuznetsov, *Elektrokhimiya*, 1970, **6**, 208.
[160] M. A. Vorotyntsev, R. R. Dogonadze, and A. M. Kuznetsov, *Doklady Akad. Nauk S.S.S.R.*, 1970, **195**, 1135.

consider in the formal treatment of the energy levels of these species. More needs to be done. However, the basic form of the theory of the energy levels of solvated molecular species has been developed. It is clear that the further formal development of this particular aspect of the overall theory of electron-transfer reactions will be stimulated by the necessity to resolve specific computational difficulties as they arise.

The theory of the electron-transfer and electrochemical processes can be of great use in the near future. The usefulness of the theory, however, rests upon its definitive test against experiment. As indicated, a crucial test is the experimental demonstration of the predicted non-linearity of the Tafel behaviour. In addition, there is room for the continuing assessment of activation energies. The formal aspects of the theory have developed to the point where now it is mandatory that consideration be given to the development of a computational theory. Because it is now possible to consider detailed quantum-chemical calculations for quite complicated molecular systems, there is every reason to expect that in the next few years extensive tests comparing theoretical and experimental activation energies will be made. Tests have been made in the past.[161,162] Hale's calculations, in particular, are based upon the use of the simplest ionic model representations, *viz.* the Born model, and upon other more or less straightforward parameterizations. The results of his calculations, when compared against experimental values for the activation energies, are good. There is certainly the suggestion in Hale's work, as well as in the calculations of others,[111] that ultimately it may not be necessary to use the full set of equipment of modern quantum chemistry in order to obtain meaningful and useful predictions of activation energies and other quantities. If this is to be the case, it will be proved by comparison against detailed, rigorous calculations.

It is necessary that the theory should account adequately for the experimental observations already made. This goal seems visible at this time. It is also necessary that the theory should provide a fabric which weaves a unification of diverse observations. This objective also seems within reach. Finally, it is desirable, but not mandatory, that the theory should be capable of some extent of suggesting new, hitherto undetected, phenomena, and that it should be predictive. That the theory will be predictive in the sense of providing activation energies, *etc.* for new reaction systems seems likely. To expect the theory to predict new phenomena may be asking too much. In the end, the theory may prove to be rather mundane, and limited only to the electron-transfer event. It is entirely possible, although not yet demonstrated, that much of the complicated behaviour, particularly of electrode systems, can be understood entirely in terms of current mass-transport analyses, *etc.* Even if this turns out to be the case, the electron-transfer theory can still play a major part in the overall analysis of electron-transfer phenomena.

Acknowledgements

Work on Part I and on the initial stages of this Part was generously assisted in part by a grant from the SRC and by a Fulbright Senior Research Scholarship.

It is a pleasure to be able to thank numerous friends and colleagues for discus-

---

[161] N. Sutin, *Ann. Rev. Phys. Chem.*, 1966, **17**, 119.
[162] J. M. Hale, in 'Reactions of Molecules at Electrodes', ed. N. S. Hush, J. Wiley and Sons, New York, 1971, pp. 229—255.

sions of various aspects of the electron-transfer theory over the years. In particular, I wish to express my thanks to Professors Marcus, Dogonadze, Fleischmann, Anson, and Fischer, and to Drs. Kornyshev, Ulstrup, Kharkats, and Schmickler. The opinions expressed in these two chapters, however, are entirely mine.

I owe an apology to Professor Hush. In Part I his initial assessment of the usefulness of the potential-energy-surface concept was quoted as 'illusory'. I neglected to refer the reader to his subsequent works, in which this point of view was substantially amended. Hush's current views on the potential-energy-surface concept definitely are no longer reflected in the statement quoted.

Author Index

Abbott, E. M., 11
Abe, T., 61
Adam, N. K., 69
Adams, R. N., 52
Adamson, A. W., 66, 68
Adrian, F. J., 161
Aitkin, A. C., 212
Akhmedov, A. M., 60
Aki, O., 25
Alexander, A. E., 66
Aliev, I. Ya, 44
Alt, G. H., 51
Alt, H., 15
Alumyan, Zh. R., 60
Aminova, F. Kh., 30
Ammar, F., 8
Anantharaman, P. N., 1
Anderson, L. B., 33
Anderson, T. N., 65, 76
Anderson, W. J., 65
Andreoli, R., 27
Andrieux, C. P., 8
Andruzzi, R., 16
Aniskova, L. V., 45, 47
Anson, F., 238
Arbuzov, B. A., 24
Arenhövel, H., 196
Arfken, G., 165
Argade, S. D., 65
Armand, J., 17
Armstrong, N. R., 9
Armstrong, R. D., 103, 106, 108, 109, 111, 115, 188, 121
Asahi, Y., 25
Ashurov, D. A., 60
Asirvatham, M. R., 16
Astruc, A., 28
Astruc, M., 28
Avagyan, S. P., 26, 27
Avrutskaya, I. A., 1, 14, 18, 25, 64
Aylmer-Kelly, A. W. B., 34

Baba, H., 61
Babievski, K. K., 14
Bailey, G. L. J., 69
Bailey, W. F., 24
Baizer, M. M., 2, 5, 6
Balagopalan, S., 14
Balakrishnan, T. D., 14
Balashova, N. A., 65
Banda, F. M., 47
Bangham, D. H., 68
Barbey, G., 54, 55
Bard, A. J., 4, 15
Barradas, R. G., 25

Barry, J. E., 49
Batley, G. E., 15
Batrakov, V. V., 65
Battiotuzzi, G., 27
Bauer, H. H., 66
Baumgärtel, H., 57
Baxter, S., 69
Beach, K. W., 71
Becher, J., 14
Bechgaard, K., 43
Beck, F., 1
Beck, T. R., 71
Becker, J. Y., 34, 63
Begley, W. J., 18
Behret, H., 40
Bell, M. F., 109, 115
Bellamy, A. J., 11
Belnikov, V. M., 14
Benedetti, L., 27
Benninger, S., 60, 62
Berdnikov, E. A., 24, 28
Berenblit, V. V., 46
Berge, H., 52
Berne, B. J., 163
Berenblit, V. V., 46, 61
Berezina, S. I., 6
Bernett, M. K., 73
Bewick, A., 34
Bewig, K. W., 73
Bezuglyi, V. D., 87
Bighi, C., 27
Bikerman, J. J., 68, 69
Biolchini, G., 23
Bingulae, S. P., 113
Bison, G., 26
Bixon, M., 183, 222
Blackburn, G. M., 39
Bloomfield, J. J., 48
Blount, H. N., 39
Bobilliart, F., 34
Bockris, J. O'M., 65, 71
Bodé, D. D., jun., 76
Bonnemay, M., 65, 71
Bontempelli, G., 12, 58, 64
Booth, F., 168
Born, M., 135, 151
Busch, D. H., 33
Butin, K. P., 26, 32
Butler, J. A. V., 66
Brenner, A., 67
Breslow, R., 2
Brettle, R., 36, 47
Britton, W. E., 24, 43
Bronoël, G., 71
Brooks, C. J., 47
Brown, O. R., 22, 26
Brun, A., 43

Bruni, P., 16
Bykhovskaya, T. N., 21
Byrd, L. R., 63
Byzov, B. A., 61

Campbell, C. R., 4
Campbell, D. S., 67
Campbell, J. A., 8
Canfield, N. D., 59
Cantrill, P. R., 34
Cape, J. N., 29
Carelli, I., 16
Carr, J. P., 91, 92, 93
Casalbore, G., 32, 60
Casanova, J., 24
Case, B., 168
Cassie, A. B. D., 69
Cataliotti, M. L., 27
Caullet, C., 9, 11, 53, 54, 55, 59, 62
Cauquis, G., 54
Cekovic, Z., 57
Cernak, J., 7
Chabaud, B., 54
Chambers, J. Q., 28, 29, 59
Chechina, O. N., 46
Chekir, K., 17
Chernyk, I. N., 22
Chiba, T., 4
Chikayuki, K., 46
Child, M. S., 153, 186
Chizmadzhev, Y. A., 237
Chkir, M., 45
Chong Min Pak, 13
Christov, S. G., 148
Churina, D. Kh., 64
Clavilier, J., 65
Coe, P. J., 47
Cogoni, G., 12, 64
Cole, R. H., 106
Coleman, J. P., 34, 39, 47
Colon, F., 101
Colonna, M., 16
Comminellis, Ch., 62
Corbon, P., 54
Costa, G., 32
Covitz, F. H., 1
Cox, J. A., 28
Cramer, J., 15
Croxton, C. A., 163
Curphey, J. J., 9
Cwiklinski, C., 35
Czochralska, B., 16

Dahms, H., 71
Dalla Betta, R. A., 72

Author Index

Damaskin, B. B., 65
Damjanovic, A., 71
Damoni, Sh., 57
Dana, G., 8
Danilov, S. D., 56
Danly, D. E., 4
Danos, M., 191, 196
Darchen, A., 14
Darlewski, W., 14
Davies, T. J., 68
Davydov, G. A., 61
De Battisti, A., 27
de Laplace, P. S., 66
de Levie, R., 104
Dean, S. W. jun., 68
Dedek, V., 10
Degner, D., 10
Delahay, P., 66, 98, 113
Dennery, P., 171
Deshler, L., 8
Desideri, P. G., 50
Despic, A. R., 113
Dettre, R. H., 69, 72
Devanathan, M. A. V., 65, 66
Devynck, J., 60
Dickerson, D. L., 15
Divisek, K., 20
Dogonadze, R. R., 129, 130, 133, 150, 160, 207, 208, 213, 222, 233, 234, 237, 239
Dolgopolskii, I. M., 61
Dolin, P., 98
Dolnakov, Yu. P., 61
Dorofeenko, G. N., 30
Doughty, A. G., 60
Doupeux, H., 29
Drakesmith, F. G., 60
Drossbach, P., 102
Drury, R. F., 2
Dryhurst, G., 52, 57
Dunsch, L., 50
Dupré, A., 54, 67
Duty, R. C., 23

Eaborn, C., 31
Eberson, L., 37
Edmondson, K., 118
Efrima, S., 183, 222
Eglstaff, P., 163
Eick, J. D., 69
Eisenbach, W., 64
Eisenberg, J. M., 129
El'tsov, A. V., 12
Elving, P. J., 11, 17
Englman, R., 208
Epelboin, I., 111, 115
Epling, G. A., 24
Erb, R. A., 72
Ercoli, R., 11
Erfurt, G., 23
Erimbetov, A. K., 35
Ershler, A. B., 32
Ershler, D., 98
Evans, D. H., 5, 7, 22
Evstifeev, M. M., 30
Evstigneeva, R. P., 18
Eyring, H., 76

Fagioli, F., 27
Faingold, N. I., 46
Fano, U., 195
Falck, J. R., 39, 40
Farmery, K., 33
Farnier, M., 21

Fedoronko, M., 1
Fedorova, G. T., 32
Feldman, A., 72
Feoktistov, L. G., 25
Fetter, A. L., 149
Feynman, R. P., 148
Fierz, M., 203
Filardo, G., 11
Finkelstein, M., 49
Fioshin, M. Ya, 1, 14, 18, 25, 35, 45, 47, 64
Firman, R. E., 111, 115, 121
Fischer, H., 71, 133, 135
Fleischmann, M., 1, 60
Fletcher, S., 26
Florence, T. M., 15
Flugge, F., 203
Fodiman, Z. I., 15
Fong, F. K., 129
Foulatier, P., 9
Fowkes, F. M., 72
Frazier, H. W., 51
Fredlein, R. A., 71
Freed, K. F., 218
Fremery, M., 34
Friedlin, G. N., 46
Friedman, H. L., 207
Fritz, H. P., 34
Froeling, A., 10
Frumkin, A., 70
Frumkin, A. N., 65, 66, 105
Fry, A. J., 2, 24
Fues, E., 187
Fujii, M., 2
Fuks, N. Sh., 10

Gabrielli, C., 115
Galluzzo, M., 11
Gammel, D., 13
Genies, M., 54
Genshaw, M. A., 93
Gerischer, H., 99, 110, 111
German, E., 207
Getmanskaya, Z. I., 46
Ghe, A. M., 2, 3
Giannici, B., 11
Gibbs, J. W., 66
Gilbert, J. C., 15
Gileadi, E., 65
Gilman, S., 65
Glasiewicz, Z. M., 173
Glueckauf, E., 161, 168
Gluzman, S. S., 46
Glybovskaya, V. A., 35
Gokhshtein, A. Ya, 71
Goldin, M. M., 25, 52
Goldschmidt, V. M., 161
Good, R. J., 69
Goodin, R. C., 15
Gore, E. S., 33
Gorodetskaya, A., 70
Gostisa-Mihelcic, B., 20
Goubeau, D., 28
Gourary, B. S., 161
Gourey, J. G., 26, 59
Gouy, G., 66
Grachev, V. I., 61
Grahame, D. C., 66, 99, 111
Grandi, G., 27
Greiner, W., 129, 196
Grimshaw, J., 18, 21
Gromova, E. V., 25
Gubenko, I. I., 18
Gudeleva, N. N., 4
Guillanton, G. Le, 10, 30, 31
Gulick, W. M., 13

Hale, J. M., 240
Halpern, J., 234
Hamano, Y., 2
Hamelin, A., 65
Hammerlich, O., 38, 39
Hampson, N. A., 91, 92, 93
Hand, R. L., 50
Hanschmann, H., 11
Hansen, W. N., 65
Harado, K., 10, 19
Harkins, W., 72
Harned, H. S., 207
Harp, G. D., 163
Harrison, J. A., 26, 121
Hatayama, T., 2
Hauschild, J., 23
Hawley, M. D., 16
Hay, J. V., 8
Hayter, B., 66
Hayward, D. O., 92
Hazard, R., 57
Heckel, W. A., 4
Heilmann, S. M., 22, 34, 48
Heimler, D., 50
Heitz, E., 34
Helgee, B., 37
Henderson, M., 111
Henrix, J., 173
Henze, G., 55
Herlem, M., 34
Herring, C., 67
Hibino, K., 45
Hiller, S., 1
Hoare, J. P., 65
Hoever, H., 34
Hoffman, R. W., 67
Holley, S. N., 91
Holman, R. J., 5
Holstein, T., 139
Holub, K., 113
Hori, F., 64
Horner, L., 10, 13
Houtala, R., 72
Howard, M. B., 60
Hub, C., 95
Hush, N. S., 130
Husousky, A. A., 104

Ibrisagic, Z., 9
Ichikawa, M., 34
Ihenoya, S., 56
Inesi, A., 23, 24
Ingall, G. B., 22
Ingard, K. U., 193
Inoüe, G., 2
Irish, D. E., 222
Itoh, M., 64
Ivanova, T. A., 64
Ivanova, V. Kh., 15
Iverson, P. E., 2, 11, 13, 14, 27
Iwaski, T., 19

Jackson, R. A., 31
Jaeger, P., 14
Jaenchen, M., 33
Jakopcic, K., 57
Jason, M. F., 24
Javet, Ph., 62
Jeffreys, B. S., 235
Jeffreys, H., 235
Jeminet, G., 26, 59
Jensen, B. S., 2
Jensen, N. J., 25
Jeroschewski, P., 52
Johnson, C. A., 68

Author Index

Johnson, D. A., 15
Johnson, R. E., jun., 68, 69, 72
Jonville, P. J., 71
Jortner, J., 129, 139, 181, 208, 218
Jovanovic, D. R., 113
Jubault, M., 10
Jugelt, W., 54, 58
Junghans, K., 3

Kabanoff, B. N., 70
Kabanov, B., 70
Kadis, V., 1
Kaiser, U., 34
Kamada, K., 71
Kanakam, R., 14
Kanakam, R. I., 4
Kaplan, G. I., 23
Kapovits, I., 29
Kargin, Yu. M., 6, 28
Kargina, N. M., 6
Kasafirek, E., 26
Kastening, B., 20
Kato, S., 18, 57
Katoh, Y., 71
Katovic, V., 33
Katz, M., 36
Katzer, E., 47
Kazarinov, V. E., 65
Keddam, M., 111, 115
Keller, E., 55
Kenkel, J. V., 4
Kenyhercz, T. M., 33
Kerr, J., 11
Kerrinnes, H. J., 33
Kestner, N. R., 139
Kharkats, Yu. I., 129, 150, 168, 207, 233, 234
Khmelnitskaya, E. Yu, 21
Kimmerl, P., 64
Kinza, H., 34
Kirilyus, I. V., 10
Kitaev, Yu. P., 6, 15
Klemm, L. H., 13
Klimovitskii, E. N., 24
Klug, J. T., 49
Klyuev, B. L., 10
Knittel, D., 20
Knunyants, I. L., 44
Koch, D., 35
Koch, R. W., 8
Kodaira, K., 61
Kollar, V., 7
Kopachevskii, E. N., 35
Korczynski, A., 25
Korienko, A. G., 45
Kornyshev, A. A., 150, 160
Korshikov, L. A., 87
Kotlyarevskaya, B. M., 18
Kovac, J., 7
Kovsman, E. P., 46
Krasnokutskaya, D. M., 52
Kratzer, A., 187
Krayanskii, O. B., 63
Kropotova, N. V., 10
Kruger, J., 65
Krysin, E. P., 18
Kryuchkova, E. I., 35
Krzywicki, A., 171
Kubo, R., 128, 133
Kuhlmann-Wilsdorf, D., 67
Kumar, A., 59
Kurkina, L. G., 25
Kushin, A. N., 32
Kutowy, O., 25
Kuznetsov, A. M., 133, 150, 208, 222, 239

Kyazmov, Sh. K., 60

Lacan, M., 1, 9, 57
Laidler, K. J., 161
Lambert, F. L., 22
Lamm, B., 27
Lamy, E., 6
Landau, L. D., 134, 203
Landsberg, R., 41
Langbein, U., 33
Laurent, A., 42
Laurent, E., 42
Layloff, T. 9
Lax M., 222
Le Guillanton, G., 10, 30, 31
Lehmkuhl, H., 64
Leikis, D. I., 65
Lelandais, D., 45
Lepri, L., 50
Lester, J. H., 4
Lettau, H., 23
Levart, E., 71
Levich, V. G., 129, 130, 150, 154, 233
Levin, A. I., 46
Levin, E. S., 15
Levy, M., 5, 9
Lexa, D., 33
Lharitonov, N. P., 32
Lhoste, J. M., 33
Liao, Chi-Sheng, 29
Libert, M., 11, 53, 59, 62
Lifshitz, E. M., 203
Likhachev, V. A., 6
Lin, K. F., 71
Lin, S. H., 212
Lindhard, J., 173
Line, L. L., 51
Lines, R., 47
Linford, R. G., 66
Ling, I., 68
Lippmann, G., 66
Liska, F., 10
Llopis, J., 101
Lloyd Jones, H., 49
Logan, J., 20
Loginova, N. F., 26, 28
Lomax, A., 4
Longchamp, S., 53
Lorenz, W., 105, 107, 119
Loshkarev, Yu. M., 17
Louisell, W., 180
Lovecchio, F. V., 33
Luknitskii, F. I., 64
Lumry, R. W., 225
Lund, H., 1, 3, 13, 25, 27, 29

Madsen, J. Ø., 11
Madumarov, A. V., 234
Madumarov, M. V., 233
Magno, F., 58
Mahisko, R. F., 24
Mairanovskii, V. G., 18, 26, 28
Malakhova, T. A., 64
Mamaev, V. M., 18
Mann, C. K., 49
Marcoux, L., 52
Marcus, R. A., 128, 130, 148, 217
Marinova, R. I., 18
Margel, S., 5
Mark, H. B., 33
Mark, H. B., jun., 217
Martini, T., 62
Mason, M., 24

Mason, R., 103
Masson, J. P., 60
Mastragostino, M., 60
Mathews, W., 23
Mathieu, P., 58
Matschiner, H., 21, 23
Matsumoto, K., 16
Matsumoto, S., 10
Matsumura, Y., 45
Matsuoka, M., 14, 16
Matuschke, R., 23
Mayeda, E. A., 49
Mazzocchin, G. A., 58
McAllister, D. L., 52
McKinley, J. M., 158
Mehl, W., 111
Melik-Gaikazyan, V. I., 105
Mellor, M. E., 64
Melnikov, I. A., 21
Melnikov, N. N., 21
Meshitsaka, S., 34
Metcalfe, A. A., 121
Michaels, A. S., 68
Michel, I. A., 72
Michel, M.-A., 3, 29
Mikhailov, V. S., 23
Miller, L. L., 34, 39, 40, 41, 45, 63
Millington, J. P., 34
Mills, K. L., 60
Mil'man, V. I., 63
Mirkind, L. A., 45, 47
Mirzoyan, V. L., 10
Misono, A., 2
Mispeltier, J., 33
Mitui, Yu. V., 59
Miyawaki, S., 45
Miyoshi, M., 16
Möckel, F., 107
Möller, G., 70
Mohilner, D. M., 66
Momenteau, M., 33
Mora, R., 21
Morcos, I., 71, 90, 95
Moreira, H., 104
Morel, Ph., 115
Morimoto, A., 25
Moritake, M., 2
Morris, M. D., 1
Morse, P. M., 187, 193
Moses, P. R., 28
Mousset, G., 29
Müller, S., 41
Mueller, W. H., 4
Muirhead-Gould, J. S., 161
Mulina, T. E., 64
Mullins, W. M., 67
Muravich-Aleksandr, Kh. L., 12
Murbach, H. P., 67
Murgia, S. M., 15
Musha, S., 17
Myers, H., 225

Nadjo, L., 6
Nagase, S., 61
Naidin, V. A., 64
Naimishu, A. I., 24
Nakagawa, Y., 45
Nakamura, Y., 71
Nakaya, J., 18
Nasui, M., 56
Nechaev, B. P., 32
Nechepurnoi, V. D., 10
Nekrassov, L. N., 14
Nekrassov, N., 70

Nelson, R. F., 50, 51
Nemec, M., 10
Nesbet, R. K., 169
Neubert, G., 49
Neumann, A. W., 69, 72, 86
Nguyen-Van-Huong, G., 65
Nichikova, P. R., 46
Nikulin, V. N., 6
Nitzan, A., 181, 208
Nohe, H., 25, 46
Novak, L., 26
Nygard, B., 27

O'Brien, R. N., 93
O'Rourke, R. C., 212
Ochiai, M., 25
Odham, G., 46
Oelschlaeger, H., 18
Ogliaruso, M. A., 8
Ohme, R., 64
Ohmori, H., 56
Ohta, M., 60
Okada, T., 25
Okamoto, T., 48
Okazaki, K., 61
Okazaki, T., 2
Olekhanovich, E. P., 30
Omel'chenko, V. A., 17
Ono, M., 14
Oppenheimer, J. R., 135
Oriani, R. A., 66
Orgel, L. E., 234
Orlova, L. N., 15
Osa, T., 2
Osteryoung, R. A., 49
Osuda, T., 63
Owen, B. B., 207
Owen, D. M., 47
Ovchinnikov, A. A., 217
Ovchinnikova, M. Ya, 217
Ozment, C. L., 28

Paliani, G., 15, 27
Palint, I. N., 52
Palmquist, U., 40, 43
Panitkova, E. S., 46
Paprotskaya, V. A., 60
Pardini, V. L., 11
Parker, V. D., 2, 38, 39, 40, 43
Parker, V. D., 2
Parsons, R., 66
Paul, R. A., 60
Payne, R., 65
Pearce, R., 31
Pease, D. C., 69
Pederson, C. T., 1
Pedler, A. E., 47
Peltier, D., 10, 14
Perichon, J., 35
Person, M., 21
Persson, B., 27
Petrii, O. A., 65
Petterson, B., 46
Pfister-Guillouzo, G., 28
Pickett, D., J., 25
Pilla, A. A., 102
Pinson, J., 17, 20
Piszczek, L., 25
Pitaevskii, L. P., 162
Plashkin, V. S., 62
Plattner, E., 62
Pletcher, D., 1, 34, 60
Plzak, V., 36
Poling, M., 57
Polynikova, T. K., 14

Ponomarev, A. M., 26
Ponomarev, G. V., 18
Pople, J. A., 169
Portis, L. C., 49
Pospisil, L., 104
Power, E. A., 160
Pragst, F., 54, 58
Prater, K. B., 49
Puglisi, V. J., 4
Pulidori, F., 27
Pushkina, L. N., 62
Puxedda, A., 32

Quinn, R. K., 9

Rabai, J., 29
Racah, G., 195
Race, W. P., 103, 106
Ramachandran, V., 45
Ramakrishnaiah, K., 65
Rampasso, L., 23, 24
Randin, J. P., 90
Randles, J. E. B., 98
Rangarajan, S. K., 98, 113, 119
Raoult, E., 10
Rayleigh, Lord, 203
Razouk, R. J., 68
Rehm, D., 223
Reisenhofer, E., 32
Renaud, R. N., 22
Renzow, D., 72
Reynolds, R., 51
Reynolds, W. L., 225
Rich, R. L., 225
Richter, C., 23
Rickayzen, G., 208
Rideal, E. K., 68
Rifi, M. 1
Rifkin, S. C., 7
Rinishi, K., 46
Rodina, L. L., 58
Roeder, J., 13
Rogers, H. R., 24
Rogers, J. W., 15
Rogic, O., 57
Rogic, V., 57
Roman, P., 148
Rondarev, D. S., 46
Ronlan, A., 38, 39, 40, 43
Ross, S. D., 49
Rossett, R., 43
Rozhkov, I. N., 44
Rozin, Yu. I., 60
Ruedenberg, K., 170
Ruehlen, F. N., 4
Rukhadze, A. A., 173
Rusina, A., 7
Ryan, M. D., 5
Rybalka, K. V., 65
Rysakov, A. A., 17

Sadler, D. L., 4
Sadykh-zade, S. I., 60
Saito, Z., 7
Salaün, J. P., 9
Samoshina, T. F., 46
Sandstede, G., 40
Santhanam, K. S. V., 11, 17
Saotome, K., 3
Sass, V. P., 46, 61
Sato, M., 63
Satonaka, H., 7
Saveant, J. M., 6, 8, 11, 20
Savin, V. I., 15
Saxena, R. S., 59
Saygin, Oe, 36

Sayo, H., 56
Sazonova, T. V., 64
Schäfer, H., 35
Schaeffer, W., 33
Schcherbak, S. N., 16
Scheller, F., 33
Schiavon, G., 12, 64
Schilling, H., 21, 23,
Schmakel, C. O., 11, 17
Schmickler, W., 129, 217, 238
Schmid, C. M., 93
Schmidt, P. P., 128, 129, 149, 157, 158, 171, 174, 180, 182, 186, 207, 217, 220, 233
Schrader, M. E., 73
Schroeder, K. L., 11
Schuebel, H., 26
Schuhmann, D., 102, 115
Schultz, J., 102
Schuttenberg, A., 2
Schwarzlose, G., 25, 34
Schwertfeger, I., 54
Scriven, L. E., 95
Seebar, R., 58
Seko, N., 10
Seleznev, L. G., 64
Semenov, V. A., 24
Semenova, A. D., 10
Senderoff, S., 67
Senguen, F. I., 18
Senoo, S., 3
Senyushov, L. N., 61
Sevastyanov, E. S., 65
Shang, D. T., 39
Shapovalov, A. A., 56
Shimokawa, M., 71
Shimura, T., 7
Shinozaki, K., 25
Shoesmith, D. W., 25
Shono, T., 45
Shugar, D., 16
Shuttleworth, R., 67, 69
Shvetsov-Shilovskii, N. I., 56
Siebrand, W., 223
Siefke, B., 54
Siegel, T. M., 34
Silin, V. P., 173
Silverstone, H., 165, 171, 192
Simonet, J., 26, 27, 29, 59
Simonov, V. V., 24
Sing, R., 59
Sirohi, R. S., 93
Skyrme, T. H. R., 190
Slater, J. C., 166
Sluyters, J. H., 98, 100, 102
Sluyters-Rehbach, M., 98, 100
Smirnov, K. M., 25
Smirnov, S. K., 4
Smirnov, Yu. D., 4, 60
Smirnov, V. A., 63
Smith, D. E., 98
Smith, L. R., 51
Smolders, C. A., 71
Smolentseva, L. G., 6
Sneider, H., 36
Sokolov, S. V., 46, 61, 62
Sokol'skii, D. V., 10
Solomin, A. V., 35
Sommerfeld, A., 193
Sopilnyak, D. S., 16
Sorriso, S., 15
Sosonkin, I. M., 14

Author Index

Spaans, A., 27
Sparnaay, M. J., 66, 71
Steckhan, E., 35
Steelink, C., 43
Stenhagen, E., 46
Stermitz, F. R., 39, 40
Sternerup, H., 35, 37
Stewart, J. S., 4
Stokes, P. D., 26
Stoney, G. G., 67
Stradins, J., 1
Strelets, V. V., 32
Strogov, G. N., 14
Stüwe, A., 57
Sturostenko, E. P., 25
Sturostenko, N. F., 25
Subbiah, P., 15
Subramanian, G. S., 14
Sugawara, M., 63
Su Khac Binh, 11
Sukhmaneva, L. M., 64
Sundholm, F., 64
Sundholm, G., 64
Supin, G. S., 21, 23, 56
Susbielles, G. G., 113
Suter, H., 46
Sutin, N., 240
Sutton, J. R., 36
Sutula, C. L., 72
Suvorova, L. B., 64
Suzuki, A., 4, 64
Suzuki, M., 16
Svanholm, U., 43
Swiderski, J., 25
Syrkin, Y. K., 207

Tabakovic, I., 1, 57
Taguchi, T., 64
Takahashi, Y., 10, 64
Takamura, T., 71
Takata, Y., 4
Takvoryan, N., 33
Takenouti, H., 115
Tallec, A., 57
Tamera, K., 34
Tanaka, H., 48
Tanaka, Y., 18
Tanner, W., 72
Tantasheva, F. R., 28
Taran, L. A., 6
Tardevil, R., 42
Tarkhanov, G. A., 46
Tatlow, J. C., 47
Taube, H., 225
Taylor, K., 22, 91, 92
Tedoradse, G. A., 60
Tellert, K., 52
Tember, G. A., 46
ter Haar, D., 137
Tessari, G., 113
Thangavelu, S., 14
Thibault, C., 58

Thiebault, A., 34
Thijsse, H., 27
Thirsk, H. R., 22, 106, 111
Thirunavukkarasu, P., 15
Thomas, H. G., 47
Tikhonova, L. S., 12
Tilak, B. V. K. S. R. A., 66
Timofeeva, Z. N., 12
Tiwari, S. K., 59
Todres, Z. V., 15, 26
Tomilov, A. P., 4, 10, 22, 25, 60
Torii, S., 48
Toshiro, I., 46
Touboul, E., 8
Toure, V., 9
Toyozawa, Y., 133
Turner, D. R., 89
Tuxford, A. M., 34
Tyssee, D. A., 4, 5, 6
Trapnell, B. M. W., 92
Trautner, K., 21, 23
Trazza, A., 16
Tremillon, B., 60
Trivedi, L. D., 9
Trocha-Grimshaw, J., 18, 21
Troll, T., 2
Tsar'kova, J. G., 18
Tverdovskii, I. P., 70

Udo, K., 2
Udupa, H. V., 4, 14
Udupa, H. V. K., 1, 15
Udupa, K. S., 14, 15
Ulery, H. E., 21
Ulstrup, J., 129, 207, 234
Urabe, N., 64
Urushadze, Z. D., 213
Usatenko, Yu. I., 16
Utley, J. H. P., 5, 47

van der Helm, D., 57
Van der Meer, D., 27
Van Duyne, R. P., 133
Vainshtein, Y. I., 52
Valcher, S., 2, 3, 32, 60
Valette, G., 65
Vanden Born, H. W., 22
Vanderborgh, N. E., 9
Vargaftik, M. N., 207, 232
Vargalyak, V. F., 17
Varshavskii, S. L., 25
Vasileff, H. D., 212
Vasilev, Yu. B., 46
Veksler, M. A., 64
Veinberg, A. Ya., 26
Venkatachalapathy, M. S., 4, 14
Vermaak, J. S., 67
Vertier, H., 11
Vielstich, W., 129

Vincent, C. A., 29
Voigtlaender, R., 21
Volke, J., 7
Vol'kenshteyn, M. V., 233
Volodkovich, S. D., 23
Vorotyntsev, M. A., 133, 150, 208, 234, 239
Vovchenko, G. D., 10

Wagenknecht, J. H., 19, 51
Walecka, J. D., 149
Warren, C. B., 48
Watanabe, A., 71
Watkins, B. F., 41
Wawzonek, S., 22, 34, 48
Weber-Schäfer, M., 57
Weedon, B. C. L., 47
Weisskopf, V. F., 175
Weller, A., 223
Wendt, H., 36
Wenisch, F., 46
Wenzel, R. N., 68
White, M. L., 72
Wiberg, K. B., 24
Wigner, E., 175
Wilcock, J. D., 67
Wild, W., 174
Wilke, G., 64
Will, J. P., 39
Wilman, H., 67
Wladislaw, B., 11
Wolf, J. F., 8
Wrona, M., 16
Wuerminghausen, T., 34
Wulff, G., 67

Yagupol'skii, L. M., 62
Yakhontov, L. N., 52
Yamamoto, F., 10
Yamamoto, H., 63
Yamamoto, T., 2, 128
Yao, T., 17
Yap, K. S., 25
Yeager, E., 90
Yonekura, M., 61
Yontanza, A., 10
Young, T., 66, 67
Yuldasheva, L. K., 24
Yzermans, A. B., 102

Zakarina, N. A., 4
Zakumbaeva, G. D., 4, 64
Zaporozhets, E. V., 14, 18
Zapevalova, N. P., 59
Zecchin, S., 12, 64
Zeeh, B., 14
Zener, C. A., 134
Zettlemoyer, A. C., 72
Zhurinov, M. Zh., 35
Zisman, W. A., 73
Zuman, P., 8, 9

JUN 29 1978